A Comprehensive Guide to HSMM

Series Editor
Nikolaos Limnios

A Comprehensive Guide to HSMM

Theory, Software, and Advanced Extensions

Edited by

Nathalie Peyrard
Benoîte de Saporta

iSTE

WILEY

First published 2026 in Great Britain and the United States by ISTE Ltd and John Wiley & Sons, Inc.

ISTE Ltd
27-37 St George's Road
London SW19 4EU
UK

www.iste.co.uk

John Wiley & Sons, Inc.
111 River Street
Hoboken, NJ 07030
USA

www.wiley.com

Library of Congress Control Number: 2025944984

British Library Cataloguing-in-Publication Data
A CIP record for this book is available from the British Library
ISBN 978-1-83669-035-1

The manufacturer's authorized representative according to the EU General Product Safety Regulation is Wiley-VCH GmbH, Boschstr. 12, 69469 Weinheim, Germany, e-mail: Product_Safety@wiley.com.

Contents

Introduction

Hidden Markov models (HMMs) were introduced in the 1960s by Baum and Petrie (1966), first in the case of discrete observations. They have become classical tools to analyze time series whose dynamics can be explained by those of a hidden process. The model is composed of two sets of random variables that are two linked dynamical stochastic processes: one hidden and one observed. This family of models was originally popularized by applications in speech recognition (Baker 1975) and later in other domains like genomics (Churchill 1989). In parallel, developments were achieved in computational statistics to make statistical estimation possible. At the beginning of the 1980s, it was observed that the Markovian assumption on hidden state dynamics was not satisfied in the context of speech recognition. Thus, different relaxations of this assumption were proposed, leading to explicit duration HMMs (Ferguson 1980, ED-HMMs), reformulated shortly after in their modern, more parsimonious form by Russell and Moore (1985). More general semi-Markov models were introduced by Murphy (2002), as an extension of so-called segment HMMs, which are themselves generalizations of ED-HMMs.

Concomitantly with those generalizations of HMMs leading progressively to HSMMs, focusing on refinements on how state and sojourn durations at the current jump depend on the same quantities as the previous jump, HMMs underwent developments on modeling dependencies between several interacting chains, which is the subject of Chapter 3 in this book. These developments were once again motivated by applications in speech or video processing and oriented toward two directions: representing coupling of (shared) hidden states through observations (Ghahramani and Jordan 1997) or directly coupling through states, keeping observations conditionally independent given their unique associated hidden state (Brand et al. 1997).

Chapter written by Benoîte DE SAPORTA, Jean-Baptiste DURAND, Alain FRANC and Nathalie PEYRARD.

Markov processes are also a key concept in control problems, whenever stochastic effects have an impact on the dynamics of the system states, or on the efficiency of actions taken to control the system, or even on both aspects. To account for such stochastic effects, Markov decision processes were introduced in the 1950s (Bellman 1958). In cases where the system state cannot be observed directly, but only indirectly through stochastic functions of the state process as in HMMS, the appropriate formalism to address control is that of partially observed MDPs (POMDPs), introduced by Åström (1965). Actually, in such models, "states" in the sense of HMMS may never be observed but POMDPs encompass states and observations into a broader concept of partially observed states. Some extensions of control problems to partially observed semi-Markov dynamics were introduced in the 1990s (Puterman 1994); they are presented in this book under the more general framework of impulse controlled piecewise-deterministic Markov processes.

In parallel with these theoretical developments, the application fields of HSMMs have largely expanded. Speech recognition is less present nowadays but HSMMs have successfully been applied in numerous other domains. This is well illustrated in the introduction of the book by Yu (Yu 2016), where the author presents an overview of the main areas of application up to 2016, covering almost 40 fields. In recent years, HSMMs have been applied to activities recognition for humans (van Kuppevelt et al. 2019; Cavallo et al. 2022; Thornton et al. 2023) and animals (Ruiz-Suarez et al. 2022; Koslik et al. 2023), as well as eye-movement analysis (Olivier et al. 2022; Gao et al. 2023). Seismology (Pertsinidou et al. 2017), epidemiology (Touloupou et al. 2020), occupational health (Haji-Maghsoudi et al. 2021), cardiovascular disease screening (Oliveira et al. 2018), neurophysiology (Chakravarty et al. 2019), plant growth (Mészáros et al. 2020; Labadie et al. 2023) and ecology (Nicol et al. 2022) also exploited the flexibility of HSMMs. Controlled versions of HMM or HSMM have been recently applied to reliability and safety (Srinivasan and Parlikad 2014; Zhang and Revie 2016), path planning for automated vehicles (Bravo et al. 2019), healthcare (Skandari and Shechter 2021; Fatemi et al. 2022; Liu et al. 2022; Garcia et al. 2024; de Saporta et al. 2024) and health economics (Cao et al. 2016; Mohammadi et al. 2023).

This book is written by a consortium of French researchers as part of the project *Hidden Semi-Markov Models: INference, Control and Applications* (HSMM-Inca) funded by the French *Agence Nationale de la Recherche* (ANR, grant ANR-21-CE40-005). This consortium is a unique group of researchers with a long experience of statistics, probability, inference and control for temporal and spatial processes. It also has expertise on the modeling and inference of hidden dynamic processes in health, ecology and natural risks.

We have written this book to offer an accessible introduction to the framework of HSMMs, covering the main methods and theoretical results for maximum likelihood estimation in HSMMs, together with an opening onto new, less classical related topics such as multichain HSMM and controlled HSMM. This book unifies and generalizes existing results. It is also complementary to more detailed textbooks such as Barbu and Limnios (2009) and Yu (2016) that go deeper into technical details but stay focused on classical HSMM.

This book is primarily intended for master and PhD students, researchers and academic faculty in the fields of statistics, applied probability, graphical models, computer science and connected domains. It is also meant to be accessible to practitioners involved in modeling, analysis or control of time series in the fields of reliability, theoretical ecology, signal processing, finance, medicine, epidemiology, etc.

The book is organized as follows. Chapter 1 introduces the general HSMM formalism and well-known particular cases. Then, it introduces maximum likelihood estimation (MLE) for HSMMs: likelihood expression and evaluation, asymptotic properties of the MLE and EM algorithm for MLE computation. Finally, it presents recent results on the definition of reliability indicators and an extensive review of how to introduce mixed effects into HSMM components.

Chapter 2 provides an overview of the packages and software (primarily in R and Python) dedicated to the estimation, simulation and application of HSMMs and other types of Markovian models that are central to the themes of the subsequent chapters.

Chapter 3 defines the framework of multichain HMM that enables us to present in a unified way existing models from literature and also generalize them. It illustrates how such models can be used to model dynamics in ecology and epidemiology. Then, it discusses inference of multichain HMM in the context of the EM algorithm.

Chapter 4 deals with the introduction of the semi-Markov assumption in multichain HMMs. It proposes a sound formalization of two classes of models that extend standard and general semi-Markov models to the multichain setting. Then, it considers the hidden framework and builds various classes of multichain HSMMs that generalize some MHMM structures defined in Chapter 3.

Chapter 5 focuses on the evaluation of the time complexity of marginal inference calculated with the forward-backward algorithm. It presents an algebraic formalism that leads to writing the forward-backward algorithm and calculating its complexity by counting the multiplications, with the same approach for several models studied in the previous chapters. It also shows how the sparsity of the transition matrices leads to a reduction of this complexity.

Chapter 6 focuses on controlled HSMMs. The chapter starts with a step-by-step introduction to the Markov decision processes (MDPs) formalism and progressively introduces partial observation semi-Markov assumption and continuous time to present the current available methods for controlled HSMMs.

This book presents distinct definitions, models and issues, which despite their differences share some common features. To guide the reader through this diversity of approaches, two simple case studies have been selected as toy examples, which are used during the course of the different chapters to illustrate the different notions and approaches, enabling an intuitive understanding before entering into the technique. They illustrate distinct extensions of HSMMs, like inference with covariates, multichain setting and linking with decision-making. The first one is referred to as *Squirrel* and the second as *Deer*. Both are models in behavioral ecology and introduced in Chapter 1. Chapters 2 and Chapter 6 present numerical illustrations of the two toy examples and provide the code for reproducing it.

Throughout the book, we also present open questions on the different aspects of HSMMs covered in the chapters. We hope we have created a source of inspiration for future research.

I.1. References

Åström, K.J. (1965). Optimal control of Markov processes with incomplete state information. *Journal of Mathematical Analysis and Applications*, 10(1), 174–205.

Baker, J. (1975). The DRAGON system – An overview. *IEEE Transactions on Acoustics, Speech, and Signal Processing*, 23(1), 24–29.

Barbu, V.S. and Limnios, N. (2009). *Semi-Markov Chains and Hidden Semi-Markov Models Toward Applications: Their Use in Reliability and DNA Analysis*. Springer, New York.

Baum, L. and Petrie, T. (1966). Statistical inference for probabilistic functions of finite state Markov chains. *The Annals of Mathematical Statistics*, 37(6), 1554–1563.

Bellman, R. (1958). Dynamic programming and stochastic control processes. *Information and Control*, 1(3), 228–239.

Brand, M., Oliver, N., Pentland, A. (1997). Coupled hidden Markov models for complex action recognition. In *Proceedings of IEEE Computer Society Conference on Computer Vision and Pattern Recognition*, San Juan. IEEE Press, Piscataway, 994-999.

Bravo, R.Z.B., Leiras, A., Cyrino Oliveira, F.L. (2019). The use of UAVs in humanitarian relief: An application of POMDP-based methodology for finding victims. *Production and Operations Management*, 28(2), 421–440.

Cao, Q., Buskens, E., Feenstra, T., Jaarsma, T., Hillege, H., Postmus, D. (2016). Continuous-time semi-Markov models in health economic decision making: An illustrative example in heart failure disease management. *Medical Decision Making*, 36(1), 59–71.

Cavallo, F., Toumazou, C., Nikolic, K. (2022). Unsupervised classification of human activity with hidden semi-Markov models. *Applied System Innovation*, 5(4).

Chakravarty, S., Baum, T.E., An, J., Kahali, P., Brown, E.N. (2019). A hidden semi-Markov model for estimating burst suppression EEG. In *2019 41st Annual International Conference of the IEEE Engineering in Medicine and Biology Society (EMBC)*.

Churchill, G.A. (1989). Stochastic models for heterogeneous DNA sequences. *Bulletin of Mathematical Biology*, 51(1), 79–94.

Fatemi, M., Wu, M., Petch, J., Nelson, W., Connolly, S.J., Benz, A., Carnicelli, A., Ghassemi, M. (2022). Semi-Markov offline reinforcement learning for healthcare. In *Conference on Health, Inference, and Learning*. PMLR.

Ferguson, J.D. (1980). Variable duration models for speech. In *Symposium on the Application of Hidden Markov Models to Text and Speech*. IDA-CRD, Princeton, 143–179.

Gao, L., Wang, C., Wu, G. (2023). Hidden semi-Markov models-based visual perceptual state recognition for pilots. *Sensors*, 23(14).

Garcia, G.-G., Lavieri, M., McAllister, T., McCrea, M., Broglio, S. (2024). 8.2 design and assessment of a novel decision-theoretic approach to personalize return-to-play from concussion. *British Journal of Sports Medicine*, 58(Suppl 1), A41–A41.

Ghahramani, Z. and Jordan, M. (1997). Factorial hidden Markov models. *Machine Learning*, 29(2-3), 245–273.

Haji-Maghsoudi, S., Bulla, J., Sadeghifar, M., Roshanaei, G., Mahjub, H. (2021). Generalized linear mixed hidden semi-Markov models in longitudinal settings: A Bayesian approach. *Statistics in Medicine*, 40(10), 2373–2388.

Koslik, J.-O., Feldmann, C. C., Mews, S., Michels, R., Langrock, R. (2023). Inference on the state process of periodically inhomogeneous hidden Markov models for animal behavior. arXiv preprint arXiv:2312.14583

van Kuppevelt, D., Heywood, J., Hamer, M., Sabia, S., Fitzsimons, E., van Hees, V. (2019). Segmenting accelerometer data from daily life with unsupervised machine learning. *PLOS ONE*, 14(1), 1–19.

Labadie, M., Guy, K., Demené, M.-N., Caraglio, Y., Heidsieck, G., Gaston, A., Rothan, C., Guédon, Y., Pradal, C., Denoyes, B. (2023). Spatio-temporal analysis of strawberry architecture: Insights into the control of branching and inflorescence complexity. *Journal of Experimental Botany*, 74(12), 3595–3612.

Liu, Z., Khojandi, A., Li, X., Mohammed, A., Davis, R.L., Kamaleswaran, R. (2022). A machine learning–enabled partially observable Markov decision process framework for early sepsis prediction. *INFORMS Journal on Computing*, 34(4), 2039–2057.

Mészáros, M., Guédon, Y., KrVska, B., Costes, E. (2020). Modelling the bearing and branching behaviors of 1-year-old shoots in apricot genotypes. *PLoS ONE*, 15(7), e0235347.

Mohammadi, N., Skandari, R., and Shah, A. (2023). Efficient discovery of cost-effective policies: The case of hearing loss screening for cystic fibrosis. *SSRN*. doi: 10.2139/ssrn.4338022.

Murphy, K. (2002). Hidden semi-Markov models (HSMMs) [Online]. Available at: https://www.cs.ubc.ca/~murphyk/Papers/segment.pdf.

Nicol, S., Cros, M.-J., Peyrard, N., Sabbadin, R., Trépos, R., Fuller, R.A., Woodworth, B.K. (2022). Flywaynet: A hidden semi-Markov model for inferring the structure of migratory bird networks from count data. *Methods in Ecology and Evolution*, 14(1), 265–279.

Oliveira, J., Renna, F., Mantadelis, T., Coimbra, M. (2018). Adaptive sojourn time HSMM for heart sound segmentation. *IEEE Journal of Biomedical and Health Informatics*, 23(2), 642–649.

Olivier, B., Guérin-Dugué, A., Durand, J.-B. (2022). Hidden semi-Markov models to segment reading phases from eye movements. *Journal of Eye Movement Research*, 15(4), 1–19.

Pertsinidou, C.E., Tsaklidis, G., Papadimitriou, E., Limnios, N. (2017). Application of hidden semi-Markov models for the seismic hazard assessment of the north and south Aegean Sea, Greece. *Journal of Applied Statistics*, 44(6), 1064–1085.

Puterman, M.L. (1994). *Markov Decision Processes: Discrete Stochastic Dynamic Programming*. John Wiley & Sons, Hoboken, NJ.

Ruiz-Suarez, S., Leos-Barajas, V., Morales, J.M. (2022). Hidden Markov and semi-Markov models: When and why are these models useful for classifying states in time series data? *Journal of Agricultural, Biological and Environmental Statistics*, 27, 339–363.

Russell, M. and Moore, R. (1985). Explicit modelling of state occupancy in hidden Markov models for automatic speech recognition. In *ICASSP'85 – IEEE International Conference on Acoustics, Speech, and Signal Processing*, Tampa, FL, 5–8.

de Saporta, B., Thierry d'Argenlieu, A., Sabbadin, R., Cleynen, A. (2024). A Monte-Carlo planning strategy for medical follow-up optimization: Illustration on multiple myeloma data. *PLOS ONE*, 19(12).

Skandari, M.R. and Shechter, S.M. (2021). Patient-type Bayes-adaptive treatment plans. *Operations Research*, 69(2), 574–598.

Srinivasan, R. and Parlikad, A.K. (2014). Semi-Markov decision process with partial information for maintenance decisions. *IEEE Transactions on Reliability*, 63(4), 891–898.

Thornton, C., Kolehmainen, N., Nazarpour, K. (2023). Using unsupervised machine learning to quantify physical activity from accelerometry in a diverse and rapidly changing population. *PLOS Digit Health*, 2(4).

Touloupou, P., Finkenstädt, B., Spencer, S.E.F. (2020). Scalable Bayesian inference for coupled hidden Markov and semi-Markov models. *Journal of Computational and Graphical Statistics*, 29(2), 238–249.

Yu, S.-Z. (2016). *Hidden Semi-Markov Models Theory, Algorithms and Applications*. Elsevier, Amsterdam.

Zhang, M. and Revie, M. (2016). Continuous-observation partially observable semi-Markov decision processes for machine maintenance. *IEEE Transactions on Reliability*, 66(1), 202–218.

1

Monochain HSMM

A hidden semi-Markov model (HSMM) is an extension of an hidden Markov model (HMM) where the sojourn duration of the chain in a state is a realization of a random variable that, contrary to the Markovian case, does not necessarily have a geometric distribution. In this chapter, the general HSMM formalism and well-known particular cases are presented, starting with a gentle introduction to illustrate the main notions and notations for a toy example. Then we introduce the maximum likelihood estimation (MLE) for HSMM: the likelihood expression and evaluation, asymptotic properties of MLE and expectation-maximization (EM) algorithm for MLE computation. Finally, we present recent results on two topics seldom addressed for HSMMs: the definition of reliability indicators and introduction of mixed effects into HSMM components, with the latter presenting a thorough survey of the literature. In this chapter, the reader will also find a presentation of the two toy examples that will accompany all the book's chapters.

1.1. Introduction

As mentioned in the book's Introduction, HSMMs are versatile models that have been extensively used for diverse applications. These models are built using several *components*: first, two sets of random variables that are dynamical stochastic processes (one hidden and one observed) and second, distributions defining the joint behavior of these two processes (with possibly parameters associated with these distributions).

With this first chapter, we present the basic elements to start with HSMMs. In section 1.2, we introduce the general definition with the different elements

---

Chapter written by Jean-Baptiste DURAND, Alain FRANC, Nathalie PEYRARD, Nicolas VERGNE and Irene VOTSI.

composing the joint distribution of hidden and observed variables, as well as particular cases more often used in practice. In section 1.3 we review the results existing on the MLE asymptotic properties in HSMMs. The standard implementations of the EM algorithm (Dempster et al. 1977) for HSMM inference and their computational complexity are then discussed. We present also two toy examples (section 1.4) to illustrate the different concepts introduced and to prepare the comparison of existing packages for inference in HSMMs, which is proposed in Chapter 2.

In addition, we present less classical topics on HSMMs: the definition of reliability indicators (section 1.5) and the introduction of covariates and random effects (section 1.6). Regarding mixed effects, we present how to model them in the different components of a HSMM and we discuss specific modeling and algorithmic issues. We also propose a review of the literature on mixed models in HMM and HSMM.

1.2. HSMM framework

In order to introduce progressively the components defining a HSMM, we first provide an intuitive idea of what they may actually represent through the presentation of a fictive but realistic use case in ecology (section 1.2.1). This example will be one of the two toy examples of the book.

Then, we present formally the HSMM framework, starting from the more general definition (section 1.2.2), before continuing with the standard acceptation of HSMMs in statistics (section 1.2.3) and with the particular case of explicit duration HMM (ED-HMM) (section 1.2.4). The last example introduced is the well-known HMM.

1.2.1. *Intuitive presentation with the Squirrel toy example*

Let us imagine a squirrel that has spent the summer gathering hazelnuts for the winter and hidden them under a bed of foliage in various places in the forest known only by itself. When winter comes, it feeds on its reserves. It wanders from reserve to reserve, spending varying amounts of time on each one. When it visits a reserve, it may not use it up and it may return to the same reserve several times. The reserve where the squirrel feeds is referred to as the squirrel *state*. The sequence of states is referred to as a *chain*. Thus, this chain is a process with *jumps*, the jump being the event where the squirrel changes reserve. Then it stays there for a while (between two jumps the state does not change). The time of jump and the time spent in a reserve are stochastic. A naturalist with a passion for *Sciuridae* arrives to observe them. The squirrel is very difficult to spot, as its fur blends in with dead leaves. The state of the squirrel is then a *hidden process* for the naturalist. The naturalist therefore relies on the observation of traces left by the squirrel to guess in which reserve the squirrel is. The reserve guessed by the naturalist is the *observed process*. However, the traces are ambiguous, sometimes misinterpreted and they are the result of some probabilistic connection between the occupied reserve and traces.

This little story can be formalized in the HSMM framework as follows: the set of K reserves is numbered from 1 to K. The squirrel is in state i when it exploits resources in reserve i. The change of reserve is a jump and J_n denotes the reserve occupied by the squirrel (i.e. its state) after the nth jump. The date of the nth jump is denoted as S_n. The number of days the squirrel exploits the reserve reached after n jumps is the *sojourn duration*, denoted as $X_{n+1} = S_{n+1} - S_n$. Here, $X_{n+1} = d$ means that the squirrel exploits the reserve J_n exactly d days once entering it, then jumps to a different reserve after that duration. Chains $J_0, J_1, \ldots, S_0, S_1, \ldots$ and $X_0, X_1, \ldots$ are indexed by the jumps. They are hidden variables in the model, because they are only observed indirectly. The first assumption is made: Markovianity of (J_n, X_{n+1}). The new reserve and its duration depends only on the previous reserve and its duration (and not the entire history). Beyond the Markovian assumption, it is assumed here that the reserve reached after a jump does not depend on the time spent in the previous reserve but only on the reserve before jumping. It is also assumed that the time spent in the reserve after a jump does not depend on the previous reserve and the associated duration but only on the reserve reached after jumping. These are conditional independencies made for the Squirrel toy example, which are not necessarily true for any general HSMM. Under these assumptions, a simple way to describe the stochastic behavior of the squirrel is to define:

– the probabilities $\mathbb{P}(J_{n+1} = j \,|J_n = i)$ to jump from reserve i to reserve j, given that the squirrel finished its sojourn in reserve i; this probability explicits the way the squirrel explores its territory. For example, the geographical location of the patches may matter, and at time S_n of jump n from patch J_{n-1}, patches neighbor to J_{n-1} may be favored (for the sake of energy saving while travelling from patch to patch);

– the sojourn duration distribution $\mathbb{P}(X_{n+1} = d \,|J_n = j)$ for $d \geq 1$, given that the squirrel jumped into reserve j after nth jump.

The observations are defined at a daily time step t. Therefore, it is convenient to also have a variable representing the hidden state at time t: Z_t (to be distinguished from J_n which is indexed by the jumps). Let denote as Y_t the reserve where the naturalist believes the squirrel is from the sight of traces: the stochastic connection between Z_t and Y_t is modeled by the emission distribution $\mathbb{P}(Y_t = j \,|Z_t = i)$ for i and $j \in \{1, \ldots, K\}$. It is assumed here that, given the squirrel state at time t, the reserve guessed by the naturalist at the same time does not depend on any other random variable. Again, this choice is specific to the Squirrel toy example.

The three distributions $\mathbb{P}(J_{n+1} = j \mid J_n = i)$, $\mathbb{P}(X_{n+1} = d \mid J_n = j)$ and $\mathbb{P}(Y_t = j \mid Z_t = i)$ can be non-parametric or parametric ones. We will provide an example of parameterization for the Squirrel example in section 1.4.1.

The joint distribution of the observed process $Y_{0:T} = (Y_0, Y_1, \ldots, Y_T)$ and the hidden process $Z_{0:T} = (Z_0, Z_1, \ldots, Z_T)$ defined by the above distributions is an example of Ferguson (1980). It is a particular case of HSMM. More complex dependencies between the couple (J_n, X_{n+1}) and the couple (J_{n-1}, X_n) can be considered in the general HSMM framework and will be discussed in the next sections.

1.2.2. *General HSMM framework*

In the literature, there exist several models that are related to HSMMs. We begin with the presentation of the most general model, referred to as the general HSMM in Yu (2016). The other classes of HSMMs are derived from the general HSMM by successive introductions of additional assumptions on conditional independencies in the semi-Markov model defining the hidden process.

Let us define formally the different processes introduced in the Squirrel example and that will allow us to present the definition of the general HSMM. There are two sets of variables, one indexed on jump times and one indexed on calendar time. The first set describes the hidden process:

– $J = (J_n)_{n \in \mathbb{N}}$ belonging to the finite state space $\Omega_Z = \{1, 2, \ldots K\}$, where J_n is the *state into which the chain enters at jump* n.

– $S = (S_n)_{n \in \mathbb{N}}$ belonging to state space $\mathbb{N}$, where S_n is the *time at jump* n. By convention, one defines $S_0 = 0$. With this choice, it is assumed that at $t = 0$ the system enters a new state.

– $X = (X_n)_{n \in \mathbb{N}^*}$ belonging to state space $\mathbb{N}^*$, where $X_n = S_n - S_{n-1}$ is the sojourn duration in state J_{n-1}, that is, the time during which the chain remains in the state J_{n-1}. By convention, one defines $X_0 = 0$. The above-defined different processes are represented in Figure 1.1 (Votsi and Brouste 2019).

The second set corresponds to the values of the hidden and observed processes at each time step of observation.

– $Z = (Z_t)_{t \in \mathbb{N}}$ belonging to the finite state space $\Omega_Z = \{1, 2, \ldots K\}$ is the *state of the chain at time* t.

– $Y = (Y_t)_{t \in \mathbb{N}}$ belonging to the state space Ω_Y is the *observation at time* t.

If we introduce now $N(t) = \max_n (S_n \leq t)$, the index of the current jump, the process J and Z are linked by the following relation: $\forall t \in \mathbb{N}, Z_t = J_{N(t)}$. The random variable Y_t can be discrete or continuous but for the sake of simplicity in the following, unless otherwise stated, we use notations for Ω_Y being a finite set.

There exist two alternative notations to designate events of the type "the chain enters state i and remains in this state for exactly d time steps". The first uses indexation by jump times and relies on variables J_n and X_{n+1}: $(J_n = i, X_{n+1} = d)$ is the event "at the nth jump, the chain enters state i and remains in this state for exactly d time steps". The second notation uses calendar time and relies on the variable Z_t as follows. Let us define the random vector $Z_{t+1,t+d} = \{Z_{t+1}, \ldots, Z_{t+d}\}$. In Yu (2016), the following notations are introduced: $(Z_{t+1,t+d} = i)$ is the event "the chain is in state i at times $t+1, \ldots, t+d$, but without specification of the value of Z_t and Z_{t+d+1}"; $(Z_{[t+1,t+d]} = i)$ is the event "at time $t+1$, the chain enters in state i and it remains in this state for exactly d time steps". For instance in Figure 1.1, we have $(Z_{[S_1,S_2-1]} = j)$. Events like $(Z_{[t+1,t+d} = i)$ or $(Z_{t+1,t+d]} = i)$ are defined similarly depending on whether the date of entrance or the date of exit of state i is specified or not. This second set of notations is more convenient to express in a compact way certain events. Depending on the topics of the different book's chapters, one notation or the other will be used. Here, we will try when possible to present concepts with both notations.

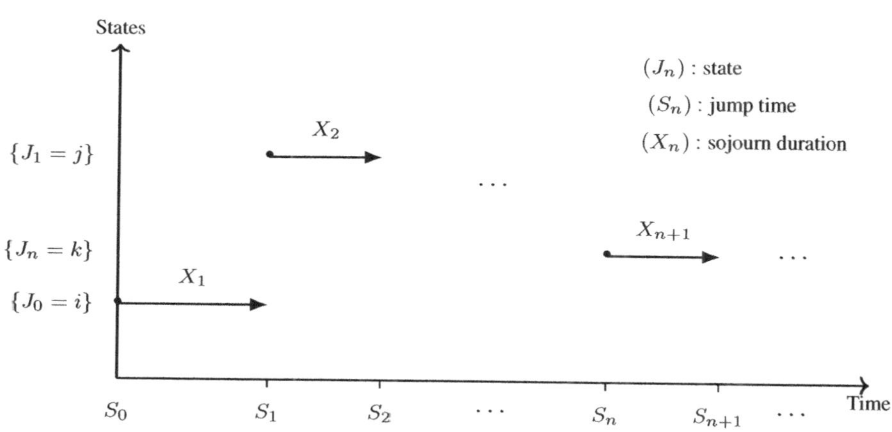

Figure 1.1. *Example of the realization of a discrete time semi-Markov process (from Votsi and Brouste 2019)*

1.2.2.1. *Hidden variables distribution*

In the general HSMM, the only assumption about the hidden process is that (J_n, X_{n+1}) is a Markov chain. It means that (J_n, X_{n+1}) is independent of $(J_{n-k}, X_{n+1-k})_{k>1}$ given (J_{n-1}, X_n).

DEFINITION 1.1.– *The chain $(Z_t)_{t\in\mathbb{N}}$, indexed by calendar time, is a general semi-Markov model if the associated chain $(J_n, X_{n+1})_{n\in\mathbb{N}}$ indexed by jump times is Markovian.*

The dynamics of a general SMM is governed by $a_{(i,d),(i,l)}$ the probability of transition from (i,d) to (j,l):

$$a_{(i,d),(j,l)} = \mathbb{P}(X_{n+1} = l, J_n = j \mid X_n = d, J_{n-1} = i), \quad \forall n.$$

This *joint transition probability* does not depend on n and it could be defined equivalently as:

$$a_{(i,d),(j,l)} = \mathbb{P}(Z_{[t+d+1:t+d+l]} = j \mid Z_{[t+1,t+d]} = i) \quad \forall t.$$

Let us also define the *initial distribution* as follows:

$$\pi_{id} = \mathbb{P}(J_0 = i, X_1 = d) = \mathbb{P}(Z_{[0:d-1]} = i).$$

1.2.2.2. *Emission distribution*

In the general HSMM, J and S are hidden, the observation at time t is denoted Y_t. While the hidden chain is in state i for d time steps, there will be d observations emitted, denoted $y_{t+1:t+d}$. It is assumed that, conditionally to $Z_{[t+1:t+d]} = i$, the sequence of observations $Y_{t+1:t+d}$ is independent of the other observations and the other hidden states.

DEFINITION 1.2.– *The chain $(Z_t, Y_t)_{t\in\mathbb{N}}$ is a general HSMM if (i) $(Z_t)_{t\in\mathbb{N}}$ is a general semi-Markov model, and (ii) conditionally to $Z_{[t+1:t+d]} = i$, the sequence of observations $Y_{t+1:t+d}$ is independent of the other observations and the other hidden states.*

The *joint emission probability* is defined as follows:

$$b_{id}(y_{t+1:t+d}) = \mathbb{P}(Y_{t+1:t+d} = y_{t+1:t+d} | Z_{[t+1:t+d]} = i).$$

Usually, when modeling applications, the following simplifying assumption is made:

H1: Conditionally to the chain $(Z_t)_{t\in\mathbb{N}}$, the Y_t are independent from each other and Y_t depends only on Z_t.

If we define the *emission probability* $b_{iy} = \mathbb{P}(Y_t = y \mid Z_t = i)$, it means that:

$$b_{id}(y_{t+1:t+d}) = \prod_{s=t+1}^{t+d} b_{iy_s}$$

The transition probabilities $a_{(i,d),(j,l)}$, the initial probabilities π_{id} and the joint emission probabilities $b_{id(y_{t+1:t+d})}$ fully define the distribution of the general HSMM.

In the following sections, we present nested subclasses of the general HSMM, corresponding to adding independence assumptions on the transition $a_{(i,d),(j,l)}$.

1.2.3. *Standard HSMM*

1.2.3.1. *Definition*

The transition probability in a general HSMM can be decomposed as follows:

$$a_{(i,d),(j,l)} = \mathbb{P}(X_{n-1} = l \mid J_n = j, X_n = d, J_{n-1} = i) \\ \times \mathbb{P}(J_n = j \mid X_n = d, J_{n-1} = i).$$

Let us consider the following assumption:

H2: The duration of the current state does not depend on the previous state nor on its duration, given the current state.

If H2 applies, we obtain the simplification:

$$a_{(i,d),(j,l)} = \mathbb{P}(X_{n+1} = l \; J_n = j) \times \mathbb{P}(J_n = j|X_n = d, J_{n-1} = i). \qquad [1.1]$$

This expression uses what is called the *distribution of sojourn duration in state* i:

$$h_i(d) = \mathbb{P}(X_{n+1} = d \,|J_n = i).$$

DEFINITION 1.3.– *The chain* $(Z_t)_{t\in\mathbb{N}}$, *indexed by calendar time, is a standard semi-Markov model if the associated chain* $(J_n, X_{n+1})_{n\in\mathbb{N}}$ *indexed by jump times is Markovian and hypothesis H2 is satisfied.*

DEFINITION 1.4.– *The chain* $(Z_t, Y_t)_{t\in\mathbb{N}}$ *is a standard HSMM if (i)* $(Z_t)_{t\in\mathbb{N}}$ *is a standard semi-Markov model (SMM), and (ii) conditionally to* $Z_{[t+1:t+d]} = i$, *the sequence of observations* $Y_{t+1:t+d}$ *is independent of the other observations and the other hidden states.*

The graphical representation of the conditional independencies for this subcase of the general SMM is displayed in Figure 1.2. The graph is build as follows: if in the expression of the joint distribution of $(J_0, X_1, J_1, X_2, \ldots, J_N, X_{N+1})$, decomposed using the chain rule, a variable appears in the conditional distribution of another; then, there is an arc from the former toward the latter. Otherwise stated, the expression of the joint distribution of $(J_0, X_1, J_1, X_2, \ldots, J_N, X_{N+1})$ can be decomposed into the product of the conditional probability of each variable in the graph, given its parents variables in the same graph.

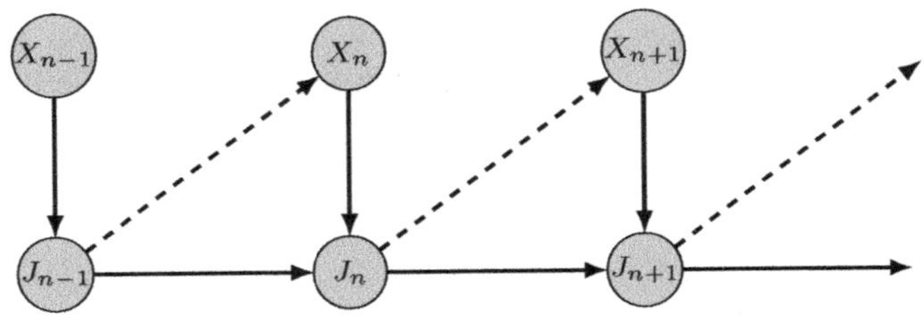

Figure 1.2. *Graphical representation of the conditional independencies in the chain $(J_n, X_n)_n$ in a standard HSMM. Solid arcs correspond to the sojourn duration distribution $h_i(d)$ and dotted ones to the transition $\mathbb{P}(J_{n+1} = j \,|X_{n+1} = d, J_n = i)$*

1.2.3.2. *Alternative definition*

In Barbu and Limnios (2008), another definition of HSMM is used based on different concepts like the Markov renewal chain and the Markov kernel. We present this definition and show that it is a standard HSMM. The above definition of a standard HSMM follows the framework introduced in Yu (2016), where computational aspects of inference in HSMMs are well developed. The content of Barbu and Limnios (2008) is complementary, it provides a statistical vision of the topic, with results on asymptotic properties of MLE. Therefore, we found it useful to introduce the two visions of HSMMs to the reader.

Let us start with the definition of the distribution of the hidden variables in Barbu and Limnios (2008), $(J_n, X_n)_{n\in\mathbb{N}}$. The following conditional independence assumption is considered:

$$\begin{aligned}\mathbb{P}(J_{n+1} &= j, S_{n+1} - S_n = d \,|J_0 = i_0, ..., J_n = i; S_0 = s_0, ..., S_n = s_n)\\ &= \mathbb{P}(J_{n+1} = j, S_{n+1} - S_n = d \,|J_n = i).\end{aligned} \quad [1.2]$$

Since knowing the sequence of jump times $S_0, ..., S_n$ is equivalent to knowing the sequence of sojourn durations $X_0, ..., X_n$ (S_0 and X_0 are fixed to 0), and vice versa, we can deduce from the above equality that:

$$\begin{aligned}\mathbb{P}(J_{n+1} &= j, X_{n+1} = d \,|J_0 = i_0, ..., J_n = i; X_0 = d_0, ..., X_n = d_n)\\ &= \mathbb{P}(J_{n+1} = j, X_{n+1} = d \,|J_n = i),\end{aligned} \quad [1.3]$$

which means that the couple (J_n, X_n) is Markovian and furthermore the new state J_{n+1} and the duration X_{n+1} of the current state depend only on the current state J_n and not on the duration X_n of the previous state. By comparison, in the general

HSMM, the Markovian assumption is on (J_n, X_{n+1}), that is, the current state and its duration.

A chain $(J_n, S_n)_n$ satisfying [1.2] is called a *Markov renewal chain*. To define the distribution of (J_n, X_n), it is necessary to define the quantity $\mathbb{P}(J_{n+1} = j, X_{n+1} = d \mid J_n = i)$, named the *semi-Markov kernel* of chain $(J_n)_n$, denoted $q_{ij}(d)$. This kernel is a key quantity from which other transition probabilities of interest can be computed, for instance, $\mathbb{P}(Z_t = j \mid Z_0 = i)$. Using Bayes theorem, we can see that the kernel can be decomposed in a product of two terms:

$$q_{ij}(d) = p_{ij} f_{ij}(d), \qquad [1.4]$$

where p_{ij} is defined by:

$$p_{ij} = \mathbb{P}(J_{n+1} = j \,|J_n = i),$$

and $f_{ij}(d)$ is the *conditional distribution of* X_{n+1}, defined by:

$$f_{ij}(d) = \mathbb{P}(X_{n+1} = d \,|J_n = i, J_{n+1} = j).$$

Note here that there is a conditioning on a future event, which is not always easy to interpret or model in applications.

Based on this decomposition of the kernel, the graphical representation of the conditional independencies in the Markov chain (J_n, X_n) is shown in Figure 1.3. We can see from this figure that (J_n, X_{n+1}) is Markovian, so this model is in the class of general SMM. Furthermore, assumption H2 is satisfied so it is actually in the class of standard SMM. Stated otherwise and borrowing from concepts of graphical models, the moralized version of the graph in Figures 1.2 and 1.3 are the same, therefore they define the same set of conditional independence assumptions (Koller and Friedman 2009).

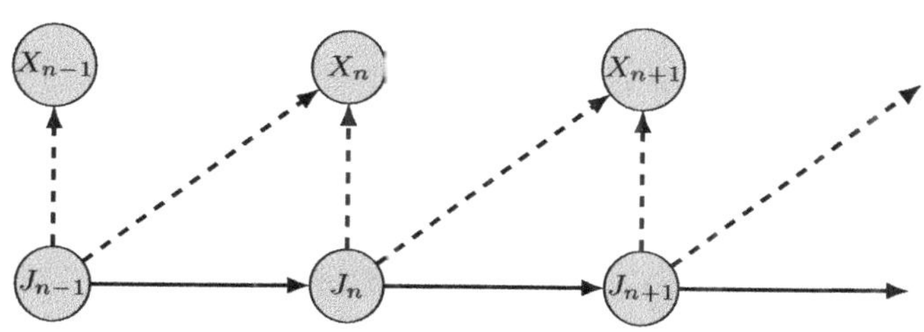

Figure 1.3. *Graphical representation of the conditional independencies in the chain (J_n, X_n) of a SMM defined by the kernel $q_{ij}(d) = p_{ij} f_{ij}(d)$. Dotted arcs correspond to the conditional distribution $f_{ij}(d)$ and solid ones to the transition p_{ij}*

Regarding now the joint emission distribution, in Barbu and Limnios (2008), assumption H1 is considered, together with more general situations where Y_t depends on the k previous observations.

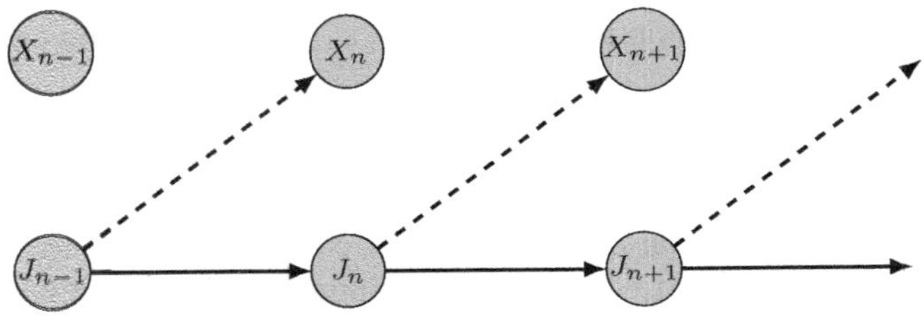

Figure 1.4. *Graphical representation of the conditional independence relationships in the chain $(J_n, X_{n+1})_n$ for the ED-MM. Dotted arcs correspond to the sojourn duration distribution $h_i(d)$ and solid ones to the transition p_{ij}*

1.2.4. *Explicit duration HMM*

The general HSMM is seldom used in applications because of its complexity. The intermediate model of the standard HSMM is sometimes considered but in many applications the explicit duration (ED-HMM) that we present now is chosen. This model is referred to as the variable duration HMM in Ferguson (1980), where it was first mentioned. It is also often referred to as the ED-HSMM (Yu 2016), since the hidden chain is a semi-Markov chain and not a Markov chain. The name ED-HMM is linked to the fact that this model was first introduced as an extension of a HMM where the distribution of the sojourn duration is explicitly modeled. This is the model we will use for the two toy examples of the book.

Starting from the standard HSMM and the associated decomposition of a in equation [1.1], the following additional assumption is made:

H3: J_n depends only on J_{n-1} and not on X_n. We obtain:

$$a_{(i,d),(j,l)} = \mathbb{P}(X_{n+1} = l | J_n = j) \times \mathbb{P}(J_n = j | J_{n-1} = i) = h_j(l) p_{ij}.$$

The graphical representation of the corresponding conditional independence relationships is shown in Figure 1.4.

DEFINITION 1.5.– *The chain $(Z_t)_{t\in\mathbb{N}}$, indexed by calendar time, is an explicit duration Markov model (ED-MM) if the associated chain $(J_n, X_{n+1})_{n\in\mathbb{N}}$ indexed by jump times is Markovian and hypotheses H2 and H3 are satisfied.*

DEFINITION 1.6.– *The chain $(Z_t, Y_t)_{t\in\mathbb{N}}$ is an ED-HMM if (i) $(Z_t)_{t\in\mathbb{N}}$ is an ED-MM, and (ii) conditionally to $Z_{[t+1:t+d]} = i$, the sequence of observations $Y_{t+1:t+d}$ is independent of the other observations and the other hidden states.*

1.2.5. *HMM*

A classical definition of an Hidden Markov Model (MacDonald and Zucchini 1997) is as follows.

DEFINITION 1.7.– *The chain $(Y_t)_{t\in\mathbb{N}}$ is a HMM if there exists some unobserved process $Z = (Z_t)_{t\in\mathbb{N}}$ such that the process Z is a Markov chain and, given Z, the observations $(Y_t)_{t\in\mathbb{N}}$ are mutually independent and Y_t depends on Z_t only.*

The three quantities necessary to define a HMM are the initial probabilities $\pi_i = \mathbb{P}(Z_0 = i)$, the transition probabilities $a_{ij} = \mathbb{P}(Z_t = j|Z_{t-1} = i)$ and the emission probabilities $b_{iy} = \mathbb{P}(Y_t = y|Z_t = i)$.

HMM have been defined before HSMM, and they are actually a particular case of ED-HMM. Indeed, in a HMM, the distribution of the sojourn duration in state i is a geometric distribution with parameter a_{ii}, which depends only on the current state. At the end of the sojourn duration, the next state j is chosen according to the a_{ij}s, which depend only on the current state i. Assumptions H2 and H3 are satisfied.

1.3. Inferential topics for HSMMs

In HSMM like in any statistical model, parameters can be considered either as deterministic (frequentist approach) or random (Bayesian approach). The frequentist approach mainly relies on the maximization of the likelihood function. In the Bayesian approach, the target function is the posterior distribution of the parameters, which depends not only on the likelihood function but also on the prior and the normalization constant. Bayesian models and estimation algorithms have been proposed in the case of ED-HMMs (Johnson and Willsky 2013; Economou et al. 2014; Hadj-Amar et al. 2023). In this chapter, we focus on the frequentist approach and the method of MLE. This is the most common estimation method for HSMMs, as demonstrated by the set of available software described in Chapter 2. Note that besides MLE, spectral methods (Melnyk and Banerjee 2017) have also been proposed for general HSMM. Moment-based methods have been developed for HMMs (Anandkumar et al. 2012) and more recently have been extended to HSMMs (Melnyk and Banerjee 2017).

In the frequentist context, the three main computational tasks to be tackled when working with HSMMs are as follows:

1) *Likelihood evaluation*: this task consists of computing the likelihood, for a given set of observations and parameters. The parameters of a model can be estimated by means of the likelihood function. Moreover, different models can be compared by means of information criteria or statistical tests that are based on the likelihood. Such criteria include the AIC (Akaike 1974) and the BIC (Schwarz 1978), which are composed of the likelihood evaluated at the MLE and a penalty term to control the

complexity. Likelihood ratio tests can also be considered. They have been introduced for SMMs by Votsi et al. (2021); Cardot and Frascolla (2024) and HMMs (e.g. Giudici et al. 2000) but not yet for HSMMs. In the context of HSMMs, the likelihood can be evaluated by means of the forward-backward algorithm.

2) *Likelihood maximization for parameter estimation*: for a given set of observations, and in order to estimate the parameters of a model, we need to maximize the likelihood function. In the context of HSMMs, this is done by using the EM algorithm Dempster et al. (1977). This algorithm is based on the complete likelihood. It computes parameter estimates that are updated iteratively and generally converge toward a local maximum of the likelihood.

3) *State restoration*: this task is also referred to as state decoding. Given an observation sequence and parameters values, the aim is generally to find the most probable state sequence, that is, the state sequence that is most probable to have given rise to the observations. An alternative is to perform the restoration independently at each time step.

In the following, in section 1.3.1 we define the likelihood and present the classical algorithm for likelihood computation, *the forward-backward algorithm*. Then, in section 1.3.2 we recall the main elements on MLE: asymptotic properties and efficient computation with the EM algorithm. In section 1.3.4, we explain how to compute smoothing and filtering probabilities for a HSMM. Finally, in section 1.3.5, we describe the two main methods for state restoration. For the sake of simplicity, unless otherwise stated, the definitions and algorithms are presented for a standard or an ED-HMM (even if they exists also for general HSMM, see Yu (2016)), and in the case where the hidden sequence enters a new state at $t = 0$ and the last state visited ends at time T, end of the observation period (absence of left and right censoring). Some comments are provided to explain which changes are to be brought when considering either censoring (meaning that the observed processes may have continued after the observer stopped recording it) or more general HSMMs.

1.3.1. *Likelihood evaluation*

In this section, we define the likelihood function for standard HMMs and ED-HMMs and we explain how to evaluate it. Extension to general HSMMs is rather straightforward, at the cost of larger computational complexity and more tedious notations.

1.3.1.1. *Likelihood function*

For the standard HSMM, in a non-parametric context (or with canonical parameterization), we denote θ as the union of $(p_{ij})_{i,j\in\Omega_Z,i\neq j}$, $(f_{ij}(d))_{i,j\in\Omega_Z,i\neq j,d\in\mathbb{N}^*}$ and $(b_{ia})_{i\in\Omega_Z,a\in\Omega_Y}$. Given a sample path of observations

$y_{0:T}$, and the associated hidden chain $Z_{0:T}$, the complete likelihood function is defined by:

$$\mathcal{L}(Z_{0:T}, y_{0:T}; \theta) = \pi_{Z_0} \left(\prod_{n=1}^{N(T)} p_{J_{n-1}J_n} f_{J_{n-1}J_n}(X_n) \right) \left(\prod_{t=0}^{T} b_{Z_t y_t} \right) \overline{H}_{J_{N(T)}}(E_T).$$

In this expression, X_n and J_n are derived from $Z_{0:T}$. The quantity $\overline{H}_i(d)$ corresponds to the survival function in state i:

$$\overline{H}_i(d) = \mathbb{P}(X_{n+1} \geq d | J_n = i).$$

In the expression of the likelihood, the survival function is evaluated at E_T, the elapsed time in the last visited state: $E_t = \min\{d \in \mathbb{N}^*, Z_{t-d} \neq Z_t\}$.

For the ED-HMM, the complete likelihood becomes:

$$\mathcal{L}(Z_{0:T}, y_{0:T}; \theta) = \pi_{Z_0} \left(\prod_{n=1}^{N(T)} p_{J_{n-1}J_n} h_{J_n}(X_n) \right) \left(\prod_{t=0}^{T} b_{Z_t y_t} \right) \overline{H}_{J_{N(T)}}(E_T).$$

To obtain the likelihood function, we need to sum over the hidden states sequence, that is:

$$\mathcal{L}(y_{0:T}; \theta) = \mathbb{P}_\theta(y_{0:T}) = \sum_{z_0} \cdots \sum_{z_T} \mathcal{L}(z_{0:T}, y_{0:T}; \theta). \qquad [1.5]$$

In a parametric context, $q_{ij}(.)$, p_{ij} and b_{zy} depend on some parameters, but the expression of the likelihood remains the same.

1.3.1.2. *Likelihood computation with the forward-backward recursions*

The likelihood $\mathcal{L}(y_{0:T}; \theta)$ cannot be computed directly from expression [1.5], since it requires the summation over K^{T+1} terms. However there exists an algorithm with quadratic time complexity that allows the evaluation of the likelihood, known as *forward-backward recursion*.

For the ED-HMM and for a fixed value of θ, the forward and backward quantities are the following. For $1 \leq j \leq K$ set:

$$\alpha^t(j) = \mathbb{P}_\theta(Z_t = j, Z_{t+1} \neq j, y_{0:t}), \quad \forall t = 0, \ldots, T,$$

$$\beta^t(j) = \mathbb{P}_\theta(y_{t+1:T} \mid Z_t = j, Z_{t+1} \neq j), \quad \forall t = T-1, \ldots, 0.$$

For α, the recursive formulas for $1 \leq j \leq K$ are (Yu 2016):

$$\alpha^t(j) = \sum_{d=1}^{+\infty} \alpha_*^{t-d+1}(j) h_j(d) u^t(j,d), \quad \text{for } t = 0, \ldots T, \qquad [1.6]$$

with

$$\alpha_*^{t+1}(j) = \mathbb{P}_\theta(Z_t \neq j, Z_{t+1} = j, y_{0:t}) = \sum_{i \neq j} \alpha^t(i) p_{i,j}, \quad \text{for } t = 0, \ldots T \qquad [1.7]$$

and

$$u^t(j,d) = \prod_{\tau=t-d+1}^{t} b_{j y_\tau}.$$

This is the joint probability of all observations available between time $t-d+1$ and t given that hidden state is j during this period. We assume that the hidden chain enters a new state at $t = 0$ and we initialize the recursion with $\alpha_*^0(j) = \mathbb{P}(Z_{-1} \neq j, Z_0 = j)$ which is equal to the initial distribution π_j and $\alpha_*^\tau(j) = 0$ for $\tau < 0$.

For β and for $1 \leq j \leq K$, we have:

$$\beta^t(j) = \sum_{j<i} p_{i,j} \beta_*^{t+1}(i), \qquad [1.8]$$

with

$$\beta_*^{t+1}(j) = \mathbb{P}_\theta(y_{t+1:T} \mid Z_t \neq j, Z_{t+1} = j) = \sum_{d \in \mathbb{N}^*} h_j(d) u^{t+d}(j,d) \beta^{t+d}(j),$$

for $t = T-1, \ldots, 0$.

We assume that the hidden chain enters a new state at $T+1$ and we initialize the recursion with $\beta^T(j) = 1$ and $\beta^\tau(j) = 0$ for $\tau > T$.

Two important probabilities can be derived from the α's and β's: (i) the joint probability that the state is j at time t and the observation sequence is $y_{0:T}$ and (ii) the likelihood.

The former is equal to:

$$\gamma^t(j) = \mathbb{P}_\theta(Z_t = j, y_{0:T}),$$

and can be computed recursively (Yu 2016, chapter 5, p. 106) through the formula:

$$\forall t > 0, \ \gamma^t(j) = \gamma^{t-1}(j) + \alpha_*^t(j)\beta_*^t(j) - \alpha^{t-1}(j)\beta^{t-1}(j). \qquad [1.9]$$

The recursion is initialized with:

$$\gamma^0(j) = \pi_j \beta_*^0(j).$$

Then the likelihood can be expressed in terms of γ's as follows:

$$\mathcal{L}(y_{0:T};\theta) = \sum_{k=1}^{K} \mathbb{P}_\theta(Z_t = k, y_{0:T}) = \sum_{k=1}^{K} \gamma^t(k). \qquad [1.10]$$

Note that other quantities can be computed from the α's and β' like:

$$\eta^t(i,d) = \mathbb{P}_\theta(Z_{[t-d+1:t]} = i, y_{0:T}) = \alpha_*^{t-d+1}(i) h_i(d) u^t(i,d) \beta^t(i)$$

or

$$\xi_t(i,j) = \mathbb{P}_\theta(Z_{t]} = i, Z_{[t+1} = j, y_{0:T}) = \alpha^t(i)\beta_*^{t+1}(j) p_{ij}.$$

1.3.2. *Asymptotic properties of the MLE*

We denote by Θ the parameter space for the set of parameters θ of the standard HSMM. The MLE of θ is defined by:

$$\widehat{\theta}(T) = \arg\max_{\theta\in\Theta} \mathcal{L}(y_{0:T};\theta).$$

To the best of our knowledge, the asymptotic properties of the MLE for a standard HSMM have been studied for the first time in Barbu and Limnios (2006, 2008) in a large sample setting. In particular, the MLE of the parameter vector θ has been shown to be strongly consistent and asymptotically normal under some regularity conditions. We first denote by θ_0 the true parameter vector, which means the parameter vector that has generated the sample path $y_{0:T}$. We further define the observed Fisher information matrix evaluated in θ_0 by:

$$\mathrm{I}_T(\theta_0) = -\mathbb{E}_{\theta_0}\left(\frac{\partial^2 \log \mathcal{L}(y_{0:T};\theta)}{\partial\theta_u \partial\theta_v}\right)_{u,v},$$

where θ_u, θ_v are the uth, vth components of θ.

To show the asymptotic properties of the MLE, we need to assume that the following conditions hold:

– A1: the conditional sojourn duration distributions $f_{ij}(.)$ have finite support $\mathcal{D} = \{1, \ldots, \mathrm{D}\} \subset \mathbb{N}$;

– A2: the semi-Markov chain is irreducible;

– A3: there exists an integer $T \in \mathbb{N}$ such that the matrix $\mathrm{I}_T(\theta_0)$ is non-singular.

Under the conditions A1 and A2, and according to Barbu and Limnios (2006), the MLE $(\widehat{q}_{ij}(d,T))_{i,j\in\Omega_Z,i\neq j,d\in\mathrm{D}}$ and $\left(\widehat{b}_{ia}(T)\right)_{i\in\Omega_Z,a\in\Omega_Y}$ of $\left(q^0_{ij}(d)\right)_{i,j\in\Omega_Z,i\neq j,d\in\mathrm{D}}$ and $\left(b^0_{ia}\right)_{i\in\Omega_Z,a\in\Omega_Y}$, respectively, are strongly consistent. Moreover, under conditions A1–A3, the random vectors $\sqrt{T}\left(\widehat{q}_{ij}(d,T)-q^0_{ij}(d)\right)_{i,j\in\Omega_Z,i\neq j,d\in\mathrm{D}}$ and $\sqrt{T}\left(\widehat{b}_{ia}(T)-b_{ia}\right)\Big)_{i\in\Omega_Z,a\in\Omega_Y}$ are asymptotically normal (Barbu and Limnios 2006).

Later on, in Trevezas and Limnios (2009), the MLE was studied for HSMMs in the case where the state space of the observations is a subset of a Euclidean space. In this work, the conditional sojourn duration distributions are considered to belong to a specific parametric family. Moreover, an observation Y_t conditioned on the underlying semi-Markov chain depends probabilistically not only on the hidden state Z_t but also on E_t, the time elapsed since entering the last visited state. This dependency enables the transformation of the general HSMM into a general HMM following Bickel and Ritov (1996) and Leroux (1992). Under some model class assumptions and some regularity and identifiability conditions, the MLE of the parameters was shown to be strongly consistent and asymptotically normal.

1.3.3. *EM algorithm*

1.3.3.1. *Principle*

The EM algorithm, rather than a specific algorithm, is strictly speaking a family of algorithms based on a common principle. It is dedicated to maximum likelihood estimation in models whose likelihood is defined through a marginalization operation on latent variables. Let $\boldsymbol{Z}, \boldsymbol{z}, \boldsymbol{Y}$ and $\boldsymbol{y}$ denote, respectively, latent (or hidden) and observed variables and a possible realization of each. $\mathcal{L}(\boldsymbol{y};\theta) = \mathbb{P}_\theta(\boldsymbol{y})$ denote the likelihood, then:

$$\mathbb{P}_\theta(\boldsymbol{y}) = \sum_{\boldsymbol{z}} \mathbb{P}_\theta(\boldsymbol{z},\boldsymbol{y}).$$

Although the notation applies to discrete-valued $\boldsymbol{z}$, this principle extends to continuous or mixed latent variables, which can be of arbitrary number. The EM algorithm is particularly adequate whenever:

$$\arg\max_\theta \mathbb{P}_\theta(\boldsymbol{z},\boldsymbol{y}), \text{ or equivalently } \arg\max_\theta \ln \mathbb{P}_\theta(\boldsymbol{z},\boldsymbol{y})$$

has a closed form or can be computed by a straightforward routine. Usually, this property will generally not apply to:

$$\arg\max_\theta \ln \mathbb{P}_\theta(\boldsymbol{y}). \qquad [1.11]$$

In this case, the EM algorithm can be obtained by replacing [1.11] with:

$$\arg\max_{\theta} \mathbb{E}[\ln \mathbb{P}_{\theta}(\boldsymbol{Z}, \boldsymbol{Y}) | \boldsymbol{Y} = \boldsymbol{y}], \tag{1.12}$$

where by definition:

$$\mathbb{E}[\ln \mathbb{P}_{\theta}(\boldsymbol{Z}, \boldsymbol{Y}) | \boldsymbol{Y} = \boldsymbol{y}] = \sum_{\boldsymbol{z}} \ln \mathbb{P}_{\theta}(\boldsymbol{z}, \boldsymbol{y}) p(\boldsymbol{z} | \boldsymbol{Y} = \boldsymbol{y}).$$

The role of θ in the equation above is quite clear: this is the dummy variable with respect to what is to be optimized. However, the definition of $p(\boldsymbol{z}|\boldsymbol{Y} = \boldsymbol{y})$ is less clear, since it also has to depend on a parameter, but which one: also θ, or the *true* θ, or another θ'? The *true* θ is not necessarily a well-defined concept and its value is not accessible in any case, moreover it can be shown that using the same θ in $\mathbb{P}_{\theta}(\boldsymbol{z}|\boldsymbol{Y} = \boldsymbol{y})$ and $\mathbb{P}_{\theta}(\boldsymbol{z}, \boldsymbol{y})$ does not reduce the complexity of the problem. In the EM algorithm, the expectation is under a current, fixed value $\theta^{(m)}$ of the parameter:

$$Q(\theta, \theta^{(m)}) = \mathbb{E}_{\theta^{(m)}}[\ln \mathbb{P}_{\theta}(\boldsymbol{Z}, \boldsymbol{Y}) | \boldsymbol{Y} = \boldsymbol{y}], \tag{1.13}$$

and the principle is to iterate by updating this parameter using:

$$\theta^{(m+1)} = \arg\max_{\theta} Q(\theta, \theta^{(m)}). \tag{1.14}$$

The name EM is a shorthand for expectation-minimization, and it comes from the two steps defining one iteration:

1) Expectation step (E step): computing the conditional expectations involved in [1.13]. This is performed via the computation of the α and β using the forward-backward recursions presented above. The required posterior marginals can be expressed in terms of these two intermediate quantities.

2) Maximization step (M step): updating the set of parameters θ thanks to the quantities found in the E Step by resolving $\theta^{(m+1)} = \arg\max_{\theta} Q(\theta, \theta^{(m)})$.

These two steps are repeated until the algorithm converges. The main generic property of the EM algorithm is that the produced sequence $(\theta^{(m)})_{m \in \mathbb{N}}$ has non-decreasing likelihoods. In addition, it converges toward a stationary point of the log-likelihood (i.e. a local maximum, a local minimum or a saddle point). Even if it is counterintuitive, choices of very specific initial points may lead to saddle or even local minimum stationary points. However, in practice, EM algorithm converges to a local maximum of the log-likelihood function (see Koller and Friedman 2009). To minimize the risk of being trapped in a local maximum, the EM algorithm can be repeated with different initial points.

1.3.3.2. *Literature*

We do not develop here the formulas of the E step and M step for a HSMM (but we provide them for a HMM in Appendix 1.10). The complexity in deriving these formulas lies in the expression of the forward-backward update formulas. They are available in detailed form in several state-of-the-art books or articles. For the reader who is interested in implementing EM for a particular HSMM, we summarize here three main resources that provide the formulas for the general HSMM, the standard HSMM and the ED-HMM.

For the general HSMM, we refer the reader to Yu (2016) where the formulas of the forward-backward algorithm are provided in Chapter 2 under different assumptions (chapters in this paragraph refer to chapters in Yu (2016)). In particular, the right-censored assumption, classically used in HSMMs, is made: $(Z_t)_t$ enters a new state at $t = 0$ and the last state ends at T or after. The sojourn duration is a bounded discrete variable. Time complexity is $\mathcal{O}(TD^2|\Omega_Z|^2)$. The model is non-parametric and the M step update formulas for the general are presented in Chapter 3. The case of the ED-HMM is also presented in Chapter 5. Finally, practical problems (typically underflow) that can occur when implementing EM for HSMMs, with solutions, are discussed in Chapter 4.

The standard HSMM is studied in Barbu and Limnios (2008). The formulas of the forward-backward algorithm are provided under the right censored assumption and for a sojourn duration variable that can be unbounded. Time complexity is cubic in T. The EM algorithm and a respective stochastic version have been proposed for the standard HSMM in Malefaki et al. (2009), in the case where the duration distributions are attached to transitions and not to states.

In Guédon (2003), the author considers the ED-HMM (with possibility of absorbing states). Sojourn duration can take only a finite number of values, and the observation is categorical. The formulas for the forward-backward algorithm are provided under the right-censored assumption. Time complexity is quadratic in the length of the sequence T and in $|\Omega_Z|$ in the worst case and space complexity is linear in T and in $|\Omega_Z|$.

As we will see in Chapter 2, several packages propose parameter estimation with EM for HSMMs, but in the restricted framework of ED-HMMs only. There are standard tools now, but the user has to be aware that such algorithms may not be suited for very long sequences due to the quadratic complexity. The three above-mentioned references also show us that there are two aspects to handle with care when implementing EM for HSMMs. The first one is the choice of the boundary conditions. In some applications, a left and right censored assumption may be the most realistic, and this choice has an impact on the expression of the initial conditions in the forward-backward algorithm. The second one is the hypothesis of unbounded sojourn duration. In this case, infinite sums appear in the EM formulas.

One solution is to truncate the sum at time T. However, it may be too coarse for some sojourn duration distributions and truncation can be made at some value t^* such that the probability of a duration value larger than t^* is reasonably low.

1.3.4. *Smoothing and filtering probabilities*

Smoothing and filtering consist of predicting states at each time step t. The difference between both is that smoothing is performed once the whole sequence is observed, while filtering relies on the only observation of $y_{0:t}$.

1.3.4.1. *Smoothing*

Smoothing probabilities correspond to the probability of variable Z_t conditionally to all the observations $y_{0:T}$:

$$\xi_t(z_t) = \mathbb{P}(Z_t = z_t | Y_{0:T} = y_{0:T}).$$

For an ED-HMM, these quantities can be computed easily once the γ^t have been computed using the α_t and β_t provided by the forward-backward algorithm (see section 1.3.1.2):

$$\xi_t(z_t) = \frac{\mathbb{P}(Z_t = z_t, Y_{0:T} = y_{0:T})}{\sum_i \mathbb{P}(Z_t = i | Y_{0:T} = y_{0:T})} = \frac{\gamma^t(z_t)}{\sum_i \gamma^t(i)}.$$

1.3.4.2. *Filtering*

Filtering probabilities are formally defined as $\mathbb{P}(Z_t = z_t | Y_{0:t} = y_{0:t})$. These are obtained as a by-product of the forward recursion only. For any time t, let E_t denote the elapsed time in current state Z_t since the last jump (including current time t), and R_t denote the remaining time until next jump ($E_t + R_t$ is equal to the sojourn duration). For an ED-HMM, we have for every possible state value z_t:

$$\mathbb{P}(Z_t = z_t, Y_{0:t} = y_{0:t}) = \sum_{k=1}^{t+1} \sum_{d=0}^{\infty} \mathbb{P}(Z_t = z_t, E_t = k, R_t = d, Y_{0:t} = y_{0:t}).$$

Since $\{Z_t = z_t, E_t = k, R_t = d\} \Leftrightarrow \{Z_t = z_t, X_{N(t)+1} = d+k, Z_{t-k]} \neq z_t\} \Leftrightarrow \{Z_{[t-k+1} = z_t, X_{N(t-k+1)+1} = d + k, Z_{t-k]} \neq z_t\}$, we have

$$\mathbb{P}(Z_t = z_t, Y_{0:t} = y_{0:t}) = \sum_{k=1}^{t+1} \sum_{d=0}^{\infty} \mathbb{P}(Z_t = z_t, X_{N(t-k+1)+1}$$

$$= d + k, Z_{t-k]} \neq z_t, Y_{0:t} = y_{0:t})$$

$$= \sum_{k=1}^{t+1} \sum_{d=0}^{\infty} \sum_{z_{t-1} \in \Omega_Z \backslash z_t} \mathbb{P}(Z_t = z_t, X_{N(t-k+1)+1} = d+k, Z_{t-k]} = z_{t-1}, Y_{0:t} = y_{0:t})$$

$$= \sum_{k=1}^{t+1} \sum_{d=0}^{\infty} \sum_{z_{t-1} \in \Omega_Z \backslash z_t} \mathbb{P}(Y_{t-k+1:t} = y_{t-k+1:t}, Z_{[t-k+1:t} = z_t, X_{N(t-k+1)+1}$$

$$= d+k \mid Z_{t-k]} = z_{t-1}) \mathbb{P}(Z_{t-k]} = z_{t-1}, Y_{0:t-k} = y_{0:t-k})$$

$$= \sum_{k=1}^{t+1} \sum_{d=0}^{\infty} \sum_{z_{t-1} \in \Omega_Z \backslash z_t} \mathbb{P}(Y_{t-k+1:t} = y_{t-k+1:t} \mid Z_{[t-k+1:t} = z_t) \mathbb{P}(X_{N(t-k+1)+1}$$

$$= d+k \mid Z_{[t-k+1} = z_t) \mathbb{P}(Z_{[t-k+1} = z_t \mid Z_{t-k]} = z_{t-1})$$

$$\times \mathbb{P}(Z_{t-k]} = z_{t-1}, Y_{0:t-k} = y_{0:t-k})$$

$$= \left(\prod_{u=t-k+1}^{t} b_{z_t y_u} \right) \sum_{z_{t-1} \in \Omega_Z \backslash z_t} a_{z_{t-1} z_t} \sum_{k=1}^{t+1} \alpha^{t-k}(z_{t-1}) \sum_{d=0}^{\infty} h_{z_t}(d+k),$$

where the infinite sum actually is $1 - \mathbb{P}(X_{N(t)+1} < k \mid J_{N(t)} = z_t)$. Eventually,

$$\mathbb{P}(Z_t = z_t \mid Y_{0:t} = y_{0:t}) = \frac{\mathbb{P}(Z_t = z_t, Y_{0:t} = y_{0:t})}{\sum_i \mathbb{P}(Z_t = i, Y_{0:t} = y_{0:t})}.$$

A different definition of filtering probabilities is provided in Yu (2016):

$$\mathbb{P}(Z_{t]} = z_t | Y_{0:t} = y_{0:t}) = \frac{\mathbb{P}(Z_{t]} = z_t, Y_{0:t} = y_{0:t})}{P(Y_{0:t} = y_{0:t})}$$

$$= \frac{\alpha^t(z_t)}{\mathbb{P}(Y_{0:t} = y_{0:t})},$$

where $\mathbb{P}(Y_{0:t} = y_{0:t}) = \sum_{i=1}^{K} \gamma^t(i)$.

1.3.5. *State restoration*

Two approaches are usually used for state restoration. The first one restores states at each time step t separately by maximizing the smoothing probabilities:

$$\hat{z}_t = \arg \max_{z_t \in \Omega_Z} \mathbb{P}(Z_t = z_t | Y_{0:T} = y_{0:T})$$

$$= \arg \max_{z_t \in \Omega_Z} \mathbb{P}(Z_t = z_t, Y_{0:T} = y_{0:T}) = \arg \max_{z_t \in \Omega_Z} \xi_t(z_t)$$

It has the advantage of maximizing the average number of correct states and of directly using the forward-backward algorithm (see section 1.3.1.2). It has the drawback of producing sequences $\hat{z}_{0:T}$ that may have null joint probability, since constraints due to possibly null transition probabilities are not directly accounted for in the criterion.

The second approach seeks the most probable hidden states sequence knowing the sequence of observations. It relies on a joint maximization:

$$\hat{z}_{0:T} = \arg \max_{z_{0:T} \in \Omega_Z^{T+1}} \mathbb{P}(Z_{0:T} = z_{0:T} | Y_{0:T} = y_{0:T}). \qquad [1.15]$$

This joint maximization is achieved by means of the Viterbi algorithm, as described in the next section.

1.3.5.1. *Viterbi algorithm*

The Viterbi algorithm, provided in Yu (2016), proceeds with two recursions. Under the same assumptions as in section 1.3.1, the forward recursion relies on:

$$\delta_t(j, x) = \max_{z_{0:t-x}} \mathbb{P}(Z_{0:t} = z_{0:t}, Z_{[t-x+1:t]} = j, y_{0:t}),$$

which satisfies

$$\delta_t(j, x) = \max_{i \in \Omega_Z \setminus \{j\}, x' \in \mathbb{N}^*} \left[\delta_{t-x}(i, x') a_{ij} h_j(x') b_{jt}(y_{t-x+1:t})\right], \qquad [1.16]$$

reminding that $b_{jt}(y_{t-x+1:t}) = \prod_{\tau=t-x+1}^{t} b_{j,y_\tau}$. By denoting:

$$\psi_t(t, j, x) = (t - x, \arg \max_{i \in \Omega_Z \setminus \{j\}, x' \in \mathbb{N}^*} [\delta_{t-x}(i, x') a_{ij} h_j(x') b_{jt}(y_{t-x+1:t})]),$$

the maximal probability P^* and the whole optimal sequence can be retrieved by the following backward recursion:

$$P^* = \max_{t \geq T, i \in \Omega_Z, x \geq t-T+1} \delta_t(i, x)$$

$$(t_1, j_1^*, x_1^*) = \arg \max_{t \geq T, i \in \Omega_Z, x \geq t-T+1} \delta_t(i, x)$$

$$\forall n,\ (t_n, j_n^*, x_n^*) = \psi(t_{n-1}, j_{n-1}^*, x_{n-1}^*).$$

This reconstructs the optimal backward sequence with the (J_n, X_n) representation of states, meaning that j_n^* is the first sojourn duration, x_n^* the first state, j_1^* is the last sojourn duration and x_1^* the last state. The sequence of states $\hat{z}_{0:T}$ is deduced from these quantities. In the Viterbi algorithm, the optimization problem [1.15] is solved by recursively breaking the problem into subproblems, which are solved by maximizing over just two variables, j_n and x_n. Then solutions are combined to produce the globally optimal solution. This kind of approach to solve optimization problems is referred to as *dynamic programming*. Up to some differences in indexing variables, the forward recursion of the Viterbi algorithm [1.16] is formally the same as the forward recursion for likelihood computation in section 1.3.1, where sums are replaced by maximizations. As a result, the time complexity of both algorithms is the same, $\mathcal{O}(TD^2|\Omega_Z|^2)$. Usually, implementations rely on the logarithm of [1.16] to avoid underflow.

An algorithm with lower complexity was provided by Pertsinidou and Limnios (2015) with time complexity $\mathcal{O}(T|\Omega_z|^2)$, which is the same as in HMMs. Their model is the standard HSMM defined in section 1.2.3. Their approach is based on a dynamic programming approach by maximizing sequentially on states $z_0, \ldots, z_T$ and thus avoids the double maximization on j and x in [1.16].

1.4. Two toy examples reappearing throughout the book

In this section, we complete the description of the Squirrel toy example introduced in section 1.2.1. The Squirrel model is an ED-HMM and we provide here numerical values for the probabilities defining it (section 1.4.1). They will be exploited in Chapter 2 to illustrate in practice parameter estimation with the EM algorithm.

A second toy example is introduced, with an ED-HMM specification linking deers behavior with accelerometry data (section 1.4.2). This second example allows us to present a model with real-valued observations, and will also be exploited in Chapter 2.

More generally, these two toy examples will be present throughout the book's chapters to illustrate the different notions introduced.

1.4.1. *Squirrel toy example*

We propose here concrete values for the conditional probabilities defining the ED-HMM of the Squirrel toy example: p_{ij}, $h_i(d)$ and b_{iy}. We consider $K = 3$ reserves and a maximum of four time steps for the sojourn duration in a reserve. The transition matrix, which terms are the p_{ij}, is given as:

$$\begin{pmatrix} 0 & 0.5 & 0.5 \\ 0.3 & 0 & 0.7 \\ 0.4 & 0.6 & 0 \end{pmatrix}.$$

The three sojourn duration distributions $h_i(d)$ are given as:

$$\begin{pmatrix} 0.1 & 0.005 & 0.005 & 0.89 \\ 0.1 & 0.005 & 0.89 & 0.005 \\ 0.1 & 0.89 & 0.005 & 0.005 \end{pmatrix}$$

and the emission matrix, which terms are the b_{iy}, is:

$$\begin{pmatrix} 0.8 & 0.1 & 0.1 \\ 0.1 & 0.8 & 0.1 \\ 0.1 & 0.1 & 0.8 \end{pmatrix}.$$

the meaning of which is given in Table 1.1.

In Chapter 2, a Jupyter Notebook is proposed that enables simulation of realizations of this model and estimation of the parameters from simulated observations.

Probability	**Name in notebook**
$\mathbb{P}(J_{n+1} = j \,\vert J_n = i)$	`hsmm_class.tmat`
$\mathbb{P}(X_{n+1} = d \,\vert J_n = j)$ for $d \geq 1$	`hsmm_class.dur`
$\mathbb{P}(Y_t = j \,\vert Z_t = i)$	`hsmm_class.emit`

Table 1.1. *Conditional probabilities for the Squirrel toy example and correspondence with the jupyter notebook*

1.4.2. *Deer toy example*

High-frequency sensor data (biologging), notably accelerometers, are used to infer the behavior of animals over long durations and in conditions that cannot be directly observed, allowing the study of individual behavorial tactics. These data are also useful to study potential differences in behaviors depending on the environmental conditions, like different levels of anthropization of the area. With HSMMs, we can use estimation and restoration to bring insight on these questions.

The Deer toy example we present now is inspired from the study of deers behavior. Accelerometers are installed in a collar on the animal neck. Accelerometer data are three-dimensional but are often transformed either into a scalar variable (e.g. Euclidean norm) or in a set of more interpretable variables. For the sake of simplicity, we will assume that the latter information is a scalar, and we have selected the variable *roll angle* (used to infer the tilt of the animal) since its distribution is contrasted between behaviors. We consider five different behaviors (foraging head-down, grooming, running, unmoving and walking head-up) and $\Omega_Z = \{1, \ldots, 5\}$. Variable Z_t is behavior at time t. Observation $Y_t \in \Omega_Y = \mathbb{R}$ is the scalar information extracted from the accelerometer at time t. The time step is

0.125 s, corresponding to an 8 Hz frequency. We propose here an ED-HMM to model the dynamics of (Z_t, Y_t) as follows. The distribution $h_i(d)$ of the sojourn duration in behavior i is Poisson with parameter λ_i. The emission distribution is Gaussian: b_{iy} is the density of a Gaussian distribution with mean m_i and variance σ_i^2 computed at value y.

In order to present realistic values of the p_{ij} and the other model parameters, we used an experimental data set (not available for diffusion, Benoit 2022) to estimate them. In this data set, the hidden chain (behaviors sequence) is available via annotated videos. The obtained transition matrix (the p_{ij}s) is given as:

$$\begin{pmatrix} 0.00 & 0.025 & 0.028 & 0.744 & 0.203 \\ 0.028 & 0.00 & 0.014 & 0.888 & 0.070 \\ 0.080 & 0.028 & 0.00 & 0.256 & 0.636 \\ 0.279 & 0.269 & 0.032 & 0.0 & 0.420 \\ 0.236 & 0.043 & 0.071 & 0.650 & 0.00 \end{pmatrix},$$

and the parameters for the sojourn duration distribution and the emission distribution are given in Table 1.2.

State i	λ_i	m_i	σ_i^2
1	98.02	20.41	12.04
2	67.42	-14.98	14.45
3	37.69	-12.88	9.01
4	103.64	-24.60	14.61
5	46.13	-4.32	8.61

Table 1.2. *Parameters for the sojourn duration distribution and the emission distribution for the deer ED-HMM model*

As for the Squirrel example, in Chapter 2, a Jupyter Notebook is proposed that enables simulation and illustrates parameter estimation from data simulated with this parameterization.

Note that a standard HSMM, where the next state depends not only on the current one but also on its duration, could be relevant for a modeling of more complex dynamics between the behaviors. For instance, if current state Z_t is walking head-up with a long sojourn duration, the most probable next behavior may be unmoving, while with a short sojourn duration the most probable behavior may be *foraging* (the animal is searching for more food elsewhere).

1.5. Reliability

Reliability indicators are crucial measures that can be used to evaluate operational performance and to manage the maintenance of systems that can be described by different stochastic models. Such indicators have been widely studied for semi-Markov processes and chains by Barbu and Limnios (2008); Limnios and Oprisan (2001). However, the study of such indicators in a hidden Markov and hidden semi-Markov context is rather limited (Gámiz et al. 2023; Votsi 2019; Votsi and Limnios 2015). The observations or both observations and hidden states could be partitioned into up and down states. When observations are considered to correspond (or not) to failures (down states), the failures are observed but still rely on the underlying, non-observable Markov chain. In another context, failures can correspond to both observations and hidden states (Gámiz et al. 2023). We summarize here the available results on reliability indicators for HSMMs.

1.5.1. *Rate of occurrence of failures*

One of the most important reliability indicators for random systems that can experience many failures (repairable) is the Rate of OCcurrences Of Failures (ROCOF). This indicator represents the mean number of failures of a system over time. It enables the identification of opportunities for improving reliability performance, which is of special interest for real-life applications. An analysis of potential failures could help practitioners focus on and understand the impact of potential process or product risks and failures.

Concerning HSMMs, ROCOF was first studied in Votsi and Limnios (2015), in a particular case where observations are made only at jump times. Each observation corresponds to the realization (or not) of a failure. In such a context, the observational space Ω_Y is partitioned into the subspaces D and U including failures (or down states) or operations (or up states), respectively. In other words, $\Omega_Y = U \cup D$.

DEFINITION 1.8.– *ROCOF is the probability of moving from an up observation to a down observation at time* $t \in \mathbb{N}^*$, *that is:*

$$r(t) = \mathbb{P}(Y_{t-1} \in U, Y_t \in D).$$

Let us define $E'_t = E_t - 1$, known as the backward recurrence time. The stochastic process $(Z_t, E'_t)_t$ is a homogeneous Markov chain governed by its initial distribution, $\tilde{\boldsymbol{\pi}} = (\tilde{\pi}(i,0); i \in \Omega_Z)$, where $\tilde{\pi}(i,0) = \mathbb{P}(Z_0 = i, E'_0 = 0)$, and its transition matrix $\tilde{\mathbf{P}} = (\tilde{p}_{(is_1)(js_2)}; (i,s_1),(j,s_2) \in \Omega_Z \times \mathbb{N})$, where:

$$\tilde{p}_{(is_1)(js_2)} = \begin{cases} q_{ij}(s_1+1)/\overline{H}_i(s_1+1), & \text{if} \quad s_2 = 0, \\ \overline{H}_i(s_1+2)/\overline{H}_i(s_1+1), & \text{if} \quad i = j, s_2 - s_1 = 1, \\ 0, & \text{elsewhere.} \end{cases}$$

Definitions of variable E_t and function $\overline{H}_i$ are given in section 1.3.1.1. An explicit formula to evaluate ROCOF at time $t \in \mathbb{N}^*$ in the case where observations are made at each time $t \in \mathbb{N}^*$ (Votsi 2019) is:

$$r(t) = \sum_{y_1 \in U} \sum_{y_2 \in D} \sum_{i \in \Omega_Z} \sum_{j \in \Omega_Z} \sum_{s_1 \in T_{t-1}} \sum_{s_2 \in T_t} b_{jy_2} \tilde{p}_{(is_1)(js_2)} b_{iy_1} (\tilde{\boldsymbol{\pi}} \tilde{\mathbf{P}})^{t-1}_{is_1},$$

with $T_t = \{0, \ldots, t\}$ and $(\tilde{\boldsymbol{\pi}}\tilde{\mathbf{P}})^{t-1}_{is_1}$ the $(i, s_1)-$th element of the vector $(\tilde{\boldsymbol{\pi}}\tilde{\mathbf{P}})^{t-1}$.

1.5.2. *Mean time to failure*

The mean time to failure (MTTF) is one of the most significant indicators in reliability theory, particularly in the study of non-repairable systems. This indicator quantifies the mean time during which a system operates without experiencing failures. Therefore, it serves as an indirect measure of the reliability of a system. Let us assume that the dates of observation and the jump times coincide, that is, observations are made only at jump times. Then the discrete-time process $\left(J_n, S_n, Y_n\right)_{n \in \mathbb{N}}$, is a *hidden Markov renewal chain* (Votsi and Limnios 2015, HMRC).

The partition of the transition matrix $\mathbf{P}$ of the HMRC into up and down states is given by:

$$\mathbf{P} = \begin{matrix} & U & D & \\ & \left(\begin{matrix} \mathbf{P}_{11} \\ \mathbf{P}_{21} \end{matrix}\right. & \left.\begin{matrix} \mathbf{P}_{12} \\ \mathbf{P}_{22} \end{matrix}\right) & \begin{matrix} U \\ D \end{matrix} \end{matrix}.$$

The elements of the transition kernel of the HMRC are given by:

$$P_{(i,t_1,r)(j,t_2,m)} = q_{ij}(t_2 - t_1) b_{jm},$$

for any $(i, t_1, r), (j, t_2, m) \in \Omega_Z \times \mathbb{N} \times \Omega_Y$.

DEFINITION 1.9.– *If we denote by* $\mathrm{T}_{(\mathrm{D})}$ *the first passage time in the subset D, that is:*

$$\mathrm{T}_{(\mathrm{D})} = \inf \{n : (J_n, S_n, Y_n) \in D\} \quad \textit{and} \quad \inf \emptyset = \infty,$$

then the MTTF is defined by:

$$MTTF = \mathbb{E}(\mathrm{T}_{(\mathrm{D})}).$$

It is expressed by:

$$MTTF = \boldsymbol{\pi}^U (\mathbf{I} - \mathbf{P}_{11})^{-1} \mathbf{1}^\top,$$

with $\mathbf{P}_{11}$ and π^U the transition matrix of the HMRC and its initial law restricted to U, and $\mathbf{1}^\top$ the column-vector of ones whose length is equal to the cardinality of U.

Let us illustrate the MTTF notion on the Squirrel toy example. A HSMM is attached to a squirrel, the hidden states of which are the locations of the reserves of hazelnuts that the squirrel visits. But a squirrel is a dizzy animal. It sometimes forgets where it has buried its hazelnuts and may make some errors in its wandering. The sets U and D are a partition of the patches, with real patches for the set U (patches with buried hazelnuts) and false patches for the state D (patches where no hazelnuts have been buried). One may be interested in estimating the mean time that the squirrel spends before visiting a false reserve (MTTF). As far as formalism with HMRC is concerned, patches in D are all the triplets (J_n, S_n, Y_n) for which J_n is a false reserve. On the other side, patches in U are all the triplets (J_n, S_n, Y_n) for which J_n is a patch where the squirrel has buried some hazelnuts.

1.6. Introducing mixed effects into HSMMs

This section addresses extensions of HSMMs obtained by introducing both fixed and random effects into their components, which all are probability distributions: initial state distribution, sojourn duration and emission distributions and transition matrix in the ED-HMM case. Since the literature in this field is somewhat rare, we also extend our focus to HMMs. In the previous sections, the elementary distributions used to define the model were assumed to be independent from any variable but states. State and observed variables are intrinsic to the HSMMs definition, as opposed to fixed and random effects, which are *extrinsic*.

Two cases are distinguished regarding extrinsic variables, depending on whether these are measured (covariates W) or hidden (random effects ξ). Models containing both covariates and random effects are referred to as mixed effects HSMMs or mixed HSMMs. The purpose of using covariates is generally straightforward: modelers are interested in assessing the impact of those covariates on the HSMM components, or sometimes in removing the impact of covariates to make experimental conditions comparable between different HSMM trajectories and compute corrected means (Chaubert-Pereira et al. 2009). The purpose of using random effects may just be to account for the impact of variables that cannot be directly measured, but that modelers would not want to ignore, since they induce dependencies within some groups of observations and some variability from one group to another. Examples are provided below.

The relationship between HSMM parameter and covariates involves specific parameters, which are referred to as fixed effects, and a design matrix used to associate subsets of parameters with subsets of covariates and observations. This principle also holds for random effects but to keep the presentation accessible for readers, we will only consider simple models that do not require to make design matrices explicit.

The difficulties associated with mixed HSMMs are twofold. First, modelers have to decide whether it is relevant or not to introduce them in all the four HSMMs components. This induces specific model selection issues. Second, estimation is made more difficult by mixed effects, especially the E step of the EM algorithm, due to continuous hidden variables (though the M step may already be turned intractable due to the presence of fixed effects).

In past publications, mixed HSMMs include the effects of W and ξ by resorting to the composition of a possibly nonlinear function with a linear function of some additional parameters, that is, added to the usual set of HSMM parameters. To make parameter inference more tractable, it is assumed that these functions ensure that the HSMM components belong to the exponential family of distributions. We rely on the two toy examples to describe step-by-step how to introduce mixed effects in the different components of a HSMM. We insist on the nature and interpretation of random variables and/or parameters introduced in such models. In both cases, the introduction of mixed HSMMs is motivated by practical needs emerging from accounting for the effects of covariates and for inter-individual variability. Section 1.6.1 starts with the Squirrel example. The Deer example, section 1.6.2, enables us to present the case of real-valued observations. This is followed by discussions on the case of dynamic covariates and on model selection issues (sections 1.6.3 and 1.6.4) and a review of the literature addressing this class of models (section 1.6.5).

1.6.1. *Mixed HSMMs explained with the Squirrel example*

In section 1.2.1, one single squirrel trajectory was considered. If several sequences are observed and modeled with HSMMs (e.g. when observing several squirrels acting independently), a convenient assumption is to consider the sequences as independent trajectories issued from the same HSMM. The possible limitations of this simplistic assumption are listed as follows:

– Modelers may know a priori that some measured extrinsic factors, which vary from one squirrel to another, have an effect on these trajectories.

– If the main focus of the study is to quantify how the HSMM parameters vary from one individual to another, the question cannot be addressed properly with this assumption.

To overcome the first limitation, covariates are introduced, while random effects are dedicated to overcome the second one.

When motivating the use of mixed HSMMs in the Squirrel toy example, we begin by addressing fixed effects only in the first step, so as to introduce progressively the notions of fixed effect, linear predictor and generalized linear model. Random effects are introduced later in the section.

1.6.1.1. *Fixed effect models*

Models with covariates are able to predict the effect on individual movements between reserves of, for example, seasons, which are global to all reserves, and at the scale of each reserve, resource abundance if known – as well as the presence of predators (pine martens and magpies) or competitors, whose trajectories are not modeled. These covariates have an effect on initial state and transition probabilities, as well as sojourn duration distributions. For example, they can reduce sojourn durations and transition probabilities toward reserves where predators are observed.

Some other covariates affect observation noise, for example, seasons/weather conditions, which make observations more or less noisy. This is also the case for confusion between species: consider then as a covariate the abundance of another species (gray squirrel or red squirrel from North America), which would increase the level of noise. It is recalled from section 1.4.1 that observation variability within each state is represented as an emission matrix; initially the probability that the naturalist believes rightly that the squirrel is in reserve i is b_{ii}, and thus, non-diagonal coefficients are probability of localization errors. As a result, weather and abundance of confusing species could increase the non-diagonal coefficients and decrease the diagonal coefficients of the emission matrix.

Such models including covariates are referred to as *fixed effect* models. The above examples show that fixed effects may have an impact on both state dynamics (initial state and transition probabilities, sojourn duration distributions) and emission probabilities.

It is assumed that C non-interacting individuals are observed and that three covariates are available at each time step. The first one is season $\varsigma \in \{1, 2, 3, 4\}$. For the sake of simplicity, sequences are assumed to be shorter than 1 year and time t is relative to season ς, so that the whole set of observations can be reindexed as $(Z^c_{\varsigma,t}, Y^c_{\varsigma,t})_{t\geq 1,\varsigma\in\{1,2,3,4\},1\leq c\leq C}$ where $Y^c_{\varsigma,t}$ is the guessed reserve at season ς at time t for individual c. The purpose of reindexing observations in such a way is to facilitate model specification using ς as covariate. This notation makes it explicit that several observations t are available in each season ς and extends to random variables $J_{\varsigma,N(t)}$ and $X_{\varsigma,N(t)}$. The other covariates are resource abundance $W_{\varsigma,t,1}$ and abundance of a predator $W_{\varsigma,t,2}$. A convenient way to let initial state distributions $\mathbb{P}(Z^c_{\varsigma,0} = i)$ for $i \in \{1, \ldots, K\}$ depend on the available covariates is using generalized linear models (GLMs, McCulloch et al. 2008). GLMs assume that probabilities depend on covariates through linear functions (referred to as linear predictor), under the form $\beta + \gamma W_{\varsigma,0,1} + \delta W_{\varsigma,0,2}$, where β, γ and δ are unknown parameters, to be estimated. To ensure that such quantities are admissible as probabilities, these are mapped into $[0, 1]$ using some bijection from $\mathbb{R}$ (referred to as the *link function*), for example, $\exp(\cdot)/(1 + \exp(\cdot))$.

In the case of the initial state distribution, we have K probabilities to model, in practice $K - 1$ given the constraint that their sum is 1, each with its own linear predictor with specific parameters, ensuring the possibility, for example, that $\mathbb{P}(Z^c_{\varsigma,0} = 2|W_{0,1}, W_{\varsigma,0,2})$ increases with $W_{0,1}$, $\mathbb{P}(Z^c_{\varsigma,0} = 3|W_{0,1}, W_{\varsigma,0,2})$ decreases with $W_{0,1}$ and $\mathbb{P}(Z^c_{\varsigma,0} = 1|W_{0,1}, W_{\varsigma,0,2})$ does not depend on $W_{0,1}$. Since the collection of linear predictors has to be mapped into the standard simplex in $\mathbb{R}^K$, we use a slightly more general link function than above, which yields (for $i < K$):

$$\log \frac{\mathbb{P}(Z^c_{\varsigma,0} = i|W_{0,1}, W_{\varsigma,0,2})}{\mathbb{P}(Z^c_{\varsigma,0} = K|W_{0,1}, W_{\varsigma,0,2})} = \beta^{(1)}_{\varsigma,i} + \gamma^{(1)}_{\varsigma,i} W_{\varsigma,0,1} + \delta^{(1)}_{\varsigma,i} W_{\varsigma,0,2},$$

where $\beta^{(1)}_{\varsigma,i}, \gamma^{(1)}_{\varsigma,i}$ and $\delta^{(1)}_{\varsigma,i}$ are fixed parameters associated with the effect of season ς, resource abundance and abundance of a predator (interacting with ς), respectively, on initial state $i < K$. Since we have β, γ and δ parameters in several distributions, they are distinguished by their exponent: $\beta^{(1)}$ for initial state, $\beta^{(2)}$ for transitions, $\beta^{(3)}$ for sojourn duration and $\beta^{(4)}$ for emission probabilities.

Such modeling principles could be extended to transition matrices, sojourn duration and emission distributions. Since the formal expressions of conditional probabilities are essentially the same when dealing with either fixed or random effects (each is an additional term in a linear predictor), we do not distinguish between both cases and consider directly mixed effects in what follows.

1.6.1.2. *Mixed effect models*

If observing several independent sequences of squirrel behaviors (more specifically, independent given the available covariates), one could ask whether each of them has a specific behavior, or whether they can be considered as independent trajectories issued from the same HSMM. Since behaviors are summarized by model parameters (assuming a common family of models), this is equivalent to asking whether HSMM parameters are individual-specific or common. If modelers are focusing on studying the specific behavior of each individual c, they will introduce a set of parameters to be estimated for each of them: $\beta^{(1)}_{\varsigma,i,c}$ for initial state probability, and in a similar way, $\beta^{(\kappa)}_{\varsigma,i,c}$ for each type κ of HSMM parameters. If, on the contrary, their aim is to quantify parameter variability between individuals, the question has to be addressed using random variables $\xi^{(\kappa)}_c$, so that it is meaningful to estimate their variances. Linear predictors at time t are now under the general form $\beta^{(\kappa)} + \gamma^{(\kappa)} W_{\varsigma,t,1} + \delta^{(\kappa)} W_{\varsigma,t,2} + \xi^{(\kappa)}_c$, where $\xi^{(\kappa)}_c$ is an additional random variable with mean 0 (otherwise the mean could not be distinguished from $\beta^{(\kappa)}$) and variance τ^2_κ, which is the parameter of primary interest. It is sufficient to introduce one random variable $\xi^{(\kappa)}_c$ per individual c and type κ of HSMM parameters, and to estimate its variance. Individual random effects are to be distinguished from randomness in emission distributions, since in the latter randomness is represented by independent random variables at each time step, while individual random effects $\xi_c^{(\kappa)}$ are shared in different time steps by a given individual c. Their distributions are usually assumed to be Gaussian for computational convenience. This family of regression models is referred to as generalized linear mixed models (McCulloch et al. 2008, GLMMs,), due to the presence of both fixed and random effects (hence the term *mixed*). The full model for initial probabilities now becomes ($\forall i < K$):

$$\log \frac{\mathbb{P}(Z^c_{\varsigma,0} = i | W_{0,1}, W_{\varsigma,0,2})}{\mathbb{P}(Z^c_{\varsigma,0} = K | W_{0,1}, W_{\varsigma,0,2})} = \beta^{(1)}_{\varsigma,i} + \gamma^{(1)}_{\varsigma,i} W_{\varsigma,0,1} + \delta^{(1)}_{\varsigma,i} W_{\varsigma,0,2} + \xi^{(1)}_c, \qquad [1.17]$$

where $(\xi^{(1)}_c)_{c=1,\ldots,C}$ are i.i.d. $\mathcal{N}(0, \tau^2)$. In the case where the question is to assess the part of variability in initial probabilities that depends on season, we have to let $\xi^{(1)}_c$ and τ^2 depend on ς, yielding the following refinement introduced in the right-hand side of [1.17]:

$$\beta^{(1)}_{\varsigma,i} + \gamma^{(1)}_{\varsigma,i} W_{\varsigma,0,1} + \delta^{(1)}_{\varsigma,i} W_{\varsigma,0,2} + \xi^{(1)}_{\varsigma,c},$$

with $\mathrm{var}(\xi^{(1)}_{\varsigma,c}) = \tau^2_\varsigma$. Further extensions consist of letting these quantities depend on i or on both i and ς.

Actually, letting initial probabilities vary with respect to individuals is not likely to bring much relevant information, since initial state occurs just once for each individual. The question is more relevant for parameters associated with events occurring several times for each individual, such as state transitions. For transition probabilities, random effects $\xi^{(2)}_{i,j,c}$ are introduced:

$$\log \frac{\mathbb{P}(J^c_{\varsigma,N(t)+1} = j | J^c_{\varsigma,N(t)} = i, W_{\varsigma,t,1}, W_{\varsigma,t,2}, \xi^{(2)}_{i,j,c})}{\mathbb{P}(J^c_{\varsigma,N(t)+1} = K | J^c_{\varsigma,N(t)} = i, W_{\varsigma,t,1}, W_{\varsigma,t,2}, \xi^{(2)}_{i,j,c})}$$
$$= \beta^{(2)}_{\varsigma,i,j} + \gamma^{(2)}_{\varsigma,i,j} W_{\varsigma,t,1} + \delta^{(2)}_{\varsigma,i,j} W_{\varsigma,t,2} + \xi^{(2)}_{i,j,c},$$

where $\xi^{(2)}_{i,j,c}$'s may be assumed to be Gaussian and independent with distributions $\mathcal{N}(0, \tau^2_{i,j})$. Such a model would reflect the assumption that within individual trajectories, the different incoming states reached along all transitions from the same outgoing state i are dependent. However, given random effects and outgoing states being i, these incoming states are independent. This is in contrast with standard HSMMs, which assume that transitions are independent given the outgoing state i. For example, $\xi^{(2)}_{i,j,c} > 0$ would mean that individual c has a preference for transitions from i to j compared to average squirrels. It could also be assumed that either $\xi^{(2)}_{i,j,c}$ or its variance depend on season ς. As above, $\beta^2_{\varsigma,i}$, $\gamma^{(2)}_{\varsigma,i}$ and $\delta^{(2)}_{\varsigma,i}$ are fixed parameters associated with the effects of season ς, resource abundance and abundance of a predator (interacting with ς), respectively, on transition probabilities.

Sojourn duration distributions can be modeled by GLMMs as well if the distribution of $X^c_{N(t)+1,\varsigma}$ given $J^c_{N(t),\varsigma}, W_{\varsigma,t,1}, W_{\varsigma,t,2}, \xi^{(3)}_{i,j,c}$ is assumed to be in the exponential family, in which case the model is defined by:

$$g\left(\mathbb{E}[X^c_{\varsigma,N(t)+1}|J^c_{N(t),\varsigma} = i, W_{\varsigma,t,1}, W_{\varsigma,t,2}, \xi^{(3)}_{i,c}]\right) \qquad [1.18]$$

$$= \beta^{(3)}_{\varsigma,i} + \gamma^{(3)}_{\varsigma,i} W_{\varsigma,t,1} + \delta^{(3)}_{\varsigma,i} W_{\varsigma,t,2} + \xi^{(3)}_{i,c},$$

where g is the link function connecting the conditional mean of $X^c_{N(t)+1,\varsigma}$ with the linear predictor in the right-hand side of [1.18]. Note that whenever there is no covariate, we retrieve the usual definition of sojourn duration distribution h_i. Let $\tau^{(3)}_i$ denote the standard deviation of $\xi^{(3)}_{i,c}$ in [1.18]: the model assumes that this standard deviation is the same whatever the season and the state. A more general model would assume a standard deviation written as $\tau^{(3)}_{\varsigma,i}$, depending on both state and season.

Emission distributions can be modeled following the same principle as transition probabilities:

$$\log \frac{\mathbb{P}(Y^c_{\varsigma,t} = j|Z^c_{\varsigma,t} = i, W_{\varsigma,t,1}, W_{\varsigma,t,2}, \xi^{(4)}_{i,j,c})}{\mathbb{P}(Y^c_{\varsigma,t} = K|Z^c_{\varsigma,t} = i, W_{\varsigma,t,1}, W_{\varsigma,t,2}, \xi^{(4)}_{i,j,c})}$$

$$= \beta^{(4)}_{\varsigma,i,j} + \gamma^{(4)}_{\varsigma,i,j} W_{\varsigma,t,1} + \delta^{(4)}_{\varsigma,i,j} W_{\varsigma,t,2} + \xi^{(4)}_{i,j,c}.$$

1.6.2. *Mixed effects for real-valued observations with the Deer example*

Covariates can be introduced to represent the effect of human presence or environment (open area versus forest, etc.) on animal behavior. Indeed, a deer may shift more often to the “running” behavior when humans are present than otherwise, and the signal emitted by the accelerometer when a deer is foraging may be different in a noisy area than in a quiet area because the animal will be tense. Let us assume once again that C non-interacting individuals are observed and that two covariates are available at each time step t: season $\varsigma \in \{1, 2, 3, 4\}$ and environment $e \in \{0, 1\}$ where 0 stands for an open and 1 for a closed environment. The whole set of observations can be reindexed as $(Z^c_{\varsigma,e,t}, Y^c_{\varsigma,e,t})_{t\geq 1, e\in\{0,1\}, \varsigma\in\{1,2,3,4\}, 1\leq c\leq C}$ where $Y^c_{\varsigma,e,t}$ is the accelerometer data at season ς and in environment e at time t for individual c. This notation extends to random variables J and X.

Similarly as in section 1.6.1, the effect of ς and e on emission distributions can be accounted for by the following Gaussian linear mixed model (LMM):

$$Y^c_{\varsigma,e,t} = \beta^{(4)}_{\varsigma,e,Z^c_{\varsigma,e,t}} + \xi^{(4)}_{c,\varsigma,e,Z^c_{\varsigma,e,t}} + \varepsilon^c_{\varsigma,e,t,Z^c_{\varsigma,e,t}}, \qquad [1.19]$$

where $\beta^{(4)}_{\varsigma,e,i}$ is a fixed parameter associated with the effect of season ς interacting with environment e and state i on $Y^c_{\varsigma,e,t}$ and $\xi^{(4)}_{c,e,\varsigma,i}$ is a Gaussian random effect distributed as $\mathcal{N}(0,(\tau^{(4)}_{\varsigma,e,i})^2)$ representing the variability of accelerometer data between individuals within state i for season ς and environment e, and $\varepsilon^c_{\varsigma,e,t,Z^c_{\varsigma,e,t}}$ is a Gaussian noise distributed as $\mathcal{N}(0,\sigma_i^2)$ given $Z^c_{\varsigma,e,t} = i$. The latter assumption reflects that the level of variability of $Y^c_{\varsigma,e,t}$, once fixed and random effects are accounted for, depends on the hidden state $Z^c_{\varsigma,e,t} = i$.

Now emission distributions are no longer represented by a matrix, but by a probability density function b and [1.19] is a compact way to write

$$b(Y^c_{\varsigma,e,t}|Z^c_{\varsigma,e,t} = i, \xi^{(4)}_{c,e,\varsigma,i}) = \mathcal{N}(\mu^c_{\varsigma,e,i}, \sigma_i^2),$$

with $\mu^c_{\varsigma,e,i} = \beta^{(4)}_{\varsigma,e,i} + \xi^{(4)}_{c,\varsigma,e,i}$.

The fact that observations are a sum of parameters and random variables in [1.19] is specific to the Gaussian assumption on $Y^c_{\varsigma,e,t}$, which leads to a LMM given the state variable.

Similarly to section 1.6.1, state transition probabilities are modeled here as GLMMs:

$$\log \frac{\mathbb{P}(J^c_{\varsigma,e,N(t)+1} = j \mid J^c_{\varsigma,e,N(t)} = i, \xi^{(2)}_{i,j,c})}{\mathbb{P}(J^c_{\varsigma,e,N(t)+1} = K \mid J^c_{\varsigma,e,N(t)} = i, \xi^{(2)}_{i,j,c})} = \beta^{(2)}_{\varsigma,e,i,j} + \xi^{(2)}_{i,c},$$

where i and $j < K$ are possible state values, $\xi^{(2)}_{i,j,c}$ is a random effect (centered random variable with variance $(\tau^{(2)}_{i,\varsigma,e})^2$) and $\beta^{(2)}_{\varsigma,e,i,j}$'s are unknown, fixed parameters.

Similar mixed models can be derived for sojourn duration distributions by adapting [1.18] to the case of deer (precisely, to include e as a covariate instead of W).

1.6.3. *Dynamic covariates: toward an alternative representation of HSMMs*

A step further in covariate inclusion is to consider a time-varying covariate w_t that modulates sojourn duration once a new state is entered. In both toy models of Deer and Squirrel, such a covariate could account for observable events (e.g. presence of wanderers and weather or seasonal effect) that may shorten or extend sojourn durations. However, the definition of (hidden) semi-Markov models based on sojourn duration X_n, whose distribution is determined at the time S_{n-1} when a new state is entered, is not adequate. This limitation can be circumvented using an alternative definition of semi-Markov models based on the discrete hazard rate. This point is detailed in Chapter 4, section 4.3.5, as a semi-Markov formalism based on hazard rate is developed to model coupling of multichain semi-Markov models. The principle and advantages of resorting to hazard rates to model the effect of dynamic covariates is explained informally hereafter.

The fundamental principle underlying the discrete hazard rate is some equivalence between the distributions of a finite discrete random variable and of a sequence of binary variables. A non-negative finite discrete random variable X can be defined through a series of binary questions:

– $X = 0$?;

– if not, given $X \geq 1$, then $X = 1$?;

– and so on.

If at step k (numbered from $k = 0$), the answer to all previous questions is "no", then the current question becomes "Given $X \geq k$, then $X = k$?". Let V_k denote the answer to kth question, where "no" is coded as 0 and "yes" as 1, then the value of X is the index k of the first V_k equal to 1. The values of V_ℓ for $\ell > k$ are arbitrarily set to 0. The equivalence between both representations, X and $(V_\ell)_{\ell \geq 0}$, is achieved if the probability of $V_k = 1$ given all previous values are 0 is set to the probability $\lambda(k)$ that $X = k$ given $X \geq k$. This is exactly what the discrete hazard rate λ of the distribution of X is.

This framework can be used to represent sojourn durations $(X_n)_{n \geq 0}$. The main adaptation to be achieved is that at time step t, $V_t = 1$ if and only if there is a state jump at t; in this case, the sojourn duration is one plus the number of zeros since the previous jump. In can be proved straightforwardly that there is a one to one mapping between $(X_n)_{n \geq 0}$ and $(V_t)_{t \geq 0}$. Resorting to the V_t representation has the advantage of introducing sojourn durations as the by-product of binary decisions at the same time step t as states Z_t and observations Y_t. This makes this representation adequate to handle dynamic covariates. The second advantage is that the canonical parameters in this representation become the values of $\lambda(k)$ for every $k \in \mathbb{N}$. Consequently, covariates are introduced in the model by letting $\lambda(k)$ depend on these covariates.

If one prefers to avoid handling dynamic covariates in hazard rates, care will need to be taken to ensure that covariates are constant between state jumps, which is approximately the case if the time step of the model is negligible compared to the time step between possible changes in the covariate. In this case, modelers will use GLMMs as in [1.18] to introduce mixed effects into sojourn duration distributions.

1.6.4. *Model selection issues for fixed and random effects*

In mixed HSMMs, a recurrent question is to identify which covariates, among the whole considered set, actually have an effect on initial distributions, transition probabilities, sojourn duration and emission distributions. To achieve this task, test and model selection procedures have to be carried out.

For example in the Squirrel toy example, the hypotheses of a test on the effect of resource abundance on sojourn duration distribution in state i being the same or not in every season would write as $(\forall \varsigma_1, \varsigma_2, \gamma^{(3)}_{\varsigma_1,i} = \gamma^{(3)}_{\varsigma_2,i})$ regarding the null hypothesis, and $(\exists \varsigma_1, \varsigma_2, \gamma^{(3)}_{\varsigma_1,i} \neq \gamma^{(3)}_{\varsigma_2,i})$ regarding the alternative hypothesis. A test on the effect of dependencies between successive sojourn durations $(X^c_{n,\varsigma})_{n\in\mathbb{N}}$ in state i (or the absence thereof), due to the fact that they are achieved by the same individual c, would write as $(\tau^{(3)}_i = 0)$ regarding the null hypothesis, and $(\tau^{(3)}_i > 0)$ regarding the alternative hypothesis. In practice, this decision problem would rather be addressed by optimizing an information criterion computed under both the null and alternative models.

Modelers could also wonder whether the behaviors of the C squirrels are homogeneous, or if some individual differences can be observed. To address this question, two models will be compared: one with individual random effects in the three types of parameters, another without random effects, and both will be assessed by an information criterion, such as BIC. This would require redefinition of BIC (Schwarz 1978) in this model, and the study of its properties in model selection for mixed HSMMs.

1.6.5. *Mixed models in the HMM/HSMM literature*

Mixed HSMMs present in the literature can be categorized by their level of generality: Do they include both kinds of effects (fixed and random), are they related to emission distributions only or to both emissions and state dynamics (i.e. transitions or sojourn duration distributions), are they specific to Gaussian observations or are they compatible to non-Gaussian data (i.e. counts or categorical data)? The categories are used to organize this section, as well as the approximation strategies used in references for parameter estimation.

1.6.5.1. *Main advances*

The first HMM was proposed by Altman (2007), with covariates and random effects introduced through GLMMs. Estimation was achieved using the EM algorithm. The intractable integrals due to random effects were addressed either by Gaussian quadrature methods or by Monte Carlo simulation, yielding an MCEM algorithm in the latter case.

A somewhat different model was introduced by Maruotti and Rocci (2012) together with a fastest estimation algorithm. Fixed and random effects were included at the levels of transition and initial state probabilities, using finite-valued random effects, thus leading to finite mixtures of HMMs. Since random effects had finite values in their model, no intractable integral prevented from using an EM algorithm with closed-form E step. This led to an increased complexity of the forward-backward recursion, which seemed to be multiplied by the number of values of the random effects, although this was not explicitly stated in their work.

A partial extension to HSMMs of the model introduced by Altman (2007) was proposed by Chaubert-Pereira et al. (2010). They proposed some stochastic approximation EM (SAEM) algorithm to address the E step of EM. Their model was restricted to LMMs in emission distributions. By definition of LMMs, random effects were assumed to be Gaussian. Moreover they were specific to individuals, meaning that several independent trajectories were observed, each one associated with a specific random variable (no random effect being shared between trajectories). Note that Chaubert-Pereira (2008) also addressed the introduction of fixed effects within emission distributions through GLMs, as well as shared random effects in LMMs for emission distributions.

Nonlinear Gaussian models with discrete observations in HMMs were considered by Delattre and Lavielle (2012) with individual random and fixed effects. An approximate E step was achieved through a MCMC-SAEM algorithm.

A somewhat similar approach in HSMMs was developed by Haji-Maghsoudi et al. (2021), now with fixed and random effects at the emission distribution level only, using GLMMs. A Monte Carlo Newton–Raphson algorithm was used to obtain an approximate E step[1].

Fixed effects in sojourn duration distributions were introduced by Rojas-Salazar et al. (2020) in the framework of Bayesian HSMMs, estimating the posterior distribution with an MCMC algorithm involving a Viterbi recursion to sample sojourn duration – although the use of a Viterbi algorithm instead of a Gibbs sampler was not discussed.

1.6.5.2. *Some more specific contributions*

Some particular case of the model by Altman (2007) for categorical data was used by Shirley et al. (2010), but with a different parameter estimation method. Their work relied on some hybrid Metropolis within Gibbs sampler. The model was extended to multivariate categorical observations and implemented as the R mHMMBayes package by Emmeke Aarts (see also Moraga and Aarts 2023).

1 It is thus unclear why the article is entitled "A Bayesian approach".

A zero-inflated Poisson HMM (ZIP-HMM) with subject-specific random effects "that allows counts to move through the state-space according to covariate-specific transition probabilities, assuming a logit model for each row of the hidden state transition matrix" was developed by DeSantis and Bandyopadhyay (2011). Emission distributions were zero-inflated Poisson with covariates and subject-specific random effects. Parameters were estimated in a Bayesian approach through Gibbs sampling using WinBUGS.

A HMM for functional observations was developed by Zhou and Song (2023). The transition and emission probabilities depended on functional covariates and incorporated random effects introduced to describe the dependency of individual functional observations. A Bayesian approach was implemented for parameter estimation using a Dirichlet process for accommodating the unspecified distribution of the random effect and a blocked Gibbs sampler.

The R package hmmTMB was implemented by Michelot (2023); this incorporated mixed HMMs using different kinds of splines instead of GLMs in transition and emission distributions.

A HMM with variable number of states was developed by Russo et al. (2022). The number of states, transition probabilities and multivariate emission distributions depended on covariates. The model parameters were estimated within an Bayesian framework by reversible-jump MCMC. A somewhat related approach was proposed by Zou et al. (2023): their model is a Bayesian HMM with a priori unknown number of states, covariate-dependent transition probabilities and Gaussian linear emission distributions. Estimation was based on reversible-jump MCMC, including penalties for estimating the number of states.

Note that a review on HMMs with covariates (fixed effects) was proposed by Bartolucci et al. (2014). An extension of the model to mixed HMMs was then developed by a related group of authors and implemented in R as the LMest package (see Bartolucci et al. 2017). A review on HMMs with mixed effects was proposed by Maruotti (2011). The use of particular cases of HMMs with discrete random effects (referred to as semi-parametric) was discussed by the authors.

As a conclusion, mixed HMMs are a class of models for which both MLE and Bayesian algorithms are available, in a general framework allowing to include mixed effects in both emission and transition probabilities. In the HSMM setting, available methods are mostly restricted to mixed effects with individual random effects in emission distributions, mostly in the Gaussian case. Thus, obtaining the same level of generality in HSMM and HMM modeling, even in the case of static covariates in sojourn duration distributions, is still a challenging research question.

1.7. Conclusion/discussion

In this chapter, we have reviewed the main theoretical and methodological aspects of HSMM. We did not addressed some aspects that are classical for HMMs but that are not often covered in the HSMM literature and that we discuss briefly now.

In this chapter, we focus on discrete-time and finite state space HSMMs. Some works considered real-valued hidden states or continuous time. When the underlying chain is a Markov chain (instead of a semi-Markov chain), the observation and the hidden spaces are general and the time is discrete, then the corresponding models are known as "state space models". In such models, the inference is made in a Bayesian context by means of sequential Monte Carlo methods (or particle filters, see e.g., Cappé et al. 2007, and the references therein). These methods have been recently used for HSMMs in Aschermayr and Kalogeropoulos (2023).

The bibliography in parameter estimation for (continuous-time) hidden Markov processes with general state and observation spaces is rather limited. The MLE for continuous-time HMMs was studied in Chigansky (2009). Recent works include the results presented in Kutoyants (2020), where two observation models are considered: a partially observed two-dimensional Gaussian process and a telegraph process observed against the background of white Gaussian noise. The asymptotic properties of the estimators are studied in both large sample and small noise context. Note that continuous time HSMMs are addressed in Chapter 6 in a context of modeling and control (where parameters are assumed known).

Model validation and selection is often a crucial step in statistical modeling. In the case of explicit duration HMMs with finite observations, validation criteria were proposed in Guédon (1999). These criteria consist of distributions of various observed statistics (e.g. time before first occurrence of an observed value y, number of occurrences of y within a subsequence of given length and time of return for occurrence y since the last occurrence), which are compared to the corresponding empirical distributions. The same principle may be applied to hidden states, which have to be restored first in this case. Model selection addresses either the choice of the number of states K or inference of the graphical model itself, which is the conditional independence relationships between states and observations in the multiple chains. Selection of the number of states in HMMs often relies on information criteria (see the review in Boucheron and Gassiat 2005). Information criteria mostly rely on penalizing the maximum log-likelihood by subtracting a non-negative function that increases with the number of parameters. This requires us to estimate as many HMMs as the number of possible values for K. As an alternative, some approaches directly estimate K together with parameter estimation by penalizing the likelihood function (Lin and Huang 2025). In HSMMs, selection of K has not been thoroughly investigated: although direct application of information criteria are proposed in Yu (2016); in HSMMs with finite-support, non-parametric sojourn distribution, more general approaches are still to be developed.

1.8. Notations

Tables 1.3–1.5 recall the main notations for HSMM used in this chapter. In the other chapters of this book, notations conform as much as possible with these ones.

Definition	Notation	Domain
Calendar time	t	$\mathbb{N}$
Duration of the time series	$T+1$	$\mathbb{N}$
State space of hidden variable	$\Omega_Z = \{1, \ldots, K\}$	finite
Cardinal of the hidden variable state space	$K = \|\Omega_Z\|$	$\mathbb{N}^*$
State of the hidden chain at calendar time t		
Random variable	Z_t	Ω_Z
Realization	z_t	Ω_Z
Sequence of hidden states from time 0 to T		
Random variable	$Z_{0:T}$	$(\Omega_Z)^{T+1}$
Realization	$z_{0:T}$	$(\Omega_Z)^{T+1}$
Index of the current jump	$N(t)$	$\mathbb{N}$
Elapsed time	E_t	$\mathbb{N}^*$
Remaining time	R_t	$\mathbb{N}$
Index of a jump	n	$\mathbb{N}$
State into which the chain enters at jump n	J_n	Ω_Z
Time at jump n	S_n	$\mathbb{N}$
Sojourn duration in state J_{n-1}	X_n	$\mathbb{N}^*$
Transition probability of $(J_n, X_{n+1})_n$ $= \mathbb{P}(J_t = j, X_{t+1} = l \| J_{t-1} = i, X_t = d)$	$a_{(i,d)(j,l)}$	$[0, 1]$
Initial distribution of $(J_n, X_{n+1})_n$ $= \mathbb{P}(J_0 = i, X_1 = d)$	π_{id}	$[0, 1]$
State space of observed variable	$\Omega_Y = \{1, \ldots, L\}$	finite
Cardinal of the observed variable state space	$L = \|\Omega_Y\|$	$\mathbb{N}^*$
Observed variable at calendar time t		
Random variable	Y_t	Ω_Y
Realization	y_t	Ω_Y
Sequence of observations from time 0 to T		
Random variable	$Y_{0:T}$	$(\Omega_Y)^{T+1}$
Realization	$y_{0:T}$	$(\Omega_Y)^{T+1}$
Emission probability of value y conditionally to hidden state z	b_{zy}	$[0, 1]$

Table 1.3. *Notations for a General HSMM*

Definition	Notation
Semi-Makov kernel	$q_{ij}(d)$
Transition probability of $(J_n)_n$	p_{ij}
Conditional distribution of X_{n+1}	$f_{ij}(d)$
Distribution of sojourn duration in state i	$h_i(d)$
Survival function in state i	$\overline{H}_i(d)$
Initial distribution of J_0	π_i

Table 1.4. *Notations for the Standard SMM and the ED-MM*

Event	Notation
$Z_{t+1} = \cdots = Z_{t+d} = i$ and $Z_t \in \Omega_Z$, $Z_{t+d+1} \in \Omega_Z$	$Z_{t+1:t+d} = i$
$Z_{t+1} = \cdots = Z_{t+d} = i$ and $Z_t \neq i$, $Z_{t+d+1} \neq i$	$Z_{[t+1:t+d]} = i$
$Z_{t+1} = \cdots = Z_{t+d} = i$ and $Z_t \neq i$, $Z_{t+d+1} \in \Omega_Z$	$Z_{[t+1:t+d} = i$
$Z_{t+1} = \cdots = Z_{t+d} = i$ and $Z_t \in \Omega_Z$, $Z_{t+d+1} \neq i$	$Z_{t+1:t+d]} = i$

Table 1.5. *Notations of events using calendar time*

1.9. Acknowledgments

The authors acknowledge fruitful discussions with Catherine Trottier on generalized linear mixed models and thank Vlad Stefan Barbu for the numerous exchanges and his constant support throughout this work.

1.10. Appendix: EM algorithm for a monochain HMM

A general presentation of EM algorithm is provided in section 1.3.3. It relies on iterations of two steps: computation of [1.13] and resolution of [1.14] until the algorithm converges. We provide here the details of these two steps for a HMM. This gives an idea of the algorithm but without the full complexity of a general HSMM. Furthermore, EM for HMM will be useful for Chapter 3.

For a HMM, the quantity $Q(\theta, \theta^{(m)})$ can be decomposed into three terms: one depending on the initial distribution, another depending on the transition matrix and the last one depending on the emission distribution:

$$\begin{aligned}
Q(\theta, \theta^{(m)}) &= E_{\theta^{(m)}}[\ln(\mathbb{P}_\theta(Z_{0:T} = z_{0:T}, Y_{0:T} = y_{0:T}))|Y_{0:T} = y_{0:T}] \\
&= \sum_{z_0 \in \Omega_Z} \xi_0^{\theta^{(m)}}(z_0) \ln(\pi_{z_0})
\end{aligned}$$

$$+\sum_{t=1}^{T}\sum_{(z_t,z_{t-1})\in\Omega_Z^2}\gamma_t^{\theta^{(m)}}(z_{t-1},z_t)\ln(a_{z_{t-1}z_t})$$

$$+\sum_{t=0}^{T}\sum_{z_t\in\Omega_Z}\xi_t^{\theta^{(m)}}(z_t)\ln(b_{z_t y_t}),$$

where:

- $a_{ij} = \mathbb{P}_\theta(Z_t = j \mid Z_{t-1} = i)$;
- $\xi_t^{(m)}(z_t) = \mathbb{P}_{\theta^{(m)}}(Z_t = z_t | Y_{0:T} = y_{0:T})$;
- $\gamma_t^{(m)}(z_{t-1}, z_t) = \mathbb{P}_{\theta^{(m)}}(Z_{t-1} = z_{t-1}, Z_t = z_t | Y_{0:T} = y_{0:T})$.

1.10.1. *E step*

The E step consists of computing the marginal probabilities of interest appeared in the expression of $Q(\theta, \theta^{(m)})$, $\xi_t^{(m)}$ and $\gamma_t^{(m)}$. To obtain these two variables, we introduce the following intermediate variables:

$$\begin{cases} \alpha_t^{(m)}(z_t) &= \mathbb{P}_{\theta^{(m)}}(Y_{0:t} = y_{0:t}, Z_t = z_t) \\ \beta_t^{(m)}(z_t) &= \mathbb{P}_{\theta^{(m)}}(Y_{t+1:T} = y_{t+1:T} | Z_t = z_t). \end{cases} \quad [1.20]$$

They are computed recursively using the forward-backward algorithm. In the forward algorithm, we express $\alpha_t^{(m)}(z_t)$ using the following recurrence formula:

$$\forall t \geq 1, \alpha_t^{(m)}(z_t) = b_{z_t y_t}^{(m)} \sum_{z_{t-1}\in\Omega_Z} \alpha_{t-1}^{(m)}(z_{t-1}) a_{z_{t-1}z_t}^{(m)}, \quad [1.21]$$

and $\alpha_0^{(m)}(z_0) = b_{z_0 y_0}^{(m)} \pi_{z_0}^{(m)}$.

In the backward algorithm, we compute $\beta_t^{(m)}(z_t)$ using the following recurrence formula:

$$\forall t \geq 1,\ \beta_t^{(m)}(z_t) = \sum_{z_{t+1}\in\Omega_Z} \beta_{t+1}^{(m)}(z_{t+1}) b_{z_{t+1}y_{t+1}}^{(m)} a_{z_t,z_{t+1}}^{(m)}, \quad [1.22]$$

and $\beta_T^{(m)}(z_T) = 1$.

The quantities $\alpha_t^{(m)}(z_t)$ and $\beta_t^{(m)}(z_t)$ are used to compute $\xi_t^{(m)}(z_t)$ and $\gamma_t^{(m)}(z_{t-1}, z_t)$ as follows:

$$- \ \xi_t^{(m)}(z_t) = \frac{\alpha_t^{(m)}(z_t)\beta_t^{(m)}(z_t)}{\sum\limits_{z_t \in \Omega_Z} \left[\alpha_t^{(m)}(z_t)\beta_t^{(m)}(z_t)\right]}; \quad [1.23]$$

$$- \ \gamma_t^{(m)}(z_{t-1}, z_t) = \frac{a_{z_{t-1}z_t}^{(m)}\alpha_{t-1}^{(m)}(z_{t-1})B_n(z_t, y_t)\beta_t^{(m)}(z_t)}{\sum\limits_{z_t \in \Omega_Z} \left[\alpha_t^{(m)}(z_t)\beta_t^{(m)}(z_t)\right]}. \quad [1.24]$$

Proofs of [1.21]–[1.24] are provided in Bishop (2006), section 13.2. From [1.21], for every t, the time complexity of computing a single $\alpha_t^{(m)}(z_t)$ from $(\alpha_{t-1}^{(m)}(z_{t-1}))_{z_{t-1}\in\Omega_z}$ is $\mathcal{O}(K)$, thus the time complexity of computing the values of $\alpha_t^{(m)}(z_t)$ for every possible t and z_t (which is, running the complete forward recursion) is $\mathcal{O}(TK^2)$. The complete backward recursion has the same time complexity. Computing $\mathbb{P}_{\theta^{(m)}}(Z_t = z_t, Y_{0:T} = y_{0:T})$ rather than $\xi_t^{(m)}(z_t) = \mathbb{P}_{\theta^{(m)}}(Z_t = z_t|Y_{0:T} = y_{0:T})$ is slightly simpler in terms of computation and proofs; however, this alternative may cause underflow, since the involved joint probabilities tend toward zero when T tends to infinity Devijver (1985).

1.10.2. *M step*

During the M step, we resolve the maximization problem to obtain the expression of updated parameters. We express these parameters in terms of $\xi_t^{(m)}(z_t)$ and $\gamma_t^{(m)}(z_{t-1}, z_t)$ as follows:

$$- \forall z_0 \in \Omega_Z, \pi_{z_0}^{(m+1)} = \xi_0^{(m)}(z_0);$$

$$- \forall z_t \in \Omega_Z, \forall z_{t-1} \in \Omega_Z, a_{z_{t-1}z_t}^{(m+1)} = \frac{\sum\limits_{t=1}^{T} \gamma_t^{(m)}(z_{t-1}, z_t)}{\sum\limits_{t=1}^{T}\sum\limits_{z_t\in\Omega_Z} \gamma_t^{(m)}(z_{t-1}, z_t)};$$

$$- \forall z_t \in \Omega_Z, \forall y_t \in \Omega_Y, b_{z_t y_t}^{(m+1)} = \frac{\sum\limits_{t=0}^{T} \xi_t^{(m)}(z_t)\mathbb{1}_{Y_t = y_t}}{\sum\limits_{t=0}^{T} \xi_t^{(m)}(z_t)}.$$

As a conclusion, the conditional probabilities of states given all observations $\xi_t^{(m)}(z_t)$ and $\gamma_t^{(m)}(z_{t-1}, z_t)$ are all that is required to implement the M step, besides observations themselves. This is why the purpose of the E step is to provide these conditional probabilities.

1.11. References

Akaike, H. (1974). A new look at the statistical model identification. *IEEE Transactions on Automatic Control*, 19(6), 716–723.

Altman, R. (2007), Mixed hidden Markov models: An extension of the hidden Markov model to the longitudinal data setting. *Journal of the American Statistical Association*, 102, 201–210.

Anandkumar, A., Hsu, D., Kakade, S. (2012). A method of moments for mixture models and hidden Markov models. In *Proceedings of the 25th Annual Conference on Learning Theory*. PMLR, Edinburgh.

Aschermayr, P. and Kalogeropoulos, K. (2023). Sequential Bayesian learning for hidden semi-Markov models. *arXiv*:2301 10494. doi: 10.48550/ arXiv.2301.10494.

Barbu, V.S. and Limnios, N. (2006). Maximum likelihood estimation for hidden semi-Markov models. *Comptes Rendus Mathematique*, 342, 201–205.

Barbu, V.S. and Limnios, N. (2008). *Semi-Markov Chains and Hidden Semi-Markov Models Toward Applications*. Springer, New York.

Bartolucci, F., Farcomeni, A., Pennoni, F. (2014). Latent Markov models: A review of a general framework for the analysis of longitudinal data with covariates. *TEST*, 23, 433–465.

Bartolucci, F., Pandolfi, S., Pennoni, F. (2017). LMest: An R package for latent Markov models for longitudinal categorical data. *Journal of Statistical Software*, 81(4), 1–38.

Benoit, L. (2022). Gestion du compromis entre acquisition des ressources et évitement du risque chez une population de chevreuils (Capreolus capreolus) suivis dans un paysage anthropisé. PhD Thesis, Université Paul Sabatier.

Bickel, P.J. and Ritov, Y. (1996). Inference in hidden Markov models I: Local asymptotic normality in the stationary case. *Bernoulli*, 2(3), 199–228.

Bishop, C. (ed.) (2006). Sequential data. In *Pattern Recognition and Machine Learning*. Springer, New York.

Boucheron, S. and Gassiat, E. (2005). An information-theoretic perspective on order estimation. In *Inference in Hidden Markov Models*, Cappé, O., Moulines, E., Rydén, T. (eds). Springer, New York.

Cappé, O., Godsill, S., Moulines, E. (2007). An overview of existing methods and recent advances in sequential Monte Carlo. *Proceedings of the IEEE*, 95(5), 899–924.

Cardot, H. and Frascolla, C. (2024). Hypotheses testing for panels of semi-Markov processes with parametric sojourn time distributions. *Journal of Statistical Planning and inference*, 228, 59–79.

Chaubert-Pereira, F. (2008). Combinaisons markoviennes et semi-markoviennes de modèles de régression. Application à la croissance d'arbres forestiers. PhD Thesis, Université Montpellier II – Sciences et Techniques du Languedoc.

Chaubert-Pereira, F., Caraglio, Y., Lavergne, C., Guédon, Y. (2009). Identifying ontogenetic, environmental and individual components of forest tree growth. *Annals of Botany*, 104(5), 883–896.

Chaubert-Pereira, F., Guédon, Y., Lavergne, C., Trottier, C. (2010). Markov and semi-Markov switching linear mixed models used to identify forest tree growth components. *Biometrics*, 66, 753–762.

Chigansky, P. (2009). Maximum likelihood estimator for hidden Markov models in continuous time. *Statistical Inference for Stochastic Processes*, 12, 139–163.

Delattre, M. and Lavielle, M. (2012). Maximum likelihood estimation in discrete mixed hidden Markov models. *Computational Statistics and Data Analysis*, 56(6), 2073–2085.

Dempster, A.P., Laird, N.M., Rubin, D.B. (1977). Maximum likelihood from incomplete data via the EM algorithm. *Journal of the Royal Statistical Society: Series B (Methodological)*, 39(1), 1–22.

DeSantis, S. and Bandyopadhyay, D. (2011). Hidden Markov models for zero-inflated Poisson counts with an application to substance use. *Statistics in Medicine*, 30(14), 1678–1694.

Devijver, P. (1985). Baum's forward-backward algorithm revisited. *Pattern Recognition Letters*, 3, 369–373.

Economou, T., Bailey, T., Kapelan, Z. (2014). MCMC implementation for Bayesian hidden semi-Markov models with illustrative applications. *Statistics and Computing*, 24, 739–752.

Ferguson, J.D. (1980). Variable duration models for speech. In *Symposium on the Application of Hidden Markov Models to Text and Speech*. IDA-CRD.

Gámiz, M.L., Limnios, N., Segovia-García, M. (2023). Hidden Markov models in reliability and maintenance. *European Journal of Operational Research*, 304, 1242–1255.

Giudici, P., Ryden, T. Vandekerkhove, P. (2000), Likelihood-ratio tests for hidden Markov models, *Biometrics*, 56(3), 742–747

Guédon, Y. (1999). Computational methods for discrete hidden semi-Markov chains. *Applied Stochastic Models in Business and Industry*, 15, 195–224.

Guédon, Y. (2003). Estimating hidden semi-Markov chains from discrete sequences. *Journal of Computational and Graphical Statistics*, 12(3), 604–639.

Hadj-Amar, B., Jewson, J., Fiecas, M. (2023). Bayesian approximations to hidden semi-Markov models for telemetric monitoring of physical activity. *Bayesian Analysis*, 18(2), 547–577.

Haji-Maghsoudi, S., Bulla, J., Sadeghifar, M., Roshanaei, G., Mahjub, H. (2021). Generalized linear mixed hidden semi-Markov models in longitudinal settings: A Bayesian approach. *Statistics in Medicine*, 40(10), 2373–2388.

Johnson, M. and Willsky, A. (2013). Bayesian nonparametric hidden semi-Markov models. *Journal of Machine Learning Research*, 14(20), 673–701.

Koller, D. and Friedman, N. (2009). *Probabilistic Graphical Models: Principles and Techniques*. MIT Press, Cambridge, MA.

Kutoyants, Y. (2020). Parameter estimation for continuous time hidden Markov processes. *Automation and Remote Control*, 81, 445–68.

Leroux, B.G. (1992). Asymptotic normality of the maximum-likelihood estimator for general hidden Markov models. *Stochastic Processes and Their Applications*, 40, 127–143.

Limnios, N. and Oprisan, G. (2001). *Semi-Markov Processes and Reliability*. Birkhäuser, Boston.

Lin, Y. and Huang, M. (2025). Penalized composite likelihood estimation for hidden Markov models with unknown number of states. *Statistics & Probability Letters*, 216, 110247.

MacDonald, I. and Zucchini, W. (1997). *Hidden Markov and Other Models for Discrete-Valued Time Series*. Chapman, London.

Malefaki, S., Trevezas, S., Limnios, N. (2009). An EM and a stochastic version of the EM algorithm for nonparametric hidden semi-Markov models. *Communications in Statistics – Simulation and Computation*, 39(2), 240–261.

Maruotti, A. (2011). Mixed hidden Markov models for longitudinal data: An overview. *International Statistical Review*, 79, 427–454.

Maruotti, A. and Rocci, R. (2012). A mixed non-homogeneous hidden Markov model for categorical data, with application to alcohol consumption. *Statistics in Medicine*, 31(9), 871–88.

McCulloch, C.E., Searle, S.R., Neuhaus, J.M. (2008). *Generalized, Linear, and Mixed Models*, 2nd edition. Wiley, Hoboken, NJ.

Melnyk, I. and Banerjee, A. (2017). A spectral algorithm for inference in hidden semi-Markov models. *The Journal of Machine Learning Research*, 18(35), 1–39.

Michelot, T. (2025). hmmTMB: Hidden Markov models with flexible covariate effects in R. *Journal of Statistical Software*, 114(5), 1–45.

Moraga, S.M. and Aarts, E. (2023). Go multivariate: Recommendations on Bayesian multilevel hidden Markov models with categorical data. *Multivariate Behavioral Research*, 59(1), 17–45.

Pertsinidou, C. and Limnios, N. (2015). Viterbi algorithms for hidden semi-Markov models with application to DNA analysis. *RAIRO-Operations Research*, 49, 511–526.

Rojas-Salazar, S., Schliep, E.M., Wikle, C.K., Hawkey, M. (2020). A Bayesian hidden semi-Markov model with covariate-dependent state duration parameters for high-frequency data from wearable devices. arXiv preprint arXiv:2010.10739.

Russo, A., Farcomeni, A., Pittau, M.-G., Zelli, R. (2022). Covariate-modulated rectangular latent Markov models with an unknown number of regime profiles. *Statistical Modelling*, 24(4), 368–388.

Schwarz, G. (1978). Estimating the dimension of a model. *The Annals of Statistics*, 6(2), 461–464.

Shirley, K.E., Small, D.S., Lynch, K.G., Maisto, S.A., Oslin, D.W. (2010). Hidden Markov models for alcoholism treatment trial data. *The Annals of Applied Statistics*, 4(1), 366–395.

Trevezas, S. and Limnios, N. (2009). Maximum likelihood estimation for general hidden semi-Markov processes with backward recurrence time dependence. *Journal of Mathematical Sciences*, 163, 262–274.

Votsi, I. (2019). Conditional failure occurrence rates for semi-Markov chains. *Journal of Applied Statistics*, 46(15), 2722–2743.

Votsi, I. and Brouste, A. (2019). Confidence intervals for the mean time to failure in semi-Markov models: An application to wind energy production. *Journal of Applied Statistics*, 46(10), 1756–1773.

Votsi, I. and Limnios, N. (2015). Estimation of the intensity of the hitting time for semi-Markov chains and hidden Markov renewal chains. *Journal of Nonparametric Statistics*, 27(2), 149–166.

Votsi, I., Gayraud, G., Barbu, V.S., Limnios, N. (2021). Hypotheses testing and posterior concentration rates for semi-Markov processes. *Statistical Inference for Stochastic Processes*, 24, 707–732.

Yu, S.-Z. (2016). *Hidden Semi-Markov Models Theory, Algorithms and Applications*. Elsevier, Amsterdam.

Zhou, X. and Song, X. (2023). Functional concurrent hidden Markov model. *Statistics and Computing*, 33, 57.

Zou, Y., Lin, Y., Song, X. (2023). Bayesian heterogeneous hidden Markov models with an unknown number of states. *Journal of Computational and Graphical Statistics*, 33(1), 15–24.

2

Review of HSMM R and Python Softwares

2.1. Introduction

This chapter aims to provide a comprehensive review of the various software and packages available for hidden semi-Markov models (HSMMs) or with a particular focus on those related to the themes of this book: multichain models, coupled chains and the incorporation of covariates. A thorough survey has been conducted to compile both R and Python packages that are specifically designed to handle HSMMs, along with those related to these key themes, although a significant lack of dedicated packages was identified in these thematic areas. Throughout this chapter, all models are explicit duration hidden semi-Markov models, (see ED-HSMM in Chapter 1, section 1.2.4) unless otherwise specified, and the packages are presented in alphabetical order.

The comparison of these packages is based on their functionalities, particularly in relation to the modeling of sojourn duration distributions and emission distributions for HSMMs, as described in Chapter 1. This comparison highlights the diversity of options offered by the packages, which vary in their level of generality, as well as their similarities and differences.

To further illustrate the practical application of these tools, we demonstrate the use of two selected packages. A Python package is applied to discrete emission distributions in the toy model Squirrel, while an R package is employed for continuous emission distributions in the toy model Deer. These examples highlight the diversity and the range of options available when working with HSMMs across different software environments.

Chapter written by Caroline BÉRARD, Marie-Josée CROS, Jean-Baptiste DURAND, Corentin LOTHODÉ, Sandra PLANCADE, Ronan TRÉPOS and Nicolas VERGNE.

The chapter is then organized into three parts: section 2.2 presents an inventory of packages for HSMMs (in R and Python), as well as others that are not strictly HSMM-specific but thematically related (see Table 2.1); section 2.3 provides a comparison based on state duration modeling and emission distributions; and section 2.4 offers a demonstration of two selected packages using the toy models introduced in Chapter 1.

Software	Language	Section
CHMM	R	2.2.3.1
DNN-HSMM	Python	2.2.3.2
edhsmm	Python	2.2.2.1
GENSCAN	C	2.2.3.3
hhsmm	R	2.2.1.1
hmmTMB	R	2.2.3.4
hsmm	R	2.2.1.2
hsmmlearn	Python	2.2.2.2
LaMa	R	2.2.1.3
mHMMBayes	R	2.2.3.5
mhsmm	R	2.2.1.4
online_hmm	Python	2.2.2.3
PHSMM	R	2.2.1.5
pyhsmm	Python	2.2.2.5
rarhsmm	R	2.2.1.6
SequenceAnalysis	Python	2.2.2.4
signalHsmm	R	2.2.3.6
smmR	R	2.2.3.7
ziphsmm	R	2.2.1.7

Table 2.1. *List of software packages and the related section*

2.2. Software around HSMMs: state of the art

The descriptions of compared packages implementing HSMMs contain the following items: last update of the code; authors; references related to the package and model; source code URL; emission distributions; sojourn duration distributions; programming language; support of multiple sequences as opposed to a single sequence in HSMM estimation; estimation of the initial state distribution and automatic initialization of estimation algorithms. Regarding emission and sojourn duration distributions, a particular focus is given to the available parametric families and also, the possibility of using non-parametric distributions, meaning multinomial (or categorical) distributions for which each probability is a parameter. Regarding estimation of the initial law, the two following cases are distinguished: it is either fixed by the user or estimated (as the stationary distribution in some cases or a free parameter of the models in the other cases). There is no standardized format for data preprocessing across the different packages. Each package provides its own usage guide, typically through a manual in R or a README file in Python.

R packages implementing generic HSMMs are presented first, followed by the Python packages and eventually, packages implementing other models: mostly very specific HSMMs or multichain HMMs.

2.2.1. *R packages*

2.2.1.1. *hhsmm*

– Last update (on CRAN) and/or last commit: May 8, 2024 on CRAN.

– Authors (of the package): Morteza Amini (Sharif University of Technology), Afarin Bayat (University of Tehran) and Reza Salehian (Edinburgh Napier University).

– References (around the package and the theory): Amini et al. (2023).

– Code source URL if available: the CRAN page provides package code for the version 0.4.0: https://cran.r-project.org/web/packages/hhsmm/index.html and a git repository is available at: https://github.com/mortamini/hhsmm/.

– Emission distributions: the user can define customized emission distributions by providing the parameters, the density function, the generation function and the estimation function of the desired distributions. Available distributions in the package are the mixture of Gaussian linear (Markov-switching) models, mixture of multivariate normals, multinomial (or categorical), mixture of B-splines, Gaussian additive (Markov-switching) model and mixture of the robust emission (Qin et al. 2024).

– Sojourn duration distributions: the sojourn duration distributions available are the shifted negative binomial, the non-parametric, the shifted logarithmic, the shifted Poisson, the gamma, the Weibull and the (truncated) log-normal distributions.

– Programming language: the base language is R but main costly functions such as the forward backward algorithm are written in C.

– Multiple sequences + initial law: the package can handle multiple sequences and the initial law is estimated using the smoothed probabilities extracted from the forward-backward algorithm.

– Initialization EM: the package provides a function to initialize the model parameters using a k-means clustering technique (Lloyd 1982).

The package **hhsmm** involves tools for modeling multivariate and multi-sample time series by hidden hybrid Markov/semi-Markov models, introduced in Guédon (2005). A hidden hybrid Markov/semi-Markov model is a model with both Markovian and semi-Markovian states. Also, the **hhsmm** packages provide the following:

1) tools for model initialization through the k-means clustering technique (function **initial_cluster**);

2) left-to-right models in which the process goes from a state to the next state and never comes back to the previous state;

3) the ability to initialize, fit (function **hsmmfit**) and predict hidden sequences (function **predict.hsmm** using smoothed forward-backward results or the Viterbi algorithm) based on datasets containing missing values.

Moreover, the package **hhsmm** handles (1) Markov switching models and an auto-regressive hybrid hidden semi-Markov model, (2) a non-parametric estimation of the emission distribution using penalized B-splines and (3) continuous sojourn duration distributions.

2.2.1.2. *hsmm*

– Last update (on CRAN) and/or last commit: the last update on CRAN is April 25, 2013, and the package has been archived on the CRAN May 9, 2022.

– Authors (of the package): Jan Bulla, Ingo Bulla (University of Caen, France).

– References (around the package and the theory): Bulla et al. (2010).

– Code source URL if available: the CRAN archive of the package is available at: https://cran.r-project.org/web/packages/hsmm/index.html and the last version is 0.4.

– Emission distributions: according to the documentation, the Bernoulli, Normal, Poisson and Student's *t* distribution are available.

– Sojourn duration distributions: according to the documentation, non-parametric, geometric, negative binomial, logarithmic and Poisson distributions can be used.

– Programming language: the base language is R but main costly functions such as the forward backward algorithm are written in C++.

– Multiple sequences + initial law: the code can handle only one sequence and the initial law is fixed by the user.

– Initialization EM: the initial values of the parameters for the EM algorithm must be provided by the user.

The functions contained in the package address three important aspects of the HSMM:

1) the simulation of sequences of states and observations given the model specifications (sojourn duration and conditional distributions) and parameters (function **hsmm.sim**);

2) maximum likelihood estimation of the model parameters (function **hsmm**), given a sequence of observation and the model specifications; it estimates the HSMM parameters using the EM algorithm based on the right-censored approach initially described in Guédon (2003);

3) acquisition of information about the hidden layer sequence via the Viterbi algorithm (function **hsmm.viterbi**) and the smoothing probabilities **hsmm.smooth**.

2.2.1.3. *LaMa*

– Last update (on CRAN) and/or last commit: January 29, 2025 on CRAN.

– Authors: Jan-Ole Koslik (Universität Bielefeld, Germany).

– References (around the package and the theory): no.

– Code source URL if available: https://janoleko.github.io/LaMa/.

– Emission distributions: all univariate distributions in R.

– Sojourn duration distributions: HMM approximation of HSMMs using state-space augmentation and state aggregates in a dedicated way (Langrock and Zucchini 2011).

– Programming language: the base language is R but main functions are written in C++.

– Multiple sequences supported, initial law free or stationary.

– Initialization: the initial values are provided by the user for parameter estimation.

The **LaMa** package provides an implementation of hidden Markov models. HMMs are not represented through an instance of a class, but essentially through functions and the usual objects in R. As a result, users have to choose an available optimization algorithm in R and apply it to the likelihood function provided by the **LaMa** package. HSMMs are obtained by HMM approximations, implying that sojourn duration distributions are non-parametric except regarding the tail, which is geometric; the approach is described in Langrock and Zucchini (2011). Covariates can be included in HMMs, at the level of initial state probabilities, transition probabilities and emission distributions.

2.2.1.4. *mhsmm*

– Last update (on CRAN) and/or last commit: August 23, 2023 on CRAN.

– Authors: Jared O'Connell (University of Oxford, Great Britain) and Søren Højsgaard (Aarhus University, Denmark).

– References (around the package and the theory): O'Connell and Højsgaard (2011).

– Code source URL if available: the CRAN page provides package code for the version 0.4.21: https://cran.r-project.org/web/packages/mhsmm/index.html.

– Emission distributions: the user can define customized emission distributions by providing the parameters, the density function, the generation function and the estimation function of the desired distributions. Available distributions in the package are the normal, the multivariate normal and the Poisson distributions.

– Sojourn duration distributions: the sojourn duration distributions available are the shifted Poisson, the gamma and the non-parametric distributions.

– Programming language: the base language is R but main costly functions such as the forward backward algorithm are written in C.

– Multiple sequences + initial law: the package can handle multiple sequences and the initial law is estimated using the smoothed probabilities derived from the forward-backward algorithm.

– Initialization EM: the initial values of the parameters for the EM algorithm must be provided by the user.

The main features of the **mhsmm** package are as follows: Observations are allowed to be multivariate. Missing values are allowed. Observations must be recorded at equally spaced times. The package is designed to allow the specification of custom emission distributions. It is possible to have multiple sequences of data. Parameter estimation is made using EM algorithms based on the right censoring approach (Guédon (2003)). This package has the ability to estimate parameters for multiple observation sequences and is extensible because the user can create custom emission distributions. The main functions are as follows:

1) **hsmmspec** to specify a HSMM model with initial and state transition probabilities and sojourn and emission distributions;

2) **hsmmfit** to estimate the parameters using the right censoring EM approach;

3) **predict.hsmm** to predict the hidden state, given a HSMM model and observations, using the Viterbi algorithm or the smoothed probabilities extracted from the forward-backward approach.

2.2.1.5. *PHSMM*

– Last update: February 9, 2021.

– Authors: Jennifer Pohle.

– Reference: Pohle et al. (2021).

– Code source URL: https://cran.r-project.org/web/packages/PHSMM/index.html.

– Emission distributions: normal, gamma, Poisson and Bernoulli.

– Sojourn duration distributions: non-parametric. Flexible and data-driven estimation of the sojourn duration distributions without the need to make any distributional assumption.

– Programming language: R.

– Multiple sequences not supported, initial law deterministic or estimated: stationarity is assumed (default). Otherwise, the underlying state-sequence is assumed to enter a new state at time t = 1 and it is necessary to define the initial distribution delta (delta is a vector of length N containing the initial distribution).

– Initialization: estimates the parameters of a HSMM for univariate time series using numerical penalized maximum likelihood estimation. The numerical penalized maximum likelihood estimation requires starting values for each HSMM parameter. If these starting values are poorly chosen, the algorithm might fail in finding the global optimum of the penalized log-likelihood function. Therefore, it is advisible to repeat the estimation several times using different sets of starting values.

`PHSMM` (Pohle et al. 2021) provides tools for penalized maximum likelihood estimation of HSMMs with flexible state sojourn duration distributions. The computation of the likelihood is performed through a forward backward algorithm and the minimization is performed using the function `nlm` (Schnabel et al. 1985) that implements a Newton-type algorithm.

2.2.1.6. *rarhsmm*

– Last update: October 18, 2017.

– Authors: Zekun Jack Xu and Ye Liu.

– References: Xu and Liu (2021)
http://cran.nexr.com/web/packages/rarhsmm/rarhsmm.pdf.

– Code source URL: https://github.com/cran/rarhsmm.

– Emission distributions: multivariate normal.

– Sojourn duration distributions: non-parametric.

– Programming language: R.

– Multiple sequences supported, initial law deterministic.

– Initialization EM: deterministic.

`rarhsmm` is dedicated to fit Gaussian hidden Markov (or semi-Markov) models with/without autoregressive coefficients and with/without regularization. The fitting algorithm for the hidden Markov model is illustrated by Rabiner (1989). It relies on RcppArmadillo for linear algebra. An application with data from finance is presented. This package is no more maintained on CRAN.

2.2.1.7. *ziphsmm*

– Last update (on CRAN) and/or last commit: The last update on CRAN is May 22 2018, and the package has been archived on the CRAN, December 12, 2022.

– Authors: Zekun (Jack) Xu and Ye Liu.

– Reference: not applicable.

– Code source URL: https://cran.r-project.org/web/packages/ziphsmm/index.html.

– Emission distribution: Zero inflated Poisson.

– Sojourn duration distributions: logarithmic, geometric or shifted Poisson.

– Programming language: R and C++.

– Multiple sequences not supported. Stationarity is assumed.

– Initialization: the parameters estimation relies on the gradient descent of log likelihood and multiple starting values should be used to reduce the risk of convergence to local minima.

The package ziphsmm allows us to fit zero-inflated Poisson hidden (semi-)Markov models with or without covariates by directly minimizing the negative log likelihood function using the gradient descent algorithm. Gradients are numerically computed.

2.2.2. *Python packages*

2.2.2.1. *edhsmm*

– Last update and/or last commit (on GitHub): May 26, 2024. The project has been archived at this date.

– Authors (of the package): Ameer Carlo Lubang.

– References (around the package and the theory): the author relies on sections 2.2 and 3.1 of Yu (2010) to implement the fitting algorithm and on Benouareth et al. (2007) to implement the Viterbi algorithm.

– Code source URL if available: the github repository archive can be found here: https://github.com/poypoyan/edhsmm.

– Emission distributions: the distribution of observations available are the multivariate normal and the multinomial (or categorical) distributions.

– Sojourn duration distributions: the sojourn duration distributions available is the non-parametric distribution.

– Programming language: the base language is python but main costly functions such as the forward backward algorithm are written in cython.

– Multiple sequences + initial law: the package can handle multiple sequences and the initial law is estimated using the smoothed probabilities extracted from the forward-backward algorithm.

– Initialization EM: the package provides a function that initializes the model parameters by using a k-means clustering technique (Lloyd 1982), only for the model with observations based on multivariate normal distribution.

The library **edhsmm**, for explicit duration HSMM, is another implementation of the EM algorithm for HSMMs. The function **sample** allows us to simulate semi-Markov chains for a specified model. The function **fit** relies on the EM to estimate the model parameters. Note that one can use an option for the right-censoring approach and also an experimental option for a left-censoring approach. The function **score** computes the log-likelihood of a given dataset for a specified model and the function **predict** relies on the Viterbi algorithm to extract the most likely sequence of hidden state given a sequence of observations. In the given code examples, the author shows how we can parallelize the estimation using different initial sets of model parameters.

2.2.2.2. *hsmmlearn*

– Last update and/or last commit (on GitHub): August 21, 2021.

– Author: Joris Vankerschaver.

– References (around the package and the theory): No.

– Source code: https://github.com/jvkersch/hsmmlearn.

– Emission distributions: normal and multinomial (or categorical).

– Sojourn duration distributions: non-parametric distributions + custom.

– Programming language: C++/cython with python interface.

– Multiple sequences + initial law: the code can handle only one sequence and the initial law is assumed to depend on free parameters; it is estimated by the smoothed probabilities at initial time.

– Initialization EM: the initial values of the parameters for the EM algorithm must be provided by the user.

hsmmlearn is a library for unsupervised learning of HSMMs in the EDHMM family. It actually wraps the same underlying C++ code as the `hsmm` R package. HSMMLearn borrows its name and the design of its API from HMMLearn. The manual includes a tutorial for customized emission distributions. Another tutorial is available as a Python notebook: https://github.com/jvkersch/hsmmlearn/blob/master/notebooks/tutorial.ipynb.

2.2.2.3. *online_hmm*

– Last update and/or last commit (on GitHub): April 10, 2015. No more maintained.

– Author: Alberto Bietti.

– Reference: Bietti (2014); Bietti et al. (2015).

– Code source URL if available: https://github.com/albietz/online_hmm.

– Emission distributions: normal and multinomial (or categorical).

– Sojourn duration distributions: binomial and Poisson.

– Programming language: Python.

– Multiple sequences + initial law: the package cannot handle multiple sequences and the initial law is defined as a set of free parameters.

– Initialization EM: the initial values of the parameters for the EM algorithm must be provided by the user.

Description: the online_hmm Python library is dedicated to online state restoration and parameter estimation in HMMs and HSMMs, that is, when the whole observed sequence is not available at once but when new data arrive sequentially, requiring restoration and parameters to be updated regularly. The corresponding algorithms are described in Bietti (2014) and Bietti et al. (2015).

2.2.2.4. *OpenAlea: SequenceAnalysis*

– Last update and/or last commit (on GitHub): January 23, 2018/August 30, 2024.

– Authors: Yann Guédon, Jean-Baptiste Durand, Christophe Pradal, and Thomas Arsouze.

– References: Dufour-Kowalski et al. (2007).

– Code source URL if available: https://github.com/openalea/StructureAnalysis/tree/master/sequence_analysis (will not compile on November 11, 2024).

– Emission distributions: multivariate, conditionally independent observations given states. Discrete families: multinomial (or categorical), binomial, Poisson, negative binomial, and geometric Poisson. Continuous families: Gamma, zero-inflacted gamma, Gaussian, inverse Gaussian, and Von Mises.

– Sojourn duration distributions: binomial, Poisson, negative binomial, geometric Poisson, and uniform.

– Programming language: C++ with Python interface.

– Multiple sequences + initial law: the package can handle multiple sequences and the initial law is estimated using either the stationary distribution of the semi-Markov chain or using the MLE, assuming free parameters for initial law in the latter case.

– Initialization EM: the initial values of the parameters for the EM algorithm can be either provided by the user or estimated by default methods (geometric distributions for sojourn duration distributions).

Description: SequenceAnalysis is a python module dedicated to statistical sequence analysis (mainly, discrete events). It is part of the StructureAnalysis module, which implements various models and algorithms for the analysis of sequences and trees, among which hidden Markov trees, tree alignment and tree compression with directed acyclic graphs. StructureAnalysis is a component of the OpenAlea platform, which consists of using python as a gluing language to chain various function-structure plant models. SequenceAnalysis implements multiple rupture detection, variable-order (hidden) Markov chains, multiple sequences alignment and HSMMs, the latter being focused on in this section. HSMMs are of type explicit-duration HMM. All discrete distributions have shift parameters. By default, parametric families in emission and sojourn duration distributions are not fixed, they are selected by an information criterion during EM iterations. The 2018 version runs on python 2.7. Documentation is currently mainly outdated and partly inconsistent with code. This version can be downloaded on https://gitlab.inria.fr/statify_public/jemr-ema.

The code is currently being refactored to run on python 3 and the documentation is currently being added to development branches on GitHub.

2.2.2.5. *pyhsmm*

– Last update and/or last commit (on GitHub): August 01, 2020. Available on PyPi.

– Authors: Matt Johnson, Alex Wiltschko, Yarden Katz, Chia-ying (Jackie) Lee, Scott Linderman, Kevin Squire and Nick Foti.

– References: Johnson and Willsky (2013).

– Source code: https://github.com/mattjj/pyhsmm, https://pythonrepo.com/repo/mattjj-pyhsmm-python-science-and-data-analysis.

– Emission distributions: multivariate normal. It can be customized.

– Sojourn duration distributions: Poisson. It can also be customized.

– Programming language: C++/cython with python interface.

– Multiple sequences + initial law: the package can handle multiple sequences and the parameters are random variables with parameterized distributions in a Bayesian framework.

– Initialization EM: the package does not provide any functions to initialize the parameter but relies on prior distributions since it works in Bayesian framework.

This code is a Python library for approximate unsupervised inference in Bayesian hidden Markov models and explicit-duration hidden Markov models, focusing on the Bayesian non-parametric extensions, the hierarchical Dirichlet process (HDP-HMM) and HDP-HSMM, mostly with weak-limit approximations. This approach employs an HDP prior over an infinite state space that enables both inference of state complexity and Bayesian mixing over models of varying complexity. For each iteration of the loop, all the latent variables of the model will be resampled by Gibbs sampling steps, including the transition matrix, the observation means and covariances, the duration parameters and the hidden state sequence.

2.2.3. *Other relevant software*

We present here additional packages related to HSMMs that offer interesting features (either in terms of methodology or application) or that are connected to the themes of this book (e.g. `CHMM` for coupled HMMs and `smmR` for general SMMs).

2.2.3.1. *CHMM: R package*

– Last update on CRAN: September 29, 2017.

– Authors: Xiaoqiang Wang and Julie Aubert.

– References: Wang et al. (2019) https://cran.r-project.org/web/packages/CHMM/ https://hal.archives-ouvertes.fr/hal-01661257/document.

– Code source URL: https://github.com/julieaubert/CHMM.

– Emission distributions: normal.

– Sojourn duration distributions: Not concerned.

– Programming language: `R`.

– No multiple sequences, initial law defined specifically by the model.

– Initialization: for both the EM and the VEM estimation algorithm, clustering techniques (`mclust` (Fraley and Raftery 2002) or `kmeans` (Forgy 1965)) are used to initialize the model parameters.

The `CHMM` package (Wang et al. 2019) is dedicated to coupled hidden Markov model (not semi-Markovian). It manages multiple hidden Markov chain with modeled correlation between two hidden states of different chains at a given time-step. It provides an EM algorithm as well as a VEM (variationnal expectation maximization) algorithm arguing that the model is not tractable with the EM if the state space or the length of the chains is large. The leading application is to detect CNV (copy number variations) in DNA sequences.

2.2.3.2. *DNN-HSMM: Python library*

– Last update and/or last commit (on GitHub): March 14, 2021.

– Authors: Shinji Takaki.

– Reference: Tokuda et al. (2016).

– Code Source URL: https://github.com/sp-nitech/DNN-HSMM.

– Emission distributions: normal.

– Sojourn duration distributions: discrete normal.

– Programming language: Python.

– Multiple sequence + initial law: the code can handle only one sequence and the initial law is fixed, deterministic (first state is state 1) and the model is a left-right HSMM.

– Initialization gradient descent: the initial values of the parameters for the gradient descent must be provided by the user.

DNN-HSMM is a python implementation based on pytorch of a model dedicated to speech synthesis. It is thus oriented toward simulation and state restoration is not implemented. The HSMM parameters are encoded by a deep neural network (DNN).

State transitions are left-right and deterministic (only from each state i to state $i+1$), so they do not need to be encoded. The other distributions are assumed to be normal and are directly encoded as an output layer of a DNN (the authors do not mention how they constrain variance parameters to be non-negative). Parameter estimation is achieved by computing the derivatives of the HSMM log-likelihood, which are back-propagated to the neural network through pytorch.

2.2.3.3. *GENSCAN: C program*

– Last update: February 18, 2003.

– Authors: Chris Burge.

– References: Burges and Karlin (1997) https://www.alliot.fr/BIO/PDF/BurgeThesis.pdf.

– Code source: http://hollywood mit.edu/GENSCAN.html.

– Emission distributions: non-parametric.

– Sojourn duration distributions: non-parametric.

– Programming language: C.

– Multiple sequence + initial law: not detailed.

`GENSCAN` is a program only dedicated to a particular application: the identification of complete gene structures in genomic DNA. It uses HSMMs, with non-parametric distributions for emission and sojourn duration.

2.2.3.4. *hmmTMB: R package*

– Last update (on CRAN): October 24, 2023. Last commit on GitHub: January 14, 2025.

– Authors: Theo Michelot and Richard Glennie.

– References: Michelot (2023) https://cran.r-project.org/web/packages/hmmTMB/.

– Code source URL: https://github.com/TheoMichelot/hmmTMB/.

– Emission distributions: beta, (zero-inflated) binomial, multinomial (or categorical), Dirichlet, exponential, folded normal, (zero-inflated) gamma, log-normal, multivariate normal, (zero-inflated or zero-truncated) negative binomial, (zero-inflated or zero-truncated) Poisson, Student's *t*, truncated normal, Tweedie, von Mises, Weibull and wrapped Cauchy.

– Sojourn duration distributions: HMM approximation of HSMMs using state-space augmentation and state aggregates in a dedicated way (Langrock and Zucchini 2011).

– Programming language: C++ and R.

– Multiple sequences supported, initial law free or stationary.

– Initialization EM: deterministic or estimated by k-means.

hmmTMB provides an implementation of hidden Markov models based on Template Model Builder (TMB). TMB (Kristensen et al. 2016) is another R package for estimating statistical models with latent variables. HMMs may include fixed and random effects, smoothing splines and approximations of HSMMs by HMMs. Maximum likelihood and Bayesian estimation are possible. Maximum likelihood estimation relies on library optimix, not on EM. HMM approximations of HSMMs implies that sojourn duration distributions are non-parametric except regarding the tail, which is geometric; the approach is described in Langrock and Zucchini (2011).

2.2.3.5. *mHMMBayes: R package*

– Last update (on CRAN): April 4, 2024.

– Authors: Emmeke Aarts and Sebastian Mildiner Moraga.

– References: Zhang and Berhane (2014); de Haan-Rietdijk et al. (2017)
https://cran.r-project.org/web/packages/mHMMbayes/
https://cran.r-project.org/web/packages/mHMMbayes/vignettes/tutorial-mhmm.html.

– Code source URL: https://github.com/emmekeaarts/mHMMbayes.

– Emission distributions: multivariate discrete multinomial (or categorical) and Poisson or normal.

– Sojourn duration distributions: geometric only (Markov model).

– Programming language: C++ and R.

– Multiple sequences supported, initial law is supposed to be the stationary distribution and thus, is not a free parameter.

– Initialization MCMC: initial values to be provided. Priors are by default inverse Wishart for covariance matrices and multivariate normal otherwise.

mHMMBayes is an implementation of a hidden Markov model with mixed effects (so-called multilevel) using Bayesian estimation in R. The multilevel HMM is tailored to accommodate (intense) longitudinal data of multiple individuals simultaneously. Using a multilevel framework offers the possibility to account for heterogeneity in the model parameters (transition probability matrix and conditional distribution), while estimating one overall HMM. The model can be fitted on multivariate data with either a multinomial (or categorical), normal or Poisson distribution and includes individual level covariates (allowing for group comparisons on model parameters). Parameters are estimated using Bayesian estimation by means of the forward-backward recursion within a hybrid Metropolis within Gibbs sampler. Missing data (NA) in the dependent variables is treated assuming missing at random.

2.2.3.6. *signalHsmm: R package*

– Last update (on CRAN): November 15, 2018; last commit (on GitHub): May 4, 2020.

– Authors: Michal Burdukiewicz, Piotr Sobczyk and Jaroslaw Chilimoniuk.

– References: Burdukiewicz et al. (2018).

– Code source: https://cran.r-project.org/web/packages/signalHsmm/index.html https://github.com/michbur/signalHsmm.

– Emission distributions: non-parametric.

– Sojourn duration distributions: non-parametric.

– Programming language: R.

– Multiple sequence + initial law: the package cannot handle multiple sequences and the initial law is defined as a set of free parameters.

– Initialization EM: not detailed.

`signalHsmm` package predicts the presence of signal peptides in eukaryotic protein using HSMMs, with non-parametric distributions for emission and sojourn duration. It is only dedicated to this particular application.

2.2.3.7. *smmR: R package*

– Last update (on CRAN): August 3, 2021.

– Authors: Vlad Barbu, Caroline Bérard, Dominique Cellier, Florian Lecocq, Corentin Lothodé, Mathilde Sautreuil and Nicolas Vergne.

– References: Barbu et al. (2023).

– Code source URL: https://cran.r-project.org/web/packages/smmR/index.html https://plmlab.math.cnrs.fr/lmrs/statistique/smmR.

– Emission distributions: not concerned.

– Sojourn duration distributions: non-parametric or uniform, geometric, Poisson, discrete Weibull and negative binomial.

– Programming language: R.

– Multiple sequences supported, initial law deterministic or estimated.

– Initialization EM: not concerned.

`smmR` package is dedicated to estimation and simulation of semi-Markov models defined in a discrete state space (see Chapter 1, section 1.2.2, definition 1.1). Authors have considered parametric and non-parametric estimation; with and without censoring at the beginning and/or at the end of sample paths; different types of sojourn duration (depending on previous state and present state, only on previous state, only on present state, or not depending on the previous and present state), one or several independent sample paths. Reliability measures are computed: reliability, maintainability, availability and failure rates.

2.3. Comparative overview: R and Python packages for HSMM

HSMMs are widely explored stochastic models, with several libraries and tools available for their application. Each package offers specific features and approaches based on the needs and priorities of the users, particularly regarding general characteristics, sojourn duration distributions and emission distributions. In this section, we provide a comparative overview of the main Python and R libraries available for estimating HSMMs, highlighting their differences and specificities. The comparison criteria include the following:

– **General characteristics** (see Table 2.2):

- *support for multiple sequences*: the ability of the libraries to handle multiple observation sequences simultaneously;

- *missing data*: the ability to deal with missing data in observations. Some software are able to estimate missing values;

- *how to deal with initial distribution*: it can be fixed by the user and/or estimated (by stationary distribution or another method);

- *model initialization*: different methods are provided for initializing model parameters, including clustering algorithms or methods based on prior distributions;

- *comparison criteria*: some packages offer comparison criteria for model selection.

– **Sojourn duration distributions**: different available distributions for the time that is spent in a hidden state before the transition of the chain to another state (see Table 2.3).

– **Emission distributions**: different available distributions for the observations given the hidden states (see Table 2.4).

The tables in the following subsections summarize these characteristics for a selection of libraries, providing a clear and detailed overview of the available options for HSMM modeling.

2.3.1. *General comparison*

Table 2.2 provides a comparison of various HSMM software tools available in R and Python through the general characteristics listed in the previous paragraph. This table also includes the programming language used and the last update year for each software. The entries are color coded to highlight the presence (green) or absence (red) of each feature, with specific notations used for different types of initial distributions (F for fixed and E for estimated). This allows users to easily compare the software's functionalities and determine which tool best suits their needs.

Notably, tools such as mhsmm, hhsmm and mHMMBayes offer comprehensive support for these features, including automatic initialization and comparison criteria, while others like hsmm, ziphsmm and CHMM are more limited in their capabilities. The programming language and update frequency also vary, with some tools being actively maintained, such as hhsmm (2022) or mHMMBayes (2024), and others, like hsmm and online_hmm, having not been updated in several years. This summary provides a quick comparison to help users choose the appropriate software for their needs.

Software	Multiple sequences	Missing data	Initial distribution	Proposed automatic initialization	Comparison criteria	Language	Last update
CHMM	-	-	E	x	-	R	2017
DNN-HSMM	-	-	F	-	-	Python	2021
edhsmm	x	-	F	x	-	Python	2023
GENSCAN	-	-	F	-	-	C	2003
hhsmm	x	x	F	x	x	R	2022
hmmTMB	x	-	E	-	-	R	2025
hsmm	-	-	F	-	-	R	2013
hsmmlearn	-	-	F	x	-	Python	2021
LaMa	x	-	E	-	-	R	2025
mHMMBayes	x	x	F, E	x	x	R	2024
mhsmm	x	x	F,E	-	-	R	2017
online_hmm	-	-	F	-	-	Python	2015
PHSMM	-	-	F, E	x	-	R	2021
pyhsmm	x	-	E	-	x	Python	2020
rarhsmm	x	-	-	-	-	R	2018
SequenceAnalysis	x	-	F, E	x	x	Python	2016
signalHsmm	-	-	F	-	-	R	2018
smmR	x	-	F,E	-	x	R	2021
ziphsmm	-	-	F	-	-	R	2018

Table 2.2. *Features of HSMM software available in R or Python for multiple observation sequences ("Multiple sequences"), missing observations ("Missing data"), F for fixed initial distribution (by the user), E for Estimated initial distribution (by stationary distribution or another method). x is used for a software that handles the feature and - is used for a software that does not handle the feature. For a color version of this table, see www.iste.co.uk/peyrard/guidetohsmm.zip*

2.3.2. *Sojourn durations*

Table 2.3 describes the different sojourn duration distributions offered by various software packages. We can note that a large number of these packages provide non-parametric distributions. The next most commonly used distributions are the geometric distribution (which corresponds to the Markovian case) and the Poisson distribution.

Software	Non-parametric	Gamma	Geometric	Logarithmic	LogNormal	Negative Binomial	Poisson	Uniform	Discrete Weibull	Binomial	Custom
CHMM	-	-	-	-	-	-	-	-	-	-	-
DNN-HSMM	-	-	-	-	-	-	-	-	-	-	-
edhsmm	x	-	-	-	-	-	-	-	-	-	-
GENSCAN	x	-	x	-	-	-	-	-	-	-	-
hhsmm	x	x	x	shifted	x	shifted	shifted	-	x	-	-
hmmTMB	x*	-	x	-	-	-	-	-	-	-	-
hsmm	x	-	x	x	-	-	x	x	-	-	-
hsmmlearn	x	-	-	-	-	-	-	-	-	-	-
LaMa	x*	-	x	-	-	-	-	-	-	-	-
mHMMBayes	-	-	x	-	-	-	-	-	-	-	-
mhsmm	x	x	-	-	-	-	-	shifted	-	-	-
online_hmm	-	-	-	-	-	-	x	-	-	x	-
PHSMM	x	-	-	-	-	-	-	-	-	-	-
pyhsmm	-	-	-	-	-	-	x	-	-	-	x
rarhsmm	x	-	-	-	-	-	-	-	-	-	-
SequenceAnalysis	x	-	shifted	-	-	shifted	shifted	x	-	shifted	-
signalHsmm	x	-	-	-	-	-	-	-	-	-	-
smmR	-	-	x	-	-	x	x	x	x	-	-
ziphsmm	-	-	x	x	-	-	x	-	-	-	-

Table 2.3. *Sojourn duration (or sojourn time, dwell time, runlength) distribution. Shifted means an estimated shift is given. * possible extension to non-parametric with geometric tail.* x *is used for a software that handles the feature and* - *is used for a software that does not handle the feature. For a color version of this table, see www.iste.co.uk/peyrard/guidetohsmm.zip*

The packages `hsmm`, `hhsmm`, `smmR` and `SequenceAnalysis` are the ones offering the most diverse range of distributions. Additionally, some packages enable the estimation of a shift for a given distribution.

In conclusion, while there is a broad variety of distribution types available across different software packages, certain tools stand out for their flexibility in offering multiple options and specialized features, such as the estimation of distribution shifts.

2.3.3. *Observations*

Table 2.4 provides an overview of the different emission distributions offered by the listed packages, including Bernoulli, custom, normal, multivariate normal, Poisson, Student's *t*, and non-parametric (i.e. finite discrete) distributions (also referred to as either multinomial or categorical distributions).

Software	Bernoulli	Custom	Normal	Multivariate normal	Poisson	Student's t	Non-parametric
CHMM	-	-	x	-	-	-	-
DNN-HSMM	-	-	x	-	-	-	-
edhsmm	-	-	x	x	-	-	x
GENSCAN	-	-	-	-	-	-	-
hhsmm	-	x	x	x	-	-	x
hmmTMB	x	x	x	x	x	x	x
hsmm	x	-	x	-	x	x	-
hsmmlearn	-	x	x	-	-	-	x
LaMa	x	x	x	-	x	x	x
mHMMBayes	-	-	x	-	x	-	x
mhsmm	-	x	x	x	x	-	-
online_hmm	-	-	x	-	-	-	x
PHSMM	x	-	x	-	x	x	-
pyhsmm	-	-	x	x	-	-	-
rarhsmm	-	-	x	x	-	-	-
SequenceAnalysis	x	-	x	-	shifted	-	x
signalHsmm	-	-	-	-	-	-	-
smmR	-	-	-	-	-	-	-
ziphsmm	-	-	-	-	x*	-	-

Table 2.4. *Conditional emission distributions.* x *is used for a software that handles the feature and* - *is used for a software that does not handle the feature. (* Zero-inflated Poisson is used ; shifted means an estimated shift is given). For a color version of this table, see www.iste.co.uk/peyrard/guidetohsmm.zip*

Several software packages, such as `hsmm`, `hhsmm`, `mhsmm`, `LaMa` and `PHSMM`, offer a broad range of distributions, with `hmmTMB` standing out for supporting all aforementioned listed distributions. The Poisson and Normal distributions are among the most commonly supported types, followed by others such as multinomial and Student's *t*. Notably, some packages, like `mhsmm` and `SequenceAnalysis`, provide support for multivariate normal and shifted distributions, where "shifted" refers to an estimated shift applied to the distribution.

2.4. Illustration of the use of two packages for the toy examples

2.4.1. *Docker image*

Docker (Merkel 2014) is a platform designed to help developers build, share and run container applications. A container is a bundle of application software with all the dependencies required for the applications to run. For testing purposes, we built such a container with snapshots of libraries listed in this chapter. The container image can be found at: https://forgemia.inra.fr/inca-hsmm/software-review/containerregistry. The libraries installed are **edhsmm** (version 0.2.2), **hhsmmlearn** (commit version 69bc8aa), **hhsmm** (version 0.2.5) and **PHSMM** (version 1.0).

To make it work, it is first required to install **docker** (or alternatively **podman**) and then launch the container. Based on this container, one can launch a Jupyter Lab (Kluyver et al. 2016) using the following command on a Linux device:

```
docker run --rm -it -p 8888:8888   \
  registry.forgemia.inra.fr/inca-hsmm/software-review:2.0
```

Then, a localhost URL is given such as http://localhost:8888/tree that must be opened into a web browser.

Alternatively one can launch a bash into the container with the following command:

```
docker run --rm -it -p 8888:8888   \
  registry.forgemia.inra.fr/inca-hsmm/software-review:2.0   \
   bash
```

A new prompt indicates the user has entered the image. Then, we can load the multinomial HSMM formalism from the **edhsmm** python library with the following commands:

```
python3
>>> from edhsmm.hsmm_multinom import MultinomialHSMM
```

Or, if we prefer to launch an R console and load for example the **hhsmm** package, one can simply use the following commands:

```
R
> library(hhsmm)
```

In the following sections, while mobilizing this container image, we offer two examples of use of the libraries cited above to run the two toy models of Squirrel and Deer, respectively.

2.4.2. Python *package* edhsmm *on toy model Squirrel*

Reminder of the application and the model

The HSMM proposed here is an instance of the toy model Squirrel detailed in Chapter 1. During winter, a squirrel wanders from reserve to reserve where hazelnuts have been stored during the summer time. The K reserves are identified and one collects, based on a naturalist expertise, and guesses on where is the squirrel every single day along winter time. As explained in section 1.2.1, this behavior can be described by a HSMM.

Let us assume that there are $K = 3$ reserves and that we have collected 300 sequences of such observations. A sequence, which is the location of a single Markov chain in our case, is composed of 500 successive reserves to be interpreted as guesses of where is a single squirrel during winter time.

For convenience, we will simulate hidden states as well as observations using a "true" HSMM model with transition probability matrix, M.

$$M = \begin{pmatrix} 0.0 & 0.5 & 0.5 \\ 0.3 & 0.0 & 0.7 \\ 0.4 & 0.6 & 0.0 \end{pmatrix}$$

We assume that the maximal duration that a squirrel can spend in a reserve is 4 days and we use a non-parametric distribution to model the duration. In the observation model, we model the error of the naturalist guesses on the location of the squirrel by a multinomial distribution. Together, the duration and the observation models are detailed below. The notations from Chapter 1 are used. Thus, for all n and t, $\mathbb{P}(X_{n+1} = d \,|\, J_n = k)$ represents the probability that the squirrel stays in reserve k for d consecutive days once it enters k. On the other hand, $\mathbb{P}(Y_t = k_1 \,|\, Z_t = k_2)$ with $k_1 \neq k_2$ is the probability that, at day t, the naturalist has guessed the squirrel to be in reserve k_1, whereas it was in reserve k_2.

k	$\mathbb{P}(X_{n+1} = 1 \,\|\, J_n = k)$	$\mathbb{P}(X_{n+1} = 2 \,\|\, J_n = k)$	$\mathbb{P}(X_{n+1} = 3 \,\|\, J_n = k)$	$\mathbb{P}(X_{n+1} = 4 \,\|\, J_n = k)$
1	0.1	0.005	0.005	0.89
2	0.1	0.005	0.89	0.005
3	0.1	0.89	0.005	0.005

k	$\mathbb{P}(Y_t = 1 \mid Z_t = k)$	$\mathbb{P}(Y_t = 2 \mid Z_t = k)$	$\mathbb{P}(Y_t = 3 \| Z_t = k)$
1	0.8	0.1	0.1
2	0.1	0.8	0.1
3	0.1	0.1	0.8

Script and results

First of all, we give the model parameters by writing a function that specifies the HSMM parameters of a Python object proposed by the package `edhsmm`. There are the initial state distribution (vector of size $K = 3$), the non-parametric sojourn distribution (a matrix of size 3*4), the transition matrix (matrix of size 3*3) and the non-parametric emission distribution (a matrix of size 3*3).

```
import sys

import numpy as np
from edhsmm.hsmm_multinom import MultinomialHSMM
from matplotlib import pyplot as plt
from scipy import stats

def init_true_model(hsmm_class):
    hsmm_class.pi = np.array([2 / 3, 1 / 3, 0 / 3])
    hsmm_class.dur = np.array([
        [0.1, 0.005, 0.005, 0.89],
        [0.1, 0.005, 0.89, 0.005],
        [0.1, 0.89, 0.005, 0.005]
    ]) # meaning that the durations are between 1 and 4
    hsmm_class.tmat = np.array([
        [0.0, 0.5, 0.5],
        [0.3, 0.0, 0.7],
        [0.4, 0.6, 0.0]
    ])
    hsmm_class.emit = np.array([
        [0.8, 0.1, 0.1],
        [0.1, 0.8, 0.1],
        [0.1, 0.1, 0.8]
    ])
```

This HSMM model will serve as a simulator of observed trajectories. To simulate those trajectories, we can rely on the following code that simulates 300 trajectories of length 500.

```
rng_seed = 12598
squirrel_true_model = MultinomialHSMM(
  n_states = 3, n_durations = 4, random_state=rng_seed)
init_true_model(squirrel_true_model)

n_samples = 500
n_obs = 300
n_var = 1

all_obs = np.empty((n_samples * n_obs, n_var),
                   dtype=np.int64)
all_len = []
ctr_len = 0
for i in range(n_obs):
    sample_len, sample, _ = squirrel_true_model.sample(
        n_samples=n_samples, random_state=rng_seed)
    all_obs[ctr_len: ctr_len + sample_len, :] = sample
    all_len += [sample_len]
    ctr_len += sample_len
    rng_seed += 1
```

We now provide another HSMM model that will be used as an initial model for the EM algorithm for parameter estimation.

```
def init_starting_model(hsmm_class):
    hsmm_class.pi = np.array([0.5, 0.5, 0])
    hsmm_class.dur = np.array([
        [0.1, 0.2, 0.1, 0.6],
        [0.1, 0.1, 0.6, 0.2],
        [0.1, 0.6, 0.1, 0.2]
    ])
    hsmm_class.tmat = np.array([
        [0.0, 0.4, 0.6],
        [0.8, 0.0, 0.2],
        [0.6, 0.4, 0.0]
    ])
    hsmm_class.emit = np.array([
        [0.6, 0.3, 0.1],
        [0.2, 0.4, 0.4],
        [0.1, 0.2, 0.7]
    ])
```

To perform 20 iterations of the EM algorithm, we can use the following code. A tolerance level must be provided in order to define the convergence criterion.

```
np.random.seed(rng_seed)
squirrel_initial_model = MultinomialHSMM(n_states = 3,
                                        n_durations = 4)
init_starting_model(squirrel_initial_model)
squirrel_initial_model.n_iter = 20
# Estimate parameters
squirrel_initial_model.tol = 1e-2
squirrel_initial_model.fit(all_obs, all_len)
```

After seven loops, for a duration of 30 s, the convergence is reached and the transition matrix obtained is the following:

$$M = \begin{pmatrix} 0.0 & 0.41 & 0.59 \\ 0.27 & 0.0 & 0.73 \\ 0.53 & 0.47 & 0.0 \end{pmatrix}$$

The non-parametric sojourn duration distribution is the following:

k	$\mathbb{P}(X_{n+1} = 1 \,\vert\, J_n = k)$	$\mathbb{P}(X_{n+1} = 2 \,\vert\, J_n = k)$	$\mathbb{P}(X_{n+1} = 3 \,\vert\, J_n = k)$	$\mathbb{P}(X_{n+1} = 4 \,\vert\, J_n = k)$
1	5.2e-02	9.8e-05	6.8e-03	9.41e-01
2	2.3e-02	4.2e-02	8.9e-01	4.1e-02
3	3.3e-02	9.7e-01	8.3e-04	1.7e-04

The non-parametric emission is the following:

k	$\mathbb{P}(Y_t = 1 \,\vert\, Z_t = k)$	$\mathbb{P}(Y_t = 2 \,\vert\, Z_t = k)$	$\mathbb{P}(Y_t = 3 \Vert Z_t = k)$
1	0.85	0.1	0.05
2	0.1	0.8	0.1
3	0.05	0.08	0.87

Finally, the initial distribution is $\pi = (1, 0, 0)$. Except the initial distribution the other parameters can be considered as well estimated. But we notice that a large number of long sequences must be simulated in order to reach those results.

The Jupyter Notebook (Kluyver et al. 2016), in the squirrel directory of the lab, can also be found at: https://forgemia.inra.fr/inca-hsmm/software-review/-/blob/main/squirrel/squirrel.ipynb

2.4.3. R *package* hhsmm *on deers*

Reminder of the application and the model

The HSMM model proposed here is an instance of the toy model Deer detailed in Chapter 1 (section 1.4.2).

We aim to establish a relationship between accelerometer data and the activities of cervids using a HSMM and more specifically an ED-HMM (see the definition in Chapter 1, section 1.2.4). In this model, the observed variable corresponds to the accelerometer data (initially three-dimensional, here summarized by the roll angle, an indicator that varies with the activity), while the hidden variable represents the animal's activity.

We consider five different behaviors (foraging head-down, grooming, running, unmoving and walking head-up), and $\Omega_Z = \{1, \ldots, 5\}$. Variable Z_t is behavior at time t. Observation $Y_t \in \Omega_Y = \mathbb{R}$ is the scalar information extracted from the accelerometer at time t. The time step is 0.125 s, corresponding to a 8 Hz frequency.

We propose here an ED-HMM to model the dynamics of (Z_t, Y_t) as follows. The transition matrix M is estimated non-parametrically. The distribution of the sojourn duration in behavior i is Poisson with parameter λ_i. The emission distribution is normal with mean m_i and variance σ_i^2, computed for value y.

The transition matrix M, obtained from an experimental dataset with available behaviors, is:

$$M = \begin{pmatrix} 0.00 & 0.025 & 0.028 & 0.744 & 0.203 \\ 0.028 & 0.00 & 0.014 & 0.888 & 0.070 \\ 0.080 & 0.028 & 0.00 & 0.256 & 0.636 \\ 0.279 & 0.269 & 0.032 & 0.0 & 0.420 \\ 0.236 & 0.043 & 0.071 & 0.650 & 0.00 \end{pmatrix}$$

and the parameters for the sojourn duration distribution and the emission distribution are:

i	λ_i	m_i	σ_i^2
1	98.02	20.41	12.04
2	67.42	-14.98	14.45
3	37.69	-12.88	9.01
4	103.64	-24.60	14.61
5	46.13	-4.32	8.61

Script and results

First of all, we give the model parameters:

```
## Transition matrix
P <- matrix(c(
  0.00, 0.025, 0.028, 0.744, 0.203,
  0.028, 0.00, 0.014, 0.888, 0.070,
  0.080, 0.028, 0.00, 0.256, 0.636,
  0.279, 0.269, 0.032, 0.0, 0.420,
  0.236, 0.043, 0.071, 0.650, 0.00
), nrow = 5, ncol = 5, byrow = TRUE)

## Parameters from sojourn duration and emissions
E = matrix(c(
   98.02, 20.41, 12.04,
   67.42, -14.98, 14.45,
   37.69, -12.88, 9.01,
   103.64, -24.60, 14.61,
   46.13, -4.32, 8.61
), nrow = 5, ncol = 3, byrow = TRUE)
```

and we create the object `model`:

```
## Number of states
J = 5
## Initial distribution
initial = c(1,rep(0,J-1))
## Are the states semi-Markovian ?
semi = rep(TRUE,J)
## Parameters of the emission distributions
par = list(
        mu=as.list(E[,2]),
        sigma = as.list(E[,3]),
        mix.p=as.list(rep(1,J)))
## Parameters of the sojourn duration distributions
sojourn <- list(shape = E[,1], scale = rep(1,J),
              type = "gamma")
## Specification of the model object
model <- hhsmmspec(
           init = initial,
           transition = P,
           parms.emis = par,
```

```
            dens.emis = dmixmvnorm,
            sojourn = sojourn,
            semi = semi)
```

Remark that we consider a Gamma distribution with parameter $(\lambda, 1)$ instead of a Poisson distribution with parameter λ, because of the assumed particularity of this package to easier estimate the parameters of a Gamma distribution, regarding the sojourn duration. Note also that this package provides the possibility of having a mixture of normal distributions for the emissions; however, this possibility is not used here.

Then, we simulate some sequences with this model:

```
train <- simulate(model, nsim = c(100, 80, 80, 180),
         seed = 1234, remission = rmixmvnorm)
```

At last, we estimate all the parameters thanks to the `hhsmmfit` command:

```
clus = initial_cluster(train, nstate = J,
        nmix = rep(1,J), ltr = FALSE,
        final.absorb = FALSE, verbose = TRUE)
initmodel = initialize_model(clus = clus,
              sojourn = "gamma", M = max(train$N),
              semi = semi)
fit = hhsmmfit(x = train, model = initmodel,
      M = max(train$N))
```

For this example, the computation time is approximately 5 min on a standard computer (Ubuntu 22.04.5 LTS, Intel Core i5-9400K, 32 Go DDR4, NVIDIA Quadro P620).

We provide then the estimated values for all the model parameters. The transition matrix M is estimated by:

$$M = \begin{pmatrix} 0.000 & 0.028 & 0.042 & 0.029 & 0.902 \\ 0.282 & 0.000 & 0.057 & 0.257 & 0.404 \\ 0.544 & 0.027 & 0.000 & 0.267 & 0.162 \\ 0.263 & 0.171 & 0.135 & 0.000 & 0.431 \\ 0.917 & 0.018 & 0.016 & 0.048 & 0.000 \end{pmatrix}$$

The estimated parameters for the sojourn duration distribution and the emission distribution are:

i	η_i	m_i	σ_i^2
1	14.168	-24.173	14.663
2	86.791	20.369	12.137
3	12.896	-14.578	13.608
4	43.019	-4.303	8.565
5	15.722	-24.991	14.385

We observe that the estimates are not particularly accurate; therefore, it may be beneficial to use the package with initial values that are closer to the true model, rather than those estimated from a preliminary basic clustering.

At last, we estimate all the parameters thanks to the `hhsmmfit` command.

```
## Initial model is the true model !
model$mstep = initmodel1$mstep
fit = hhsmmfit(x = train , model = model ,
        M = max(train$N))
```

With initial model parameters closer to the true, `hhsmm` provides good estimators for all the model parameters in only 1 min.

The estimated transition matrix M is given by:

$$M = \begin{pmatrix} 0.000 & 0.014 & 0.043 & 0.686 & 0.257 \\ 0.035 & 0.000 & 0.035 & 0.860 & 0.070 \\ 0.000 & 0.048 & 0.000 & 0.190 & 0.762 \\ 0.254 & 0.288 & 0.034 & 0.000 & 0.424 \\ 0.171 & 0.045 & 0.090 & 0.694 & 0.000 \end{pmatrix}$$

The estimated parameters for the sojourn duration distribution and the emission distribution are given as:

i	η_i	m_i	σ_i^2
1	86.791	20.369	12.137
2	74.942	-14.974	13.762
3	40.699	-12.751	9.179
4	107.340	-24.578	14.680
5	42.915	-4.301	8.560

The `hhsmm` script can be found at: https://forgemia.inra.fr/inca-hsmm/software-review/-/blob/main/deer/hhsmm.R

2.5. Conclusion

In conclusion, while the available software packages for HSMM provide various tools for modeling, they primarily focus on the more restricted framework of ED-HMMs, where sojourn durations are only dependent on the current state. Among these, smmR stands out as the only package that incorporates more general sojourn duration distributions. This extension of the ED-HMM framework is an area that we aim to further generalize through hsmmR, a package developed by ourselves and dedicated to simulation and estimation of general hidden semi-Markov models, soon available on CRAN (https://cran.r-project.org/web/packages/smmR/index.html).

Selecting the most appropriate package for a given application remains challenging, as the differences in functionality can be subtle and context dependent. Few packages address mixed effects, random effects and the incorporation of covariates, which are essential for more complex modeling. Moreover, estimation remains a major hurdle, often requiring large datasets to achieve reliable results. This highlights the need for ongoing development in this field, with particular attention to improving model flexibility and estimation accuracy.

2.6. References

Amini, M., Bayat, A., Salehian, R. (2023). hhsmm: An R package for hidden hybrid Markov/semi-Markov models. *Computational Statistics*, 38, 1283–1335.

Barbu, V.S., Lecocq, F., Lothodé, C., Vergne, N. (2023). smmR: A semi-Markov R package. *Journal of Open Source Software*, 8(85), 4365.

Benouareth, A., Ennaji, A., Sellami, M. (2007). Arabic handwritten word recognition using HMMs with explicit state duration. *EURASIP Journal on Advances in Signal Processing*, 2008, 247354.

Bietti, A. (2014). Online learning for audio clustering and segmentation. PhD Thesis, ENS Cachan.

Bietti, A., Bach, F., Cont, A. (2015). An online EM algorithm in hidden (semi-)Markov models for audio segmentation and clustering. In *Proceedings of ICASSP*, Brisbane.

Bulla, J., Bulla, I., Nenadicc, O. (2010). hsmm – An R package for analyzing hidden semi-Markov models. *Computational Statistics & Data Analysis*, 54(3), 611–619.

Burdukiewicz, M., Sobczyk, P., Chilimoniuk, J., Gagat, P., Mackiewicz, P. (2018). Prediction of signal peptides in proteins from malaria parasites. *International Journal of Molecular Sciences*, 19(12), 3709.

Burges, C. and Karlin, S. (1997). Prediction of complete gene structures in human genomic DNA. *Journal of Molecular Biology*, 268(1), 78–94.

Dufour-Kowalski, S., Pradal, C., Dones, N., Barbier de Reuille, P., Boudon, F. et al. (2007). OpenAlea: An open-source platform for the integration of heterogeneous Review of HSMM R and Python software FSPM components. *Proceedings of the 5th International Workshop on Functional Structural Plant Models – FSPM07*, Napier, 4–9 November, 1–2.

Forgy, E. (1965). Cluster analysis of multivariate data: Efficiency versus interpretability of classifications. *Biometrics*, 21, 768–780.

Fraley, C. and Raftery, A.E. (2002). Model-based clustering, discriminant analysis, and density estimation. *Journal of the American Statistical Association*, 97(458), 611–631.

Guédon, Y. (2003). Estimating hidden semi-Markov chains from discrete sequences. *Journal of Computational and Graphical Statistics*, 12(3), 604–639.

Guédon, Y. (2005). Hidden hybrid Markov/semi-Markov chains. *Computational Statistics & Data Analysis*, 49(3), 663–688.

de Haan-Rietdijk, S., Kuppens, P., Bergeman, C.S., Sheeber, L.B., Allen, N.B., Hamaker, E.L. (2017). On the use of mixed Markov models for intensive longitudinal data. *Multivariate Behavioral Research*, 52(6), 747–767.

Johnson, M.J. and Willsky, A.S. (2013). Bayesian nonparametric hidden semi-Markov models. *Journal of Machine Learning Research*, 14(1), 673–701.

Kluyver, T., Ragan-Kelley, B., Pérez, F., Granger, B., Bussonnier, M., Frederic, J., Kelley, K., Hamrick, J., Grout, J., Corlay, S., et al. (2016). Jupyter notebooks – A publishing format for reproducible computational workflows. In *Proceedings of Positioning and Power in Academic Publishing: Players, Agents and Agendas*, Loizides, F. and Schmidt, B. (eds). IOS Press, Clifton, 87–90.

Kristensen, K., Nielsen, A., Berg, C., Skaug, H., Bell, B. (2016). TMB: Automatic differentiation and Laplace approximation. *Journal of Statistical Software*, 70(5), 1–21.

Langrock, R. and Zucchini, W. (2011). Hidden Markov models with arbitrary state dwell-time distributions. *Computational Statistics & Data Analysis*, 55(1), 715–724.

Lloyd, S. (1982). Least squares quantization in PCM. *IEEE Transactions on Information Theory*, 28(2), 129–137.

Merkel, D. (2014). Docker: Lightweight linux containers for consistent development and deployment. *Linux Journal*, 2014(239), 76–91.

Michelot, T. (2025). hmmTMB: Hidden Markov models with flexible covariate effects in R. *Journal of Statistical Software*, 114(5), 1–45.

O'Connell, J. and Højsgaard, S. (2011). Hidden semi-Markov models for multiple observation sequences: The mhsmm package for R. *Journal of Statistical Software*, 39(4), 1–22.

Pohle, J., Adam, T., Beumer, L.T. (2022). Flexible estimation of the state dwell-time distribution in hidden semi-Markov models. *Computational Statistics & Data Analysis*, 172, 107479.

Qin, S., Tan, Z., Wu, Y. (2024). On robust estimation of hidden semi-Markov regime-switching models. *Annals of Operations Research*, 338, 1049–1081.

Rabiner, L.R. (1989). A tutorial on hidden Markov models and selected applications in speech recognition. *Proceedings of the IEEE*, 77(2), 257–286.

Schnabel, R.B., Koonatz, J.E., Weiss, B.E. (1985). A modular system of algorithms for unconstrained minimization. *ACM Trans. Math. Softw.*, 11(4), 419–440.

Tokuda, K., Hashimoto, K., Oura, K., Nankaku, Y. (2016). Temporal modeling in neural network based statistical parametric speech synthesis. In *Proceedings of the 9th ISCA Speech Synthesis Workshop (SSW9)*, Sunnyvale, 106–111.

Wang, X., Lebarbier, E., Aubert, J., Robin, S. (2019). Variational inference for coupled hidden Markov models applied to the joint detection of copy number variations. *The International Journal of Biostatistics*, 15(1), 20180023.

Xu, Z. and Liu, Y. (2021). A regularized vector autoregressive hidden semi-Markov model, with application to multivariate financial data. In *The International FLAIRS Conference Proceedings*, 34.

Yu, S.-Z. (2010). Hidden semi-Markov models. *Artificial Intelligence*, 174(2), 215–243.

Zhang, Y. and Berhane, K. (2014). Bayesian mixed hidden Markov models: A multi-level approach to modeling categorical outcomes with differential misclassification. *Statistics in Medicine*, 33(8), 1395–1408.

3

Multichain HMM

Hidden Markov models (HMMs) are statistical models widely used for studying dynamic processes that cannot be observed directly or are governed by a hidden layer. In many applications, particularly those involving spatial data, the hidden chain and the observed time series are multidimensional, exhibiting structured dependencies between variables at time t and variables at time $t+1$. In this chapter, we introduce the framework of multichain HMMs (MHMMs), which provides a unified view to existing models in the literature and also allows for their generalization. We demonstrate the utility of these models in ecological dynamics modeling. We then discuss inference for MHMMs in the context of the expectation-maximization (EM) algorithm. We explain why exact inference remains tractable for some structures of MHMMs, while it is computationally intractable for others. For these latter, we review the main methods to perform approximate inference.

3.1. Introduction

In the first two chapters, we considered a single hidden chain. However, in some contexts, there is actually more than one hidden process of interest and these processes influence each other. Such interacting processes are often encountered in spatiotemporal dynamics: metapopulation dynamics on a network of patches, disease spread over a network of spatial locations, earthquake activities in neighbor seismic areas, etc. It is natural to model these dynamics as multiple hidden chains in interaction, even though this remains a vague concept that we will formalize and generalize under the name of MHMM in this chapter.

Chapter written by Hanna BACAVE, Jean-Baptiste DURAND, Alain FRANC, Nathalie PEYRARD, Sandra PLANCADE and Régis SABBADIN.

To concrete the idea of interacting dynamical processes, and motivate the need for extensions of HMMs to their multichain counterpart, we present extensions of the two toy examples presented in Chapter 1 in a context with multiple hidden chains.

For the Squirrel example, we recall that the hidden chain is the succession of reserves where it feeds over time and the observation is the reserve guessed by an observer. In practice, more than one squirrel feeds in the same area and the behaviors of these squirrels are not independent. Squirrels can interact having access to the same resource in a given reserve by a process of cooperation (common access is favored) or competition (common access is disadvantaged). A squirrel looks at the reserves on which the others feed and, knowing that, stays on the same reserve or moves to a new reserve. Furthermore, the observer must guess where each squirrel is, and the sequence of guesses for each squirrel may not be independent.

The Deer toy example can also be extended to more than one animal. Indeed, the hidden behavior of a given animal can influence that of another animal. For instance, if the two deer are close to each other and one of them starts running, it is likely that the other will start running too. It this case, the two hidden chains of behaviors are not independent. However, the observations sequence of each animal (accelerometer data) are independent.

One option could be to model each of these two situations as a HMM with a multidimensional hidden state corresponding to the vector of the hidden state of each chain. However, this would not be efficient. If there are C hidden processes, each with K possible states, the transition matrix would be of size $K^C \times K^C$ and estimation complexity would be exponential in C. More parsimonious representations can be built by exploiting conditional independencies in the transition, specific of the application. Furthermore, in some cases, these conditional independencies lead to a reduction of the computational complexity of parameter estimation from exponential to linear in C. In this chapter, we formalize the notion of MHMM and we develop these topics.

In the literature, the two most famous models for interacting hidden chains are factorial HMMs (FHMMs, Ghahramani and Jordan 1997) and coupled HMMs (CHMMs, Brand 1997; Wainwright and Jordan 2008). They represent two different ways to link the hidden chains and two different assumptions on conditional independencies. In an FHMM, there is no direct interaction between the hidden chains, and they become conditionally independent given a common observation. For instance, each chain corresponds to the evolution of the localization of an animal and the observation is the count of animals in each location, with some measurement error. In a CHMM, the hidden chains are linked by a direct interaction of a hidden chain to the dynamics of the others (like for the Squirrel or Deer examples).

These two interaction structures do not capture the whole diversity of interactions encountered in real-life applications. In this chapter, we explore different ways to make hidden and observed chains interacting that correspond to different conditional independence assumptions. We propose a unified presentation of these models in the framework of MHMMs that we define rigorously (section 3.2). We illustrate it using different cases for applications, mainly from ecology or epidemiology (sections 3.3 and 3.4). The difficulty with inference in MHMMs is that the exact EM algorithm is in general not computationally efficient. In most cases complexity is exponential in C, with some exceptions for some structures of dependencies between hidden chains and observations. We present in section 3.5 the EM equations for general MHMMs and for some particular classes, showing how the complexity can vary with the structure of interactions between the chains. This should provide the necessary material to help deriving the EM equations for any MHMM. Finally, in section 3.6, we discuss classical and more recent methods that can be mobilized to derive an approximate EM.

3.2. Different concepts of MHMM

The general concept of MHMM is rather broad. Intuitively, we think of a stochastic model representing the joint distribution of a set of several hidden and observed processes that evolve through time and satisfy some Markov property. The temporal evolution of theses chains can be linked in different ways. We first formalize the more general situation, introducing only one assumption, the Markov property satisfied by the set of hidden and observed processes. Then we describe several subclasses of interest that correspond to the introduction of additional assumptions on the conditional independencies in this general model.

3.2.1. *General MHMM*

Let us consider a multivariate hidden variable $\boldsymbol{Z}_t$ with C components (C chains): $\boldsymbol{Z}_t = (Z_t^1, \ldots, Z_t^C)$ with $Z_t^c \in \Omega_{Z^c}$. The observation $\boldsymbol{Y}_t$ is also multivariate with O components (O observations): $\boldsymbol{Y}_t = (Y_t^1, \ldots, Y_t^O)$ with $Y_t^o \in \Omega_{Y^o}$. The dimensions C and O are not required to be equal.

DEFINITION 3.1.– *The multidimensional process $(\boldsymbol{Z}_t, \boldsymbol{Y}_t)_{t\in\mathbb{N}}$ is a general HMM if $(\boldsymbol{Z}_t, \boldsymbol{Y}_t)$ satisfies the following Markov property:*

$$\begin{aligned}\mathbb{P}(\boldsymbol{Z}_{t+1} &= \boldsymbol{z}_{t+1}, \boldsymbol{Y}_{t+1} = \boldsymbol{y}_{t+1} \mid \boldsymbol{Z}_s = \boldsymbol{z}_s, \boldsymbol{Y}_s = \boldsymbol{y}_s,\ 1 \leq s \leq t) \\ &= \mathbb{P}(\boldsymbol{Z}_{t+1} = \boldsymbol{z}_{t+1}, \boldsymbol{Y}_{t+1} = \boldsymbol{y}_{t+1} \mid \boldsymbol{Z}_t = \boldsymbol{z}_t, \boldsymbol{Y}_t = \boldsymbol{y}_t) \qquad [3.1]\end{aligned}$$

Let $\boldsymbol{Y}_{0:T}$ (respectively, $\boldsymbol{Z}_{0:T}$) denote the set of all observed (respectively, hidden) variables between time 0 and time T. General MHMMs satisfy the following factorization property:

$$\begin{aligned}&\mathbb{P}(\boldsymbol{Z}_{0:T} = \boldsymbol{z}_{0:T}, \boldsymbol{Y}_{0:T} = \boldsymbol{y}_{0:T})\\&= \mathbb{P}(\boldsymbol{Z}_0 = \boldsymbol{z}_0, \boldsymbol{Y}_0 = \boldsymbol{y}_0)\\&\quad\times \prod_{t=1}^{T} \mathbb{P}(\boldsymbol{Z}_t = \boldsymbol{z}_t, \boldsymbol{Y}_t = \boldsymbol{y}_t | \boldsymbol{Z}_{t-1} = \boldsymbol{z}_{t-1}, \boldsymbol{Y}_{t-1} = \boldsymbol{y}_{t-1})\end{aligned} \qquad [3.2]$$

As opposed to the HMM case, the chain $(\boldsymbol{Z}_t)_t$ does not satisfy the Markov property in general. However, given $(\boldsymbol{Y}_t)_t$, the property is satisfied.

PROPOSITION 3.1.– *Conditionally on* $(\boldsymbol{Y}_t)_t$, $(\boldsymbol{Z}_t)_t$ *is a Markov chain.*

Indeed, for every $t > 0$:

$$\begin{aligned}&\mathbb{P}(\boldsymbol{Z}_{t+1} = \boldsymbol{z}_{t+1} | \boldsymbol{Z}_{0:t} = \boldsymbol{z}_{0:t}, \boldsymbol{Y}_{0:t+1} = \boldsymbol{y}_{0:t+1})\\&= \frac{\mathbb{P}(\boldsymbol{Z}_{t+1} = \boldsymbol{z}_{t+1}, \boldsymbol{Y}_{t+1} = \boldsymbol{y}_{t+1} | \boldsymbol{Z}_{0:t} = \boldsymbol{z}_{0:t}, \boldsymbol{Y}_{0:t} = \boldsymbol{y}_{0:t})}{\mathbb{P}(\boldsymbol{Y}_{t+1} = \boldsymbol{y}_{t+1} | \boldsymbol{Z}_{0:t} = \boldsymbol{z}_{0:t}, \boldsymbol{Y}_{0:t} = \boldsymbol{y}_{0:t})}\\&= \frac{\mathbb{P}(\boldsymbol{Z}_{t+1} = \boldsymbol{z}_{t+1}, \boldsymbol{Y}_{t+1} = \boldsymbol{y}_{t+1} | \boldsymbol{Z}_t = \boldsymbol{z}_t, \boldsymbol{Y}_t = \boldsymbol{y}_t)}{\sum_{i \in \Omega_Z} \mathbb{P}(\boldsymbol{Z}_{t+1} = i, \boldsymbol{Y}_{t+1} = \boldsymbol{y}_{t+1} | \boldsymbol{Z}_{0:t} = \boldsymbol{z}_{0:t}, \boldsymbol{Y}_{0:t} = \boldsymbol{y}_{0:t})} \text{(Markov property)}\\&= \frac{\mathbb{P}(\boldsymbol{Z}_{t+1} = \boldsymbol{z}_{t+1}, \boldsymbol{Y}_{t+1} = \boldsymbol{y}_{t+1} | \boldsymbol{Z}_t = \boldsymbol{z}_t, \boldsymbol{Y}_t = \boldsymbol{y}_t)}{\sum_{i \in \Omega_Z} \mathbb{P}(\boldsymbol{Z}_{t+1} = i, \boldsymbol{Y}_{t+1} = \boldsymbol{y}_{t+1} | \boldsymbol{Z}_t = \boldsymbol{z}_t, \boldsymbol{Y}_t = \boldsymbol{y}_t)}\\&= \frac{\mathbb{P}(\boldsymbol{Z}_{t+1} = \boldsymbol{z}_{t+1}, \boldsymbol{Y}_{t+1} = \boldsymbol{y}_{t+1} | \boldsymbol{Z}_t = \boldsymbol{z}_t, \boldsymbol{Y}_t = \boldsymbol{y}_t)}{\mathbb{P}(\boldsymbol{Y}_{t+1} = \boldsymbol{y}_{t+1} | \boldsymbol{Z}_t = \boldsymbol{z}_t, \boldsymbol{Y}_t = \boldsymbol{y}_t)}\\&= \mathbb{P}(\boldsymbol{Z}_{t+1} = \boldsymbol{z}_{t+1} | \boldsymbol{Z}_t = \boldsymbol{z}_t, \boldsymbol{Y}_t = \boldsymbol{y}_t, \boldsymbol{Y}_{t+1} = \boldsymbol{y}_{t+1}).\end{aligned}$$

This quantity does not depend on $\boldsymbol{Z}_{0:t-1}$, which completes the proof.

Note that the following alternative proof is shorter and does not imply any computation: for every $T > 0$:

$$\mathbb{P}(\boldsymbol{Z}_{0:T} = \boldsymbol{z}_{0:T} | \boldsymbol{Y}_{0:T} = \boldsymbol{y}_{0:T}) \propto q_0(\boldsymbol{z}_0) \prod_{t=1}^{T} q_t(\boldsymbol{z}_t, \boldsymbol{z}_{t-1}),$$

where q_0 and q_t are non-negative functions that depend on $\boldsymbol{y}$. This factorization shows that $p(\boldsymbol{z}|\boldsymbol{y})$ is an undirected graphical probabilistic model on a linear graph, which is a Markov chain (Koller and Friedman 2009).

The advantage of the definition of an MHMM lies in its generality. However, it is very costly in terms of space needed for representation. If $|\Omega_{Z^c}| = |\Omega_Z|$ and $|\Omega_{Y^o}| = |\Omega_Y|$, the representation of the transition probability [3.1] requires $|\Omega_Z|^{2C} \times |\Omega_Y|^{2O}$ elements. This is exponential in C and O. In real-life applications, conditional independencies may exist between $(\boldsymbol{Z}_{t+1}, \boldsymbol{Y}_{t+1})$ and $(\boldsymbol{Z}_t, \boldsymbol{Y}_t)$ that will enable us to express the global transition [3.1] as a product of conditional probabilities of smaller dimensions. This will reduce the representation cost.

We now present several examples of possible simplifications of the expression of [3.1] with the additional conditional independencies hypotheses associated.

3.2.2. *MHMM with conditional independencies*

The right-hand side term of [3.1] can be decomposed into a product of two conditional probabilities, one for $\boldsymbol{Y}_{t+1}$ and one for $\boldsymbol{Z}_{t+1}$, without any additional assumption:

$$\begin{aligned}\mathbb{P}(\boldsymbol{Z}_{t+1} &= \boldsymbol{z}_{t+1}, \boldsymbol{Y}_{t+1} = \boldsymbol{y}_{t+1} \mid \boldsymbol{Z}_t = \boldsymbol{z}_t, \boldsymbol{Y}_t = \boldsymbol{y}_t) \\ &= \mathbb{P}(\boldsymbol{Y}_{t+1} = \boldsymbol{y}_{t+1} \mid \boldsymbol{Z}_{t+1} = \boldsymbol{z}_{t+1}, \boldsymbol{Z}_t = \boldsymbol{z}_t, \boldsymbol{Y}_t = \boldsymbol{y}_t) \\ &\quad \times \mathbb{P}(\boldsymbol{Z}_{t+1} = \boldsymbol{z}_{t+1} \mid \boldsymbol{Z}_t = \boldsymbol{z}_t, \boldsymbol{Y}_t = \boldsymbol{y}_t)\end{aligned}$$

Following the terminology of HMMs, the probability $\mathbb{P}(\boldsymbol{Y}_{t+1} = \boldsymbol{y}_{t+1} \mid \boldsymbol{Z}_{t+1} = \boldsymbol{z}_{t+1}, \boldsymbol{Z}_t = \boldsymbol{z}_t, \boldsymbol{Y}_t = \boldsymbol{y}_t)$ will be referred to as the emission probability and $\mathbb{P}(\boldsymbol{Z}_{t+1} = \boldsymbol{z}_{t+1} \mid \boldsymbol{Z}_t = \boldsymbol{z}_t, \boldsymbol{Y}_t = \boldsymbol{y}_t)$ to as the transition probability.

The first natural set of conditional independencies (CI) assumptions that can be added to the general framework corresponds to a factorization assumption in each of these two probabilities. By factorization, we mean that the joint probability of a set of variables is decomposed into a product of individual probabilities. For the variable $\boldsymbol{Z}_{t+1}$, we obtain:

CI on transition:

$$\begin{aligned}\mathbb{P}(\boldsymbol{Z}_{t+1} &= \boldsymbol{z}_{t+1} \mid \boldsymbol{Z}_t = \boldsymbol{z}_t, \boldsymbol{Y}_t = \boldsymbol{y}_t) \\ &= \prod_{c=1}^{C} \mathbb{P}(Z_{t+1}^c = z_{t+1}^c \mid \boldsymbol{Z}_t = \boldsymbol{z}_t, \boldsymbol{Y}_t = \boldsymbol{y}_t)\end{aligned} \qquad [3.3]$$

It means that the values of the hidden chains at time $t+1$ are independent given the hidden and observed values at time t. For the variable $\boldsymbol{Y}_{t+1}$, we obtain:

CI on emission:

$$\mathbb{P}(\boldsymbol{Y}_{t+1} = \boldsymbol{y}_{t+1} \mid \boldsymbol{Z}_{t+1} = \boldsymbol{z}_{t+1}, \boldsymbol{Z}_t = \boldsymbol{z}_t, \boldsymbol{Y}_t = \boldsymbol{y}_t) = \prod_{o=1}^{O} \mathbb{P}(Y^o_{t+1} = y^o_{t+1} \mid \boldsymbol{Z}_{t+1} = \boldsymbol{z}_{t+1}, \boldsymbol{Z}_t = \boldsymbol{z}_t, \boldsymbol{Y}_t = \boldsymbol{y}_t) \quad [3.4]$$

It means that the values of the observed variables at time $t+1$ are independent given the hidden values at $t+1$ and the hidden and observed values at time t.

DEFINITION 3.2.– *An MHMM satisfying the two conditional independence assumptions CI on transition and CI on emission defined, respectively, by equations* [3.3] *and* [3.4] *is an MHMM-CI.*

It is possible to associate a graphical representation of the transition probability of an MHMM-CI. Indeed, an MHMM-CI (as any MHMM) is a particular case of directed probabilistic graphical models, namely, Bayesian networks (BN) also known as belief networks (see Pearl 1988; Koller and Friedman 2009; Murphy 2012). If the vector of random variables $X = (X_1, X_2, \dots X_N)$ is a BN, the joint distribution is expressed as the product of conditional probabilities of each variable (the child) given its parents variables:

$$\mathbb{P}(X_1 = x_1, X_2 = x_2, \dots X_N = x_N) = \prod_{i=1}^{N} \mathbb{P}(X_i = x_i \mid X_{Pa(i)} = x_{Pa(i)})$$

where $X_{Pa(i)}$ is the set of variables that are parents of variable X_i. The graphical representation $\mathcal{G}_d$ of a BN is an acyclic-directed graph where there is one node for each variable, and an arc is directed from a parent vertex to a child vertex.

Figure 3.1 shows the graphical representation $\mathcal{G}_d$ of the joint transition on $(\boldsymbol{Z}_t, \boldsymbol{Y}_t)$ for the most general MHMM-CI. Any graphical representation included in this one, that is, with some arcs missing is also an MHMM of type MHMM-CI. Missing arcs correspond to missing variables in the conditioning of [3.3] or [3.4].

With the two factorization assumptions *CI on transition* and *CI on emission*, there is a gain in the number of elements needed for representing the model. If $|\Omega_{Z^c}| = |\Omega_Z|$ and $|\Omega_{Y^o}| = |\Omega_Y|$, the representation of [3.3] requires $C|\Omega_Z|^{C+1} \times |\Omega_Z|^O$ terms and the representation of [3.4] requires $O|\Omega_Z|^{2C} \times |\Omega_Z|^{O+1}$ terms. However, it remains exponential in C and in O.

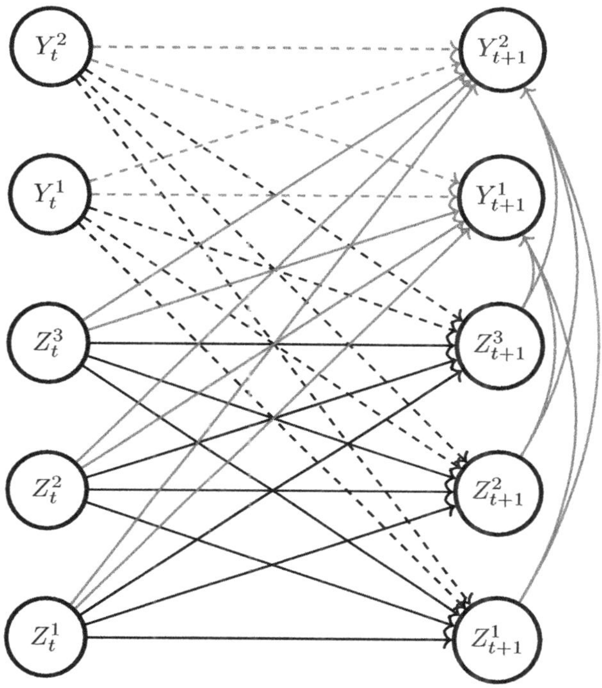

Figure 3.1. *Graphical representation $\mathcal{G}_d$ of the joint transition $\mathbb{P}(\boldsymbol{Z}_{t+1} = \boldsymbol{z}_{t+1}, \boldsymbol{Y}_{t+1} = \boldsymbol{y}_{t+1} \mid \boldsymbol{Z}_t = \boldsymbol{z}_t, \boldsymbol{Y}_t = \boldsymbol{y}_t)$ for an MHMM-CI, for $C = 3$ hidden chains and $O = 2$ observations. Full arcs (resp. dashed arcs) start from hidden (resp. observed) variables. Black arcs (resp. grey arcs) end on hidden (resp. observed) variables*

3.2.3. *Case 1 of MHMM-CI: 1to1-MHMM-CI*

An interesting subcase of the MHMM-CI definition is when $C = O$ and at each time t one observation is attached to one and only one hidden variable. We refer to this model as the 1to1-MHMM-CI.

DEFINITION 3.3.– *An MHMM-CI is a 1to1-MHMM-CI if $C = O$ and:*

$$\mathbb{P}(Y_{t+1}^c = y_{t+1}^c \mid \boldsymbol{Z}_{t+1} = \boldsymbol{z}_{t+1}, \boldsymbol{Z}_t = \boldsymbol{z}_t, \boldsymbol{Y}_t = \boldsymbol{y}_t) = \mathbb{P}(Y_{t+1}^c = y_{t+1}^c \mid Z_{t+1}^c = z_{t+1}^c, \boldsymbol{Z}_t = \boldsymbol{z}_t, \boldsymbol{Y}_t = \boldsymbol{y}_t) \quad [3.5]$$

The corresponding graphical representation is shown in Figure 3.2. Again, any graphical representation included in this one, that is, with some edges absent is also a 1to1-MHMM-CI.

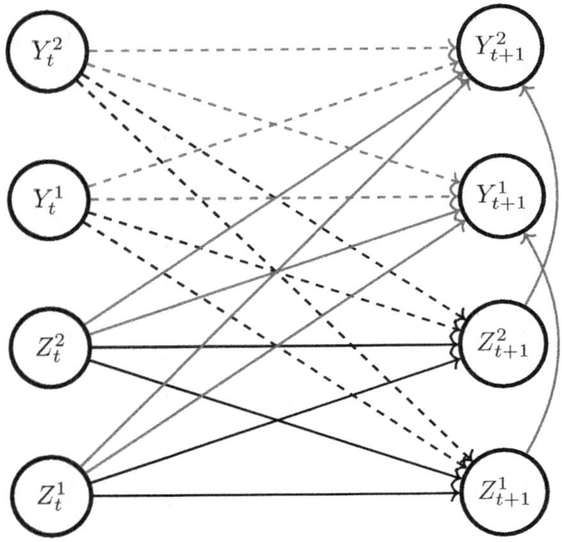

Figure 3.2. *Graphical representation $\mathcal{G}_d$ of the joint transition $\mathbb{P}(\boldsymbol{Z}_{t+1} = \boldsymbol{z}_{t+1}, \boldsymbol{Y}_{t+1} = \boldsymbol{y}_{t+1} \mid \boldsymbol{Z}_t = \boldsymbol{z}_t, \boldsymbol{Y}_t = \boldsymbol{y}_t)$ for a 1to1-MHMM-CI for $C = O = 2$. Full arcs (resp. dashed arcs) start from hidden (resp. observed) variables. Black arcs (resp. grey arcs) end on hidden (resp. observed) variables*

The joint transition [3.1] of a 1to1-MHMM-CI is therefore expressed as follows:

$$\begin{aligned}
&\mathbb{P}(\boldsymbol{Z}_{t+1} = \boldsymbol{z}_{t+1}, \boldsymbol{Y}_{t+1} = \boldsymbol{y}_{t+1} \mid \boldsymbol{Z}_t = \boldsymbol{z}_t, \boldsymbol{Y}_t = \boldsymbol{y}_t) \\
&= \prod_{c=1}^{C} \mathbb{P}(Z_{t+1}^c = z_{t+1}^c \mid \boldsymbol{Z}_t = \boldsymbol{z}_t, \boldsymbol{Y}_t = \boldsymbol{y}_t) \\
&\quad \times \prod_{c=1}^{C} \mathbb{P}(Y_{t+1}^c = y_{t+1}^c \mid Z_{t+1}^c = z_{t+1}^c, \boldsymbol{Z}_t = \boldsymbol{z}_t, \boldsymbol{Y}_t = \boldsymbol{y}_t) \qquad [3.6]
\end{aligned}$$

The case of 1to1-MHMM-CI is interesting because in many real-life applications, the observation is a noisy version or a proxy of the hidden variable, so there is exactly one observed variable associated with one hidden variable via an emission link at time t.

An example of 1to1-MHMM-CI studied in the literature is the CHMM (CHMM, Brand 1997; Wainwright and Jordan 2008) whose graphical representation is shown in Figure 3.3.

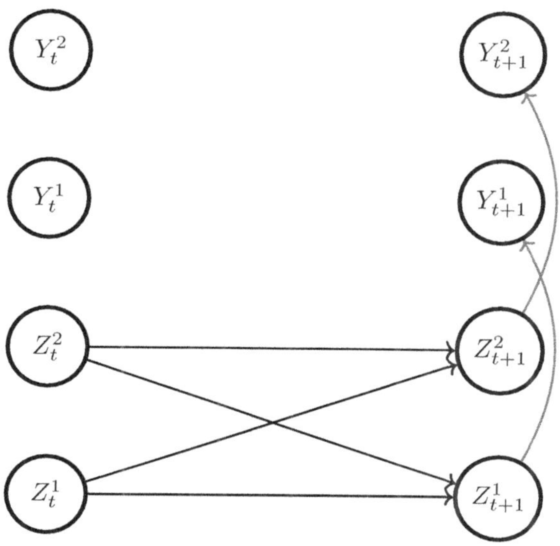

Figure 3.3. *Graphical representation $\mathcal{G}_d$ of the joint transition $\mathbb{P}(\boldsymbol{Z}_{t+1} = \boldsymbol{z}_{t+1}, \boldsymbol{Y}_{t+1} = \boldsymbol{y}_{t+1} \mid \boldsymbol{Z}_t = \boldsymbol{z}_t, \boldsymbol{Y}_t = \boldsymbol{y}_t)$ for a CHMM for $C = O = 2$. Black arcs (resp. grey arcs) end on hidden (resp. observed) variables*

DEFINITION 3.4.– *A CHMM is a 1to1-MHMM-CI such that the emission probability is reduced to:*

$$\mathbb{P}(Y_{t+1}^c = y_{t+1}^c \mid Z_{t+1}^c = z_{t+1}^c, \boldsymbol{Z}_t = \boldsymbol{z}_t, \boldsymbol{Y}_t = \boldsymbol{y}_t) = \mathbb{P}(Y_{t+1}^c = y_{t+1}^c \mid Z_{t+1}^c = z_{t+1}^c)$$

and the transition probability is reduced to:

$$\mathbb{P}(Z_{t+1}^c = z_{t+1}^c \mid \boldsymbol{Z}_t = \boldsymbol{z}_t, \boldsymbol{Y}_t = \boldsymbol{y}_t) = \mathbb{P}(Z_{t+1}^c = z_{t+1}^c \mid \boldsymbol{Z}_t = \boldsymbol{z}_t)$$

The CHMM is an example where the marginal distribution of $(Z_t^c, Y_t^c)_t$ for a given chain c is not that of a HMM. This is illustrated in Figure 3.4 with $C = 2$. In $\mathcal{G}_d$, there is a directed path from Z_{t-1}^1 to Z_{t+1}^1 via Z_t^2 and that does not go through Z_t^1. So conditionally on Z_t^1, the hidden variable Z_{t+1}^1 remains dependent on Z_{t-1}^1: the process does not satisfy the Markov property. However, the multidimensional process $(\boldsymbol{Z}_t)_t$ does.

Literature on models of the MHMM class is not always coherent in the names used. In Wainwright and Jordan (2008) and Wang et al. (2019), an MHMM is considered, where each Z_t^c is conditionally dependent on some other $Z_t^{c'}$ given $\boldsymbol{Z}_{t-1}$. Although their model is termed a CHMM, this is an MHMM but not a CHMM in the sense of the original definition.

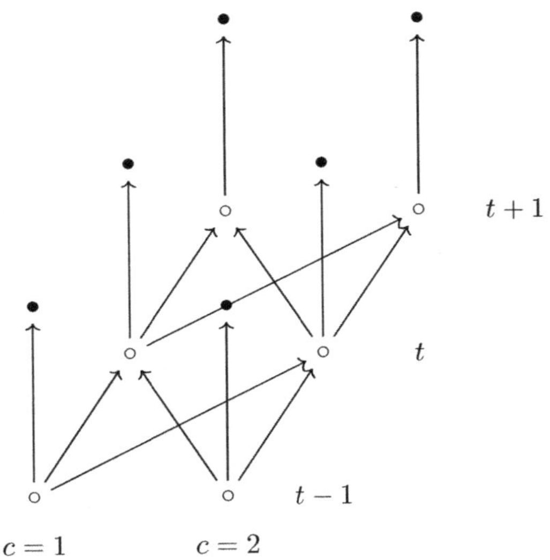

Figure 3.4. *Example of a 1to1-MHMM-CI where the marginal chains (Z_t^1, Y_t^1) and (Z_t^2, Y_t^2) are not HMMs. Hidden variables are represented by an empty bullet and observed variables by a black bullet. The structure $\mathcal{G}_d$ is that of a CHMM*

3.2.4. *Case 2 of MHMM-CI: FHMM*

Let us consider now the case $O = 1$ and assume that the unique observed variable Y_{t+1} is drawn from a distribution that depends on the C current hidden variables $\{Z_{t+1}^c\}_{1 \leq c \leq C}$. Assume furthermore that each chain $(Z_t^c)_t$ is independent of the others and is a Markov chain. This corresponds to a model known in literature as the factorial FHMM (FHMM, Ghahramani and Jordan 1997).

DEFINITION 3.5.– *An MHMM-CI is an FHMM if $O = 1$, and*

$$\mathbb{P}(Z_{t+1}^c = z_{t+1}^c \mid \boldsymbol{Z}_t = \boldsymbol{z}_t, Y_t = y_t) = \mathbb{P}(Z_{t+1}^c = z_{t+1}^c \mid Z_t^c = z_t^c) \qquad [3.7]$$

and

$$\begin{aligned}\mathbb{P}(Y_{t+1} &= y_{t+1} \mid \boldsymbol{Z}_{t+1} = \boldsymbol{z}_{t+1}, \boldsymbol{Z}_t = \boldsymbol{z}_t, Y_t = y_t) \\ &= \mathbb{P}(Y_{t+1} = y_{t+1} \mid \boldsymbol{Z}_{t+1} = \boldsymbol{z}_{t+1})\end{aligned} \qquad [3.8]$$

For an FHMM, the joint transition distribution [3.1] becomes:

$$\mathbb{P}(\boldsymbol{Z}_{t+1} = \boldsymbol{z}_{t+1}, Y_{t+1} = y_{t+1} \mid \boldsymbol{Z}_t = \boldsymbol{z}_t, Y_t = y_t) =$$

$$\mathbb{P}(Y_{t+1} = y_{t+1} \mid \boldsymbol{Z}_{t+1} = \boldsymbol{z}_{t+1}) \prod_{c=1}^{C} \mathbb{P}(Z_{t+1}^c = z_{t+1}^c \mid Z_t^c = z_t^c)$$

The graph $\mathcal{G}_d$ of the transition for an FHMM is shown in Figure 3.5.

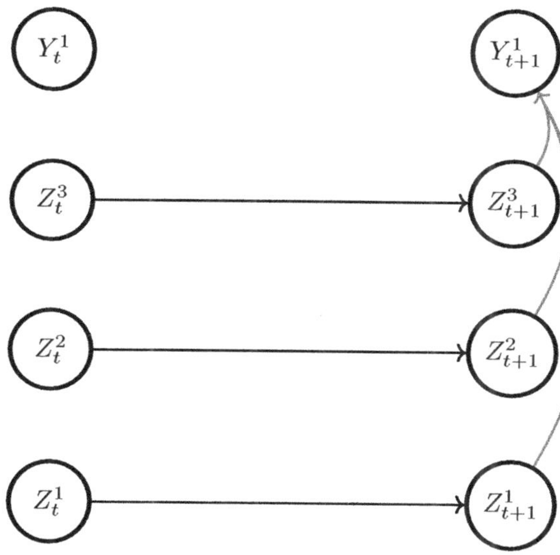

Figure 3.5. *Graphical representation $\mathcal{G}_d$ of the joint transition $\mathbb{P}(\boldsymbol{Z}_{t+1} = \boldsymbol{z}_{t+1}, \boldsymbol{Y}_{t+1} = \boldsymbol{y}_{t+1} \mid \boldsymbol{Z}_t = \boldsymbol{z}_t, \boldsymbol{Y}_t = \boldsymbol{y}_t)$ for an FHMM in the case of two chains ($C = 2$). Black arcs (resp. grey arcs) end on hidden (resp. observed) variables*

The box below summarizes the hierarchy of nested classes of MHMMs that we have defined. Each class corresponds to a new assumption of conditional independencies added to the class just above in the hierarchy. In this box, and when needed in the rest of the chapter, for sake of compactness we use the following simplification of notation for probabilities. The probability that a r.v. Z takes value z is denoted $\mathbb{P}(Z = z)$, with an upper case for the r.v., and a lower case for its realization. In case where the random variable Z is obvious from the context, we denote it more simply $p(z)$. This is naturally extended towards conditional probabilities, like transition probabilities, as in $\mathbb{P}(Z_{t+1} = z \mid Z_t = z')$, which is denoted as $p(z \mid z')$.

Nested classes of MHMMs

$\boldsymbol{Z}_t = (Z_t^1, \ldots, Z_t^C)$ and $\boldsymbol{Y}_t = (Y_t^1, \ldots, Y_t^O)$.
General MHMM: $(\mathbf{Z}_t, \mathbf{Y}_t)_{t \in \mathbb{N}}$ is a Markov chain.

∗ **MHMM-CI**: conditional independence assumptions

$$\begin{cases} Z_{t+1}^c \perp\!\!\!\perp Z_{t+1}^{c'} | (\mathbf{Z}_t, \mathbf{Y}_t) \\ Y_{t+1}^o \perp\!\!\!\perp Y_{t+1}^{o'} | (\mathbf{Z}_{t+1}, \mathbf{Z}_t, \mathbf{Y}_t) \end{cases}$$

∘ **1to1-MHMM-CI**: O=C and

$$Y_{t+1}^c \perp\!\!\!\perp (Z_{t+1}^{c'})_{c' \neq c} | (Z_{t+1}^c, \mathbf{Z}_t, \mathbf{Y}_t)$$

- **CHMM**

$$\begin{cases} Y_{t+1}^c \perp\!\!\!\perp \mathbf{Z}_t, \mathbf{Y}_t | Z_{t+1}^c \\ Z_{t+1}^c \perp\!\!\!\perp \mathbf{Y}_t | \mathbf{Z}_t \end{cases}$$

∘ **FHMM**: $O = 1$ and

$$\begin{cases} Y_{t+1} \perp\!\!\!\perp \mathbf{Z}_t, \mathbf{Y}_t | \mathbf{Z}_{t+1} \\ Z_{t+1}^c \perp\!\!\!\perp (Z_t^{c'})_{c' \neq c}, \mathbf{Y}_t | Z_t^c \end{cases}$$

3.3. Examples of models of class 1to1-MHMM-CI

The class of 1to1-MHMM-CIs is a very rich class of models. As mentioned before, in a model of the class 1to1-MHMM-CI, some variables in the conditioning of the emission probability and the transition probability can be absent. Depending on which variables actually appear in the conditioning, the resulting model corresponds to very different conditional independence assumptions. So 1to1-MHMM-CIs can be adapted to model a large variety of dynamics where one observation is attached to one hidden variable. In particular, it is adapted if the objective is to add dependencies between C bi-dimensional processes $(Z_t^c, Y_t^c)_t$, being originally modeled as mutually independent and following a HMM. We will refer to this construction as "coupling". The added dependencies may be due to spatial dependencies or social interactions. This is relevant, for instance, for the Squirrel and the Deer examples if one wants to extend the models of Chapter 1 to several individuals.

In this section, we first exhibit the four elementary 1to1-MHMM-CI structures that can be obtained by coupling. Then we describe concrete applications for which 1to1-MHMM-CIs built by coupling are well adapted.

3.3.1. *Structures obtained by coupling*

Let us consider the core directed graph $\mathcal{G}_d^*$, graphical representation of C independent HMMs. It is represented in Figure 3.6 for $C = 2$. This is a 1to1-MHMM-CI. Coupling the C chains is performed by adding arcs in $\mathcal{G}_d$ from a variable Z_t^c or Y_t^c toward a variable $Z_{t+1}^{c'}$ or $Y_{t+1}^{c'}$ with $c \neq c'$. When adding such arcs, the resulting graphical representation is still that of a 1to1-MHMM-CI since the CI on *transition condition* [3.3] and the CI on *emission condition* [3.4] as well as [3.5] are satisfied.

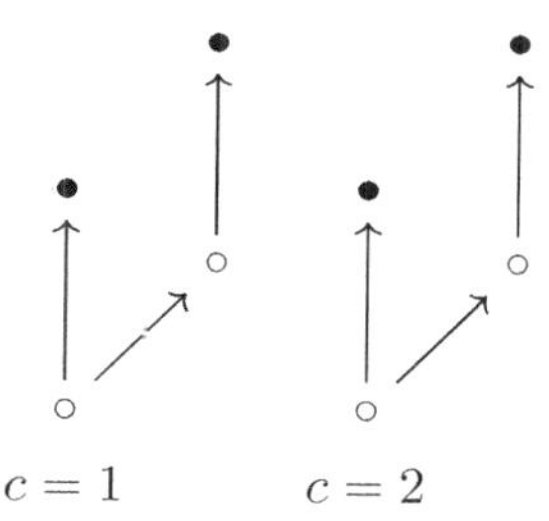

Figure 3.6. *Graphical representation $\mathcal{G}_d^*$ of a 1to1-MHMM-CI composed of 2 independent chains ($C = 2$), for two consecutive time steps. Hidden variables are represented by an empty bullet and observed variables by a black bullet*

For coupling, there are four possible types of arcs that can be added since there are two possibilities for the parent (hidden or observed) and the child (hidden or observed). In the case $C = 2$, Figure 3.7 displays the graphical representation $\mathcal{G}_d$ of the four possible elementary 1to1-MHMM-CIs obtained by coupling (in a symmetric way). The structure in Figure 3.7(d) corresponds to the CHMM presented in section 3.2.3. A 1to1-MHMM-CI obtained by coupling can combine several of each type of elementary additional arcs. An example is drawn in Figure 3.8.

Depending of the arcs added to $\mathcal{G}_d^*$, the expressions of the transition probability [3.3] and the emission probability [3.5] of the 1to1-MHMM-CI are different, since different subsets of variables are present in the conditioning. To make this more concrete, we derive these two probabilities for each structure of Figure 3.7. There are easily obtained using the semantics of the graphical representation $\mathcal{G}_d$.

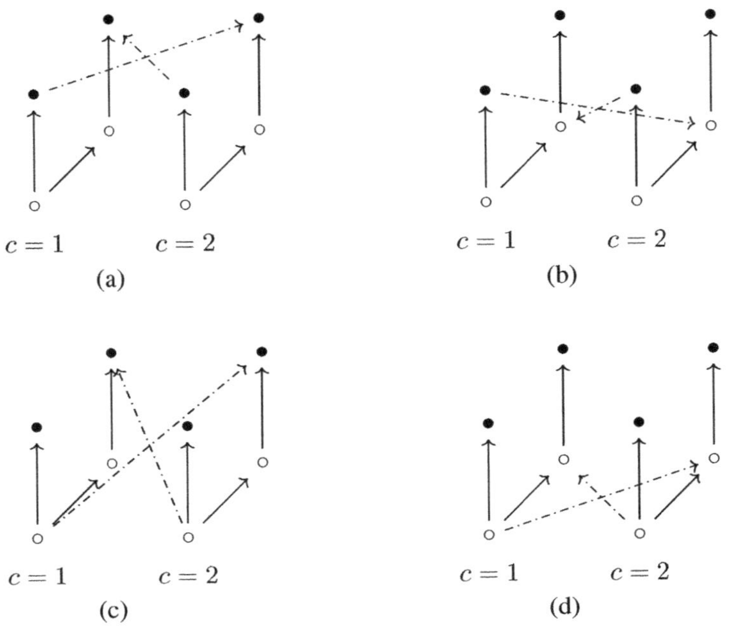

Figure 3.7. *Graphical representation $\mathcal{G}_d$ of the 4 possible elementary dependency structures in a 1to1-MHMM-CI obtained by coupling HMM structures, in the case of two chains ($C = 2$). Solid arcs are arcs of the core graph $\mathcal{G}_d^*$. Dash-dotted arcs are arcs coupling the 2 core graphs. Hidden variables are represented by an empty bullet and observed variables by a black bullet*

Coupling from observed variables toward observed variables (Figure 3.7(a)):

$$\mathbb{P}(Z_{t+1}^1 = z_{t+1}^1 \mid \boldsymbol{Z}_t = \boldsymbol{z}_t, \boldsymbol{Y}_t = \boldsymbol{y}_t) = \mathbb{P}(Z_{t+1}^1 = z_{t+1}^1 \mid Z_t^1 = z_t^1)$$

and

$$\begin{aligned}&\mathbb{P}(Y_{t+1}^1 = y_{t+1}^1 \mid Z_{t+1}^1 = z_{t+1}^1, \boldsymbol{Z}_t = \boldsymbol{z}_t, \boldsymbol{Y}_t = \boldsymbol{y}_t)\\&\quad = \mathbb{P}(Y_{t+1}^1 = y_{t+1}^1 \mid Z_{t+1}^1 = z_{t+1}^1, Y_t^2 = y_t^2)\end{aligned}$$

Coupling from observed variables toward hidden variables (Figure 3.7(b)):

$$\mathbb{P}(Z_{t+1}^1 = z_{t+1}^1 \mid \boldsymbol{Z}_t = \boldsymbol{z}_t, \boldsymbol{Y}_t = \boldsymbol{y}_t) = \mathbb{P}(Z_{t+1}^1 = z_{t+1}^1 \mid Z_t^1 = z_t^1, Y_t^2 = y_t^2)$$

and

$$\begin{aligned}\mathbb{P}(Y_{t+1}^1 &= y_{t+1}^1 \mid Z_{t+1}^1 = z_{t+1}^1, \boldsymbol{Z}_t = \boldsymbol{z}_t, \boldsymbol{Y}_t = \boldsymbol{y}_t)\\&= \mathbb{P}(Y_{t+1}^1 = y_{t+1}^1 \mid Z_{t+1}^1 = z_{t+1}^1)\end{aligned}$$

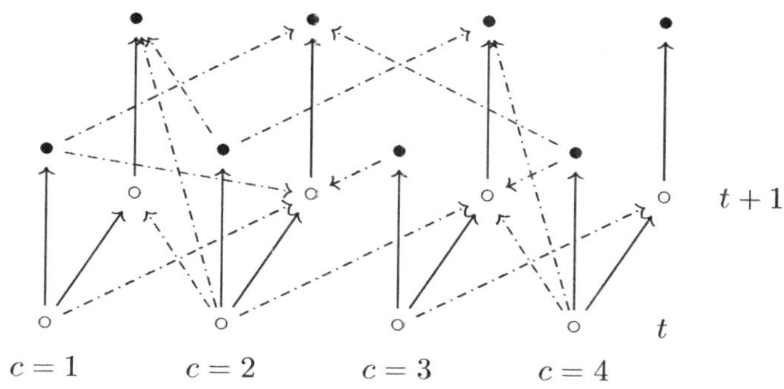

Figure 3.8. *Example of a 1to1-MHMM-CI obtained by coupling HMM structures, with $C = 4$ chains. Solid arcs are arcs of the core graph $\mathcal{G}_d^*$. Dash-dotted arcs are arcs coupling the core graphs. Hidden variables are represented by an empty bullet and observed variables by a black bullet*

Coupling from hidden variables towards observed variables (Figure 3.7(c)):

$$\mathbb{P}(Z_{t+1}^1 = z_{t+1}^1 \mid \boldsymbol{Z}_t = \boldsymbol{z}_t, \boldsymbol{Y}_t = \boldsymbol{y}_t) = \mathbb{P}(Z_{t+1}^1 = z_{t+1}^1 \mid Z_t^1 = z_t^1)$$

and

$$\mathbb{P}(Y_{t+1}^1 = y_{t+1}^1 \mid Z_{t+1}^1 = z_{t+1}^1, \boldsymbol{Z}_t = \boldsymbol{z}_t, \boldsymbol{Y}_t = \boldsymbol{y}_t)$$
$$= \mathbb{P}(Y_{t+1}^1 = y_{t+1}^1 \mid Z_{t+1}^1 = z_{t+1}^1, Z_t^2 = z_t^2)$$

Coupling from hidden variables towards hidden variables (Figure 3.7(d)):

$$\mathbb{P}(Z_{t+1}^1 = z_{t+1}^1 \mid \boldsymbol{Z}_t = \boldsymbol{z}_t, \boldsymbol{Y}_t = \boldsymbol{y}_t) = \mathbb{P}(Z_{t+1}^1 = z_{t+1}^1 \mid Z_t^1 = z_t^1, Z_t^2 = z_t^2)$$

and

$$\mathbb{P}(Y_{t+1}^1 = y_{t+1}^1 \mid Z_{t+1}^1 = z_{t+1}^1, \boldsymbol{Z}_t = \boldsymbol{z}_t, \boldsymbol{Y}_t = \boldsymbol{y}_t) = \mathbb{P}(Y_{t+1}^1 = y_{t+1}^1 \mid Z_{t+1}^1 = z_{t+1}^1)$$

We will see in section 3.5 that these four expressions of the transition and the emission probabilities do not lead to the same complexity of the task of inference from observations.

3.3.2. *Applications*

We present now how a diversity of applications where spatial or social interactions exist can be modeled by 1to1-MHMM-CIs obtained by coupling. The structures of the 1to1-MHMM-CI presented in this section follow the idea of coupling C processes

$(Z_t^c, Y_t^c)_t$ that have their own internal structure. Compared to the structures described in section 3.3.1, this internal structure can be more complex than that of a HMM and depends on the dynamics of the system under study.

Coupling from observed variables toward observed variables: temperature time series. This kind of coupling will be useful for the extension of models from the family of Markov switching auto-regressive models of order 1 (MS-AR1, Hamilton 1989) to the case where several time series are observed and are correlated. An MS-AR1 is a particular order 1 auto-regressive (AR1) model for time series where the observed AR1 process goes through different regimes, modeled by a HMM. The parameters of the AR1 model depend on the current regime. MS-AR1 have been used to model temperatures (Monbet and Ailliot 2017) or wind field time series (Bessac et al. 2016) at a given location, the hidden regime representing the weather type. Let us consider the spatial extension to C observed times series corresponding to temperatures recorded at C different locations on a regional area. Variable Y_t^c is the observed temperature at time t at location c and Z_t^c is the hidden regime at time t at location c. There is obviously a spatial correlation between temperatures, but locally each temperature (Y_t^c) dynamics has its own (independent) succession of hidden regimes (Z_t^c) that may differ due to local geographical specificities. This corresponds to a 1to1-MHMM-CI whose graphical representation $\mathcal{G}_d$ is displayed in Figure 3.9. If the different hidden regimes are correlated too, arcs from hidden state of a chain at time t to hidden states of other chains at time $t + 1$ must be added to that graph.

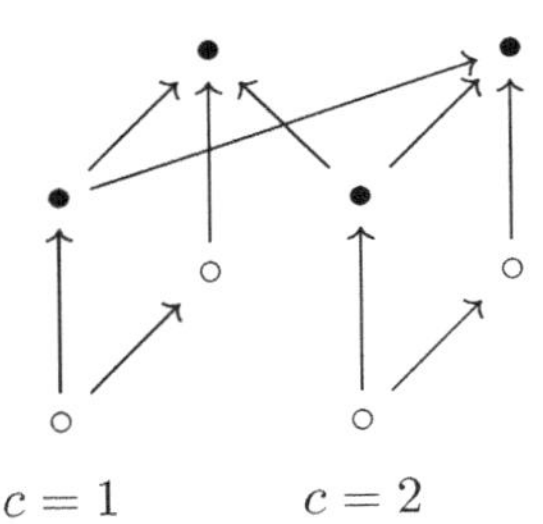

Figure 3.9. *Graphical representation $\mathcal{G}_d$ of the 1to1-MHMM-CI modeling the example of two interacting temperature time series. Each time series is an Auto-Regressive model of order 1 and has its own hidden regime as in a Markov Switching Auto-Regressive model of order 1. Hidden variables are represented by an empty bullet and observed variables by a black bullet*

Coupling from observed variables towards hidden variables: spatiotemporal dynamics of weeds. Weeds can spread between different crop fields and compete with crops, resulting in yield loss. In order to regulate the presence of weeds, it is necessary to understand the main strategies used by the species to survive from one year to another: colonization or dormancy. Indeed, seeds can be dormant and remain in the soil for several years without germinating. This reserve of seeds is called the seed bank. Its state is usually not observed. Only grown plants in the field can be observed. Grown plants produce new seeds that can enter the local seed bank or be dispersed (for instance, by wind) and colonize the seed bank of neighboring fields. To model the weeds spatiotemporal dynamics across several crop fields with an MHMM, the coupling structure of Figure 3.10 is well suited. A model with this dependency structure has been proposed in Le Coz et al. (2019). In this model, each hidden chain c correspond to a crop field. The variable Z_t^c represents evolution over time of the abundance of seeds in the seed bank of field c while Y_t^c describes the evolution of the abundance of grown weed plants in the same field. The arc from Y_t^c toward Z_{t+1}^c represents the newly produced seed that enters the soil of field c while the arc from Y_t^c toward $Z_{t+1}^{c'}$ represents the newly produced seed that enters the soil of a neighboring field c'.

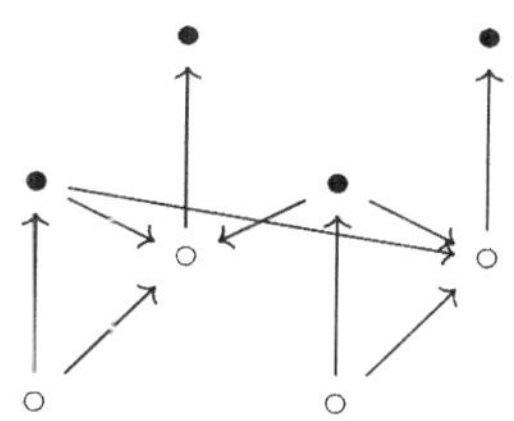

Figure 3.10. *Graphical representation $\mathcal{G}_d$ of the 1to1-MHMM-CI modeling the example of weeds dynamics in two neighbor crop fields, with possible colonization of the local seed bank by seeds produced in the other field and dispersed by wind. Hidden variables are represented by an empty bullet and observed variables by a black bullet*

Coupling from hidden variables toward observed variables. It is not as easy to find an illustrative example of this coupling as for the three others. In this type of coupling, the current hidden variable Z_t^c has an impact on the next observation of another chain c', $Y_{t+1}^{c'}$. Therefore, it could be adapted for modeling a delay in observing the effect of hidden variables on observed variables. However, the presence of arcs from Z_t^c toward Y_t^c means that an instantaneous effect also exists.

Coupling from hidden variables toward hidden variables: disease spread. This coupling corresponds to the CHMM structure (see Figures 3.3 and 3.4). It is well adapted to model the spread of a disease among individuals or among connected geographical areas when there is only imperfect information regarding the sanitary status of individuals or areas. In Touloupou et al. (2020), a CHMM is used to model the spread of infection among a population of animals in a cattle pen. There is one chain per individual, the hidden variable Z_t^c represents the sanitary status of individual c (susceptible or infected) and Y_t^c is the result of some diagnostic tests performed on individual c (with possible error). The coupling arcs represent contamination per contact between two individuals. In Peyrard et al. (2019), a CHMM model of the spread of a pest across crop fields is presented. Pest can spread from one field to another by wind or human dispersal. The hidden variable Z_t^c encodes the presence or absence of the pest in field c and Y_t^c takes value 1 if the pest was declared as present in the field and 0 otherwise. Errors of detection are possible.

3.4. Metapopulation dynamics and MHMM

In this section, we show how MHMMs offer a rich and natural framework for modeling and estimating metapopulation dynamics, where observations are often imperfect and the notion of hidden state makes perfect sense. HMMs have been used extensively in ecology, particularly behavioral ecology and population dynamics. McClintock et al. (2020) highlight the *observation process and the state process as a conceptual template for ecologists [...] for their particular systems of interest.* Here, we follow these ideas and apply them to metapopulations dynamics.

A metapopulation is a biological model of a species dynamics, consisting of a set of C connected patches, each patch hosting a local population of individuals of the species. A patch can be empty or occupied by a population, and connection is ensured by dissemination of organisms or propagules between patches. A historical reference for the metapopulation concept, is Levins (1969). This concept has become a cornerstone in ecology and population genetics, and it has been enriched from its original definition as in Ronce and Olivieri (2004).

In practice, observations are very often blurred in ecology: the presence or absence of a population in a patch at a given time can rarely be determined with certainty and observations are often subject to errors (think of difficulty to detect the presence of a fish or a frog in a pond). Therefore, as the dynamics of a single population in an isolated patch has been modeled by a HMM, we show here how naturally MHMMs can model metapopulation dynamics. We illustrate how several nested MHMMs can be built, corresponding to successive simplification assumptions on the transition matrix of the hidden chains (i.e. assumptions on the colonization paths between patches), each leading to a reduction of the number of parameters from exponential in C to constant in C.

In an MHMM of a metapopulation, the variable $\boldsymbol{Z}_t$ is the state of the metapopulation at time t, that is, the set of states of each patch c, empty or occupied, at time t: $\boldsymbol{Z}_t = (Z_t^1, \dots, Z_t^c) \in \Omega = \{0,1\}^C$. The observations at time t are denoted $\boldsymbol{Y}_t = (Y_t^1, \dots, Y_t^c)$ where $Y_t^c \in \{0,1\}$ and can be interpreted as seen/unseen.

If we make the reasonable assumption that Y_t^c depends only on Z_t^c, the emission probabilities $\mathbb{P}(Y_t^c = y_t^c \mid Z_t^c = z_t^c)$ can be modeled by a 2×2 emission matrix B, whose terms b_{ij} are the probabilities of the population to be seen/unseen knowing that the patch is empty/occupied.

The spatiotemporal dynamics of the metapopulation is modeled at the level of the hidden states. In the following, $\mathbb{P}(\boldsymbol{Z}_t = \boldsymbol{z}_t \mid \boldsymbol{Z}_{t-1} = \boldsymbol{z}_{t-1})$ is referred to as the transition matrix of the metapopulation model. Without any particular hypothesis, the number of terms needed to define all possible transitions is equal to 2^{2C}. For example, if $C = 10$, then $|\Omega| = 1,024$ and there are about 10^6 terms. As C increases, parameter estimation becomes intractable. We show now how additional assumptions on the conditional independencies between the hidden chains can lead to a reduction of this number. The first simplification is to assume that the hidden states of the patches at time $t+1$ are independent conditionally to the hidden states of the patches at time t, that is:

$$Z_{t+1}^1 \perp\!\!\!\perp \dots \perp\!\!\!\perp Z_{t+1}^C \mid (Z_t^1, \dots, Z_t^C),$$

which is equivalent to assume that:

$$p(z_{t+1}^1, \dots, z_{t+1}^C \mid z_t^1, \dots, z_t^C) = \prod_{c=1}^{C} p(z_{t+1}^c \mid z_t^1, \dots, z_t^C). \qquad [3.9]$$

This is a natural assumption since individuals colonizing patch c are individuals that have left the other patches at time t. The MHMM with this factorization of the transition probability is in the class 1to1-MHMM-CI, and more precisely, it is a CHMM: $C = O$, the transition probability and the emission probability satisfy, respectively, the condition CI on *emission condition* (see equation [3.3]) and CI on *transition condition* (see equation [3.4]), and the coupling of the hidden chains are as in Figure 3.7(d).

In a CHMM, each term $p(z_{t+1}^c \mid z_t^1, \dots, z_t^C)$ can be stored in a 2×2^c matrix with 2^{c+1} elements. Hence, there are $C\,2^{C+1}$ parameters, which is much smaller than 2^{2C}, but still huge. For example, for $C = 10$, there are $10.2^{11} \approx 2.10^4$ parameters, instead of $2^{20} \approx 10^6$, and the number of parameters still increases exponentially with C.

A second level of simplification is to assume a spatial structure on the dispersal, as for distance limited dispersal: the population of a patch disperses to its geographical neighborhood only. It means that for each patch c, $p(z_{t+1}^c \mid z_t^1, \dots, z_t^C)$ is equal to $p(z_{t+1}^c \mid \{(z_t^j)_j \ : \ j \in \mathcal{N}(c)\})$ where $\mathcal{N}(c)$ is a list of patches that can reach c by dispersal (the neighborhood of c). If each patch has d neighbors ($|\mathcal{N}(c)| = d$), the transition matrix of the metapopulation model can be represented by C local transition matrices each of size 2×2^d. The number of parameters needed to describe the transition is $C\,2^{d+1}$ and is linear with C. We refer to this model as a local CHMM (LocCHMM).

A third level of simplification is to assume that the dynamics is spatially homegeneous, that is, $p(z_{t+1}^c \mid \{(z_t^j)_j \;:\; j \in \mathcal{N}(c)\})$ is the same for every c and depends only on the values of the d neighbors. This model has 2^{d+1} parameters: it no longer depends on C and can be useful when C is huge (but not only). We refer to this model as homogeneous LocCHMM (H-LocCHMM).

We have the following nestedness of simplifications, which diminishes the number of parameters required to define the model:

$$\text{CHMM } (C2^{c+1}) \longrightarrow \text{LocCHMM } (C2^{d+1}) \longrightarrow \text{H-LocCHMM } (2^{d+1}).$$

All these simplifications assume a non-parametric definition of the transition probabilities. Another way to reduce the number of terms needed to represent $p_c(z_{t+1}^c \mid \{(z_t^j)_j \;:\; j \in \mathcal{N}(c)\})$ is to define a parametric model where the probability does not depend on the precise knowledge of the state of each neighbors, but only on a summary. The summary could be the number a^c of states $z_t^{c'} = 1$ for $c' \in \mathcal{N}(c)$ by analogy with the stochastic contact process on graphs for epidemic dynamics (Franc 2004). If we assume that only one event occurs (colonization or extinction) in a patch between time t and $t+1$, the two transition probabilities can then be defined as follows:

$$\mathbb{P}(Z_{t+1}^c = 1 \mid Z_t^c = 0, a^c) = \frac{e^{\beta a^c}}{1 + e^{\beta a^c}} \qquad \mathbb{P}(Z_{t+1}^c = 0 \mid Z_t^c = 1) = \mu,$$

where μ is the probability of extinction. Here, we have only two parameters, β and μ, to estimate.

The connection with models for disease dynamics is not coincidence. Indeed, the collection of nested CHMMs we have described for metapopulation dynamics are a natural framework for modeling in epidemiology too. They can also be used to model the two examples of disease spread dynamics presented in section 3.3.2. In epidemiology, a patch is a host, the state of which can be susceptible or infected in the simplest case, and is not observed directly but through symptoms emitted by the host. The disease is due to an infectious pathogen (a virus, a bacteria, etc.), which can disperse from host to host, in the same way that a propagule disperses between patches in a metapopulation model.

3.5. Parameter inference in MHMMs with the EM algorithm

In this chapter, we address parameter inference in MHMMs. The parameters can be considered as random, leading to Bayesian estimation, or deterministic, leading to frequentist estimation. Bayesian estimation gives access to more information based on the whole parameters distribution. However, in this section we focus on the frequentist approach, which relies on algorithms generally faster than computing or approximating posterior distributions in Bayesian estimation. In the frequentist approach, a standard method consists of computing the maximum likelihood estimate (MLE). As alternatives to MLE and Bayesian estimation, algorithms were proposed in the case of HMMs based on spectral methods (Melnyk and Banerjee 2017; De Castro et al. 2017) and moment-based methods (Anandkumar et al. 2012) but they do not seem to have been extended to MHMMs yet.

Regarding MLE computation for MHMMs, by extension of the HMM case, the most natural choice is the EM algorithm (see McLachlan and Krishnan 2008, and Chapter 1), even though alternatives to the EM algorithm were also explored, particularly quasi-Newton or Newton–Raphson algorithms (Cappé et al. 1998). The EM algorithm is strictly speaking a family of algorithms based on a common principle, rather than a specific algorithm. It is dedicated to MLE computation in models $\mathbb{P}_\theta(\boldsymbol{Y} = \boldsymbol{y})$ defined through a marginalization operation on latent variables $\boldsymbol{Z}$; in other words, from models initially defined as $\mathbb{P}_\theta(\boldsymbol{Y} = \boldsymbol{y}, \boldsymbol{Z} = \boldsymbol{z})$ with unobserved $\boldsymbol{z}$. EM proceeds iteratively by alternating between two steps, starting from an initial value θ_0 of θ: computing $\mathbb{E}_{\theta_0}[\mathbb{P}_\theta(\boldsymbol{y}, \boldsymbol{Z})|\boldsymbol{y}]$ (E step) and then, maximizing this quantity with respect to θ, providing θ_1 after the first iteration (M step), hence the name of the algorithm.

We present here how to extend the EM algorithm for HMM presented in Chapter 1 to the context of MHMMs. For the sake of keeping notation tractable, we will consider an MHMM with a symmetrical graphical representation, where chains have the same finite state space and where observations are categorical. The case of continuous observations is discussed in section 3.5.4. We consider first the case of a general MHMM, that is, with minimal conditional independence assumptions and show that the time complexity of the E step is exponential in C. Then we investigate simplifications that may occur in the EM equations that govern the E step, depending on which additional conditional independence relationships are introduced. The time complexity of the E step for general MHMMs and for the particular cases discussed in this section are reported in Table 3.1.

Model	Complexity
General MHMM	$\mathcal{O}(TK^{2C})$
1to1-MHMM-CI fig 3.7 (a)	$\mathcal{O}(CTK^2)$
1to1-MHMM-CI fig 3.7 (b)	$\mathcal{O}(CTK^2)$
1to1-MHMM-CI fig 3.7 (c)	$\mathcal{O}(TK^{2C})$
1to1-MHMM-CI fig 3.7 (d)	$\mathcal{O}(TK^{2C})$
ECHMM	$\mathcal{O}(T(C+K)K^C\binom{C+K-1}{K})$
FHMM	$\mathcal{O}(CTK^{C+1})$

Table 3.1. *Time complexity of the E step of EM for different families of MHMMs*

3.5.1. *Case of general MHMMs*

EM for a general MHMM can be handled similarly to EM for a HMM, if the hidden variables are grouped in a multidimensional variable with state space equal to the product state space $(\Omega_Z)^C$. We can therefore transpose the main formulas of the E step for HMMs provided in Chapter 1 (Appendix 1.10) to general MHMMs, up to an adaptation for the expression of $\beta_t^{(m)}$ due to the dependence of $\boldsymbol{z}_{t+1}$ on $\boldsymbol{y}_t$.

The quantities required to implement the EM algorithm are $\gamma_t^{(m)}(\boldsymbol{z}_{t-1}, \boldsymbol{z}_t) = p_{\theta^{(m)}}(\boldsymbol{z}_{t-1}, \boldsymbol{z}_t | \boldsymbol{y}_{0:T})$, which are deduced from the forward quantities $\alpha_t^{(m)}(\boldsymbol{z}_t) = p_{\theta^{(m)}}(\boldsymbol{z}_t, \boldsymbol{y}_{0:t})$ and from the backward quantities $\beta_t^{(m)}(\boldsymbol{z}) = p_{\theta^{(m)}}(\boldsymbol{y}_{t+1:T} | \boldsymbol{z}_t, \boldsymbol{y}_t)$.

The forward recursion is written as:

$$\alpha_t^{(m)}(\boldsymbol{z}_t) = \sum_{\boldsymbol{z}_{t-1} \in (\Omega_Z)^C} p_{\theta^{(m)}}(\boldsymbol{y}_t | \boldsymbol{z}_t, \boldsymbol{z}_{t-1}, \boldsymbol{y}_{t-1}) p_{\theta^{(m)}}(\boldsymbol{z}_t | \boldsymbol{z}_{t-1}, \boldsymbol{y}_{t-1}) p_{\theta^{(m)}}(\boldsymbol{z}_{t-1}, \boldsymbol{y}_{0:t-1}).$$

Thus, the time complexity to evaluate all $\alpha_t^{(m)}$ for a fixed t is $\mathcal{O}(K^{2C})$ (with K being the cardinal of ω_Z), hence the global time complexity of the whole recursion is $\mathcal{O}(TK^{2C})$. The complexity of evaluation of the $\beta_t^{(m)}$ is the same. Therefore, for a general MHMM, the time complexity of the E step is exponential in C.

3.5.2. *Case of 1to1-MHMM-CI*

Now, let us consider the case of 1to1-MHMM-CI. For a given chain c, let $\boldsymbol{Z}_t^{-c}$ denote $(Z_t^{c'})_{c' \neq c}$. Within the 1to1-MHMM-CI class, combining equations [3.2]–[3.3], [3.4] and [3.5] leads to the following expression of the log-likelihood:

$$p(\boldsymbol{z}_{0:T}, \boldsymbol{y}_{0:T}) = \prod_{c=1}^{C} [p(y_0^c | z_0^c)\, p(z_0^c)] \prod_{t=1}^{T} \prod_{c=1}^{C} [p(z_t^c | z_{t-1}^c, \boldsymbol{z}_{t-1}^{-c}, \boldsymbol{y}_{t-1}) p(y_t^c | z_t^c, \boldsymbol{z}_{t-1}, \boldsymbol{y}_{t-1})]$$

This implies the following canonical model parameterization in the case of categorical observations $\boldsymbol{y}_t$. This parameterization is referred to as “saturated”, since adding more parameters would necessarily lead to introducing additional constraints between parameters. The canonical parameters are as follows:

– initial state distributions $\mathbb{P}(Z_0^c = j) = \pi_j$;

– parameters of initial emission distributions $\mathbb{P}(Y_0^c = y_0^c | Z_0^c = j) = B_j(y_0^c)$;

– state transition probabilities:

$$\mathbb{P}(Z_t^c = j | Z_{t-1}^c = i, \boldsymbol{Z}_{t-1}^{-c} = \boldsymbol{\nu}, \boldsymbol{Y}_{t-1} = \boldsymbol{y}_{t-1}) = A_{i,\boldsymbol{\nu},\boldsymbol{y}_{t-1}}(j);$$

– parameters of non-initial emission distributions:

$$\mathbb{P}(Y_t^c = y_t^c | Z_t^c = j, \boldsymbol{Z}_{t-1} = \boldsymbol{i}, \boldsymbol{Y}_{t-1} = \boldsymbol{y}_{t-1}) = B_{\boldsymbol{i},j,\boldsymbol{y}_{t-1}}(y_t^c).$$

Let us θ denote the set of all parameters. We rewrite the log-likelihood function as follows, so that each component of parameter θ, which is a vector, appears individually and explicitly, rather than through a function of the unknown states $\boldsymbol{Z}_t$:

$$\log p_\theta(\boldsymbol{z}_{0:T}, \boldsymbol{y}_{0:T}) = \sum_{c=1}^{C} \sum_{j \in \Omega_Z} \mathbb{1}_{\{z_0^c = j\}} \log \pi_j + \sum_{c=1}^{C} \sum_{j \in \Omega_Z} \mathbb{1}_{\{z_0^c = j\}} \log B_j(y_0^c)$$

$$+ \sum_{t=1}^{T} \sum_{c=1}^{C} \sum_{i \in \Omega_Z} \sum_{j \in \Omega_Z} \sum_{\boldsymbol{\nu} \in (\Omega_Z)^{C-1}} \mathbb{1}_{\{z_{t-1}^c = i, \boldsymbol{z}_{t-1}^{-c} = \boldsymbol{\nu}, z_t^c = j\}} \log A_{i,\boldsymbol{\nu},\boldsymbol{y}_{t-1}}(j)$$

$$+ \sum_{t=1}^{T} \sum_{c=1}^{C} \sum_{\boldsymbol{i} \in (\Omega_Z)^C} \sum_{j \in \Omega_Z} \mathbb{1}_{\{z_t^c = j, \boldsymbol{z}_{t-1} = \boldsymbol{i}\}} \log B_{\boldsymbol{i},j,\boldsymbol{y}_{t-1}}(y_t^c)$$

In a 1to1-MHMM-CI, the Q function of EM is defined as:

$$Q(\theta, \theta^{(m)}) = \mathbb{E}_{\theta^{(m)}}[\log \mathbb{P}_\theta(\boldsymbol{Z}_{0:T} = \boldsymbol{z}_{0:T}, \boldsymbol{Y}_{0:T} = \boldsymbol{y}_{0:T}) | \boldsymbol{Y}_{0:T} = \boldsymbol{y}_{0:T}]$$

and written as:

$$Q(\theta, \theta^{(m)}) = \sum_{c=1}^{C} \sum_{j \in \Omega_Z} \mathbb{P}_{\theta^{(m)}}(Z_0^c = j | \boldsymbol{y}_{0:T}) \log \pi_j$$

$$+ \sum_{c=1}^{C} \sum_{j \in \Omega_Z} \mathbb{P}_{\theta^{(m)}}(Z_0^c = j | \boldsymbol{y}_{0:T}) \log B_j(y_0^c)$$

$$+ \sum_{t=1}^{T} \sum_{c=1}^{C} \sum_{i \in \Omega_Z} \sum_{\boldsymbol{\nu} \in (\Omega_Z)^{C-1}} \mathbb{P}_{\theta^{(m)}}(Z_{t-1}^c = i, \boldsymbol{Z}_{t-1}^{-c} = \boldsymbol{\nu}, Z_t^c = j | \boldsymbol{y}_{0:T}) \log A_{i,\boldsymbol{\nu},\boldsymbol{y}_{t-1}}(j)$$

$$+ \sum_{t=1}^{T} \sum_{c=1}^{C} \sum_{\boldsymbol{i} \in (\Omega_Z)^C} \sum_{j \in \Omega_Z} \mathbb{P}_{\theta^{(m)}}(Z_t^c = j, \boldsymbol{Z}_{t-1} = \boldsymbol{i} | \boldsymbol{y}_{0:T}) \log B_{\boldsymbol{i},j,\boldsymbol{y}_{t-1}}(y_t^c). \qquad [3.10]$$

The M step of the EM algorithm consists of computing:

$$\arg\max_\theta Q(\theta, \theta^{(m)}) = \theta^{(m+1)},$$

which requires the computation of

$$\delta^{(m)}(z_0^c) = \mathbb{P}_{\theta^{(m)}}(Z_0^c = z_0^c | \boldsymbol{y}_{0:T})$$

$$\gamma_t^{(m)}(z_{t-1}^c, \boldsymbol{z}_{t-1}^{-c}, z_t^c) = \mathbb{P}_{\theta^{(m)}}(Z_{t-1}^c = z_{t-1}^c, \boldsymbol{Z}_{t-1}^{-c} = \boldsymbol{z}_{t-1}^{-c}, Z_t^c = z_t^c | \boldsymbol{y}_{0:T})$$

$$\text{and } \xi_t^{(m)}(\boldsymbol{z}_{t-1}, z_t^c) = \mathbb{P}_{\theta^{(m)}}(Z_t^c = z_t^c, \boldsymbol{Z}_{t-1} = \boldsymbol{z}_{t-1} | \boldsymbol{y}_{0:T})$$

It is possible to compute these quantities by using general algorithms dedicated to computing marginal probabilities in Markov chains.

Note that $\gamma_t^{(m)}$ and $\xi_t^{(m)}$ are functions of the same variables, but separated into two or three arguments. We need to define two distinct functions since for some specific 1to1-MHMM-CIs with additional conditional independence assumptions, the set of hidden variables involved in A and B is reduced, leading to a change in $\gamma_t^{(m)}$ or $\xi_t^{(m)}$.

The property of $(\boldsymbol{Z}_t)_t$ being a Markov chain given $(\boldsymbol{y}_t)_t$ is simply the application of proposition 1, since the 1to1-MHMM-CI class is included in MHMMs. If Ω_Z and Ω_Y are both discrete (finite), one can collapse variables $\boldsymbol{Z}_t$ into a single variable $\hat{\boldsymbol{Z}}_t$ with domain size K^C and also collapse variables $\boldsymbol{Y}_t$ into a single variable $\hat{\boldsymbol{Y}}_t$. Then, quantities $\delta^{(m)}, \gamma_t^{(m)}$ and $\xi_t^{(m)}$ can be obtained by marginalization operations in Markov chains, resulting in time complexity $\mathcal{O}(TK^{2C})$, as for a general MHMM (see section 3.5.1). This is an upper bound on the complexity of computing these quantities, since in computations conditional independence relationships between random variables coming for a specific 1to1-MHMM-CI were ignored.

3.5.2.1. *Case of CHMM*

Quantities $\gamma_t^{(m)}$ and $\xi_t^{(m)}$ can be derived from $\mathbb{P}(Z_{t+1}^c = z_{t+1}^c, \boldsymbol{Z}_t = \boldsymbol{z}_t, \boldsymbol{Y}_{0:t} = \boldsymbol{y}_{0:t})$, which is the equivalent of the $\alpha_t^{(m)}$ auxiliary quantities for a HMM. In the case of a CHMM, the following specific forward recursion can be designed to compute these probabilities: for every possible value z_{t+1}^c of Z_{t+1}^c and $\boldsymbol{z}_t$ of $\boldsymbol{Z}_t$, we have:

$$\begin{aligned}
&\mathbb{P}(Z_{t+1}^c = z_{t+1}^c, \boldsymbol{Z}_t = \boldsymbol{z}_t, \boldsymbol{Y}_{0:t} = \boldsymbol{y}_{0:t}) \\
&= \mathbb{P}(\boldsymbol{Y}_t = \boldsymbol{y}_t | Z_{t+1}^c = z_{t+1}^c, \boldsymbol{Z}_t = \boldsymbol{z}_t, \boldsymbol{Y}_{0:t-1} = \boldsymbol{y}_{0:t-1}) \\
&\quad \times \mathbb{P}(Z_{t+1}^c = z_{t+1}^c, \boldsymbol{Z}_t = \boldsymbol{z}_t, \boldsymbol{Y}_{0:t-1} = \boldsymbol{y}_{0:t-1}) \\
&= \mathbb{P}(\boldsymbol{Y}_t = \boldsymbol{y}_t | \boldsymbol{Z}_t = \boldsymbol{z}_t) \mathbb{P}(Z_{t+1}^c = z_{t+1}^c | \boldsymbol{Z}_t = \boldsymbol{z}_t, \boldsymbol{Y}_{0:t-1} = \boldsymbol{y}_{0:t-1}) \\
&\quad \times \mathbb{P}(\boldsymbol{Z}_t = \boldsymbol{z}_t, \boldsymbol{Y}_{0:t-1} = \boldsymbol{y}_{0:t-1}) \\
&= \mathbb{P}(\boldsymbol{Y}_t = \boldsymbol{y}_t | \boldsymbol{Z}_t = \boldsymbol{z}_t) \mathbb{P}(Z_{t+1}^c = z_{t+1}^c | \boldsymbol{Z}_t = \boldsymbol{z}_t) \\
&\quad \times \sum_{\boldsymbol{z}_{t-1} \in (\omega_Z)^C} \mathbb{P}(\boldsymbol{Z}_t = \boldsymbol{z}_t, \boldsymbol{Z}_{t-1} = \boldsymbol{z}_{t-1}, \boldsymbol{Y}_{0:t-1} = \boldsymbol{y}_{0:t-1})
\end{aligned}$$

$$
\begin{aligned}
&= \mathbb{P}(\boldsymbol{Y}_t = \boldsymbol{y}_t | \boldsymbol{Z}_t = \boldsymbol{z}_t)\mathbb{P}(Z^c_{t+1} = z^c_{t+1} | \boldsymbol{Z}_t = \boldsymbol{z}_t) \\
&\quad \times \sum_{\boldsymbol{z}_{t-1} \in (\omega_Z)^C} \mathbb{P}(\boldsymbol{Z}_t = \boldsymbol{z}_t \; \boldsymbol{Z}_{t-1} = \boldsymbol{z}_{t-1})\mathbb{P}(\boldsymbol{Z}_{t-1} = \boldsymbol{z}_{t-1}, \boldsymbol{Y}_{0:t-1} = \boldsymbol{y}_{0:t-1}) \\
&= \left[\prod_{c=1}^{C} \mathbb{P}(Y^c_t = y^c_t | Z^c_t = z^c_t)\right] \mathbb{P}(Z^c_{t+1} = z^c_{t+1} | \boldsymbol{Z}_t = \boldsymbol{z}_t) \\
&\quad \times \sum_{\boldsymbol{z}_{t-1} \in (\omega_Z)^C} \left\{ \left[\prod_{c'} \mathbb{P}(Z^{c'}_t = z^{c'}_t | \boldsymbol{z}_{t-1}) \right] \right. \\
&\quad\quad \left. \sum_{i'=1}^{K} \mathbb{P}(Z^1_t = i', \boldsymbol{Z}_{t-1} = \boldsymbol{z}_{t-1}, \boldsymbol{Y}_{0:t-1} = \boldsymbol{y}_{0:t-1}) \right\}.
\end{aligned}
$$

Here, we chose arbitrarily in the last equation to marginalize over t $Z^1_t = i'$, but marginalization over any $Z^{C_0}_t$ would be appropriate.

If we do not make any additional assumption, the time complexity of computing all values of $\mathbb{P}(Z^c_{t+1} = z^c_{t+1}, \boldsymbol{Z}_t = \boldsymbol{z}_t, \boldsymbol{Y}_{0:t} = \boldsymbol{y}_{0:t})$ at a given time t remains in $\mathcal{O}(K^{2C})$. However, if we further assume that $\mathbb{P}(Z^{c'}_t = z^{c'}_t | \boldsymbol{z}_{t-1})$ is the same for every c' (referring to this model as to exchangeable CHMMs [ECHMMs]), then the product $\prod_{c'} \mathbb{P}(Z^{c'}_t = z^{c'}_t | \boldsymbol{z}_{t-1})$ can be precomputed for every $\boldsymbol{z}_{t-1}$ and stored for every combination of states, not accounting for the order of chains. This is referred to as C-combinations with repetitions from a set of size K in combinatorics and their number is $\binom{C+K-1}{K}$. The inner sum (with complexity $\mathcal{O}(K)$) and product (with complexity $\mathcal{O}(K)$) have to be repeated K^C times and summed. Thus, quantities

$$
\sum_{\boldsymbol{z}_{t-1} \in (\omega_Z)^C} \left\{ \left[\prod_{c'} \mathbb{P}(Z^{c'}_t = z^{c'}_t | \boldsymbol{z}_{t-1}) \right] \sum_{i'} \mathbb{P}(Z^1_t = i', \boldsymbol{Z}_{t-1} = \boldsymbol{z}_{t-1}, y_0, \ldots, y_{t-1}) \right\}
$$

can be computed with time complexity $\mathcal{O}((C+K)K^C\binom{C+K-1}{K})$ and the overall complexity of the whole recursion over t is $\mathcal{O}(T(C+K)K^C\binom{C+K-1}{K})$. The backward recursion can be built on the same principle, with the same complexity. Note that in the related model by Kwon and Murphy (2000), the provided time complexity is $\mathcal{O}(TCK^{C+F-1})$ for some F, which is consistent with $\mathcal{O}(T(C+K)K^C\binom{C+K-1}{K})$.

3.5.2.2. *Coupling with conditional independence of hidden chains*

There is an obvious particular case of 1to1-MHMM-CI where the E step complexity is reduced: when the C bidimensional processes $(Z^c_t, Y^c_t)_t$ are independent HMMs. In this case, transition and emission probabilities are as follows:

$$\mathbb{P}(\boldsymbol{Z}_t = \boldsymbol{z}_t \mid \boldsymbol{Z}_{t-1} = \boldsymbol{z}_{t-1}) = \prod_{c=1}^{C} A^c(z_{t-1}^c, z_t^c),$$

$$\mathbb{P}(\boldsymbol{Y}_t = \boldsymbol{y}_t \mid \boldsymbol{Z}_t = \boldsymbol{z}_t) = \prod_{c=1}^{C} B^c(z_t^c, y_t^c).$$

Each quantities A^c and B^c, attached to the dynamics of chain c, are independent of variables from the other chains. In this case, the expression [3.10] of the Q function involves simpler definitions of the probabilities $\delta_t^{(m)}$ and $\gamma_t^{(m)}$:

$$\gamma_t^{(m)}(z_{t-1}^c, z_t^c) = \mathbb{P}_{\theta^{(m)}}(Z_{t-1}^c = z_{t-1}^c, Z_t^c = z_t^c | \boldsymbol{y}_{0:T})$$

$$\text{and } \xi_t^{(m)}(z_t^c) = \mathbb{P}_{\theta^{(m)}}(Z_t^c = z_t^c | \boldsymbol{y}_{0:T})$$

So, all quantities can be computed independently for each chain c, using forward and backward recursions involving only z_t^c (like in a monochain HMM), leading to an overall complexity in $\mathcal{O}(CTK^2)$.

In the particular case where the chains are coupled, but conditionally independent given the observed variables, such as in Figure 3.7, cases (a) and (b), inference tasks can also be performed independently in each chain. Indeed, in case (a), transition and emission probabilities are as follows:

$$\mathbb{P}(\boldsymbol{Z}_t = \boldsymbol{z}_t \mid \boldsymbol{Z}_{t-1} = \boldsymbol{z}_{t-1}) = \prod_{c=1}^{C} A^c(z_{t-1}^c, z_t^c),$$

$$\mathbb{P}(\boldsymbol{Y}_t = \boldsymbol{y}_t \mid \boldsymbol{Z}_t = \boldsymbol{z}_t, \boldsymbol{Y}_{t-1} = \boldsymbol{y}_{t-1}) = \prod_{c=1}^{C} B^c(z_t^c, y_t^c, \boldsymbol{y}_{t-1}^{-c}).$$

and in case (b):

$$\mathbb{P}(\boldsymbol{Z}_t = \boldsymbol{z}_t \mid \boldsymbol{Z}_{t-1} = \boldsymbol{z}_{t-1}, \boldsymbol{Y}_{t-1} = \boldsymbol{y}_{t-1}) = \prod_{c=1}^{C} A^c(z_{t-1}^c, z_t^c, \boldsymbol{y}_{t-1}),$$

$$\mathbb{P}(\boldsymbol{Y}_t = \boldsymbol{y}_t \mid \boldsymbol{Z}_t = \boldsymbol{z}_t) = \prod_{c=1}^{C} B^c(z_t^c, y_t^c).$$

Like for the independent case, each quantities A^c and B^c, attached to the dynamics of chain c, are independent of variables from the other chains. So for these two 1to1-MHMM-CI, when deriving the equations for the E step, the corresponding $\gamma_t^{(m)}$ and $\xi_t^{(m)}$ functions depend only on hidden variables of a single chain, c. Therefore, it is only necessary to define one set of forward and backward quantities per chains that are computed independently. The time complexity of the E step is also in $\mathcal{O}(CTK^2)$. This has been established in Le Coz et al. (2019) for the case (a) of Figure 3.7.

Another demonstration of this complexity is provided in Chapter 5, where the conditional independence of the hidden chains given the observations is linked to the possibility to decompose the transition matrix of the model as a Kronecker product of matrices, corresponding each to a single hidden chain.

3.5.3. *Case of FHMM*

In the particular case of FHMM models, Ghahramani and Jordan (1997) propose a variable elimination algorithm for performing the E step of the EM algorithm in time complexity $\mathcal{O}(TCK^{C+1})$ instead of $\mathcal{O}(TK^{2C})$ for a naive forward-backward approach.

3.5.4. *MHMM parameterization for continuous observations*

It has been assumed in this section that observations are categorical. In applications, observations may have unbounded discrete or even continuous values and this assumption has to be relaxed.

As long as hidden states remain categorical, introducing non-categorical observations only impacts model parameterization. The reason for this is that the likelihood function is still defined by marginalization of categorical variables and the same kinds of marginalizations occur, for example, in the forward and backward recursions in EM. The case of models without outgoing arcs from observations in the graphical model is handled straightforwardly and generically. Indeed, transition probabilities are written as:

$$\mathbb{P}(Z_t^c = j | Z_{t-1}^c = i, \boldsymbol{Z}_{t-1}^{-c} = \boldsymbol{\nu})$$

and do not depend on $\boldsymbol{y}_{t-1}$ anymore. This also holds for emission probabilities:

$$\mathbb{P}(Y_t^c = y_t^c | Z_{t-1}^c = j, \boldsymbol{Z}_{t-1} = \boldsymbol{j}).$$

In this case, a classical and convenient choice is to assume that these emission distributions belong to the same parametric family of distributions $(B_\eta)_{\eta \in \mathrm{H}}$:

$$\mathbb{P}(Y_t^c = y_t^c | Z_{t-1}^c = j, \boldsymbol{Z}_{t-1} = \boldsymbol{i}) = B_{\eta_{\boldsymbol{i},j}}(y_t^c).$$

Note that non-parametric estimation has also been considered in monochain HMMs (De Castro et al. 2017); since the issue is related to emission probabilities and not to the dependency structure between states, the approach should be adaptable to MHMMs.

In the case of continuous observations, if there are outgoing arcs from the observations, the model does not have a canonical parameterization anymore. The cause for this is that, for example, transition matrices now are functions of infinite (either countable or uncountable) $\boldsymbol{y}_{t-1}$. Thus, it is not possible to define them using a finite set of parameters indexed by the possible values of $\boldsymbol{y}_{t-1}$.

A solution is to resort to regression models, which are dedicated to modeling how the distribution of some random variable (here, Z_t^c) depends on other continuous or discrete variables. Generalized linear models (GLMs, McCulloch et al. 2008) are used in most publications, such as Bartolucci et al. (2014), since they have the advantages of relying on a small number of interpretable parameters and benefit from well-known estimation methods in the context of independent observations. These methods can often be adapted to the context of HMMs (see Chapter 1, section 1.6). Moreover, the asymptotic properties of parameter estimates are also well characterized for GLMMs with independent observations. In the case of transition probabilities, GLM-based models are written as:

$$g_1\left[(A_{i,\boldsymbol{\nu},\boldsymbol{y}_{t-1}}(j))_j\right] = \left(<\boldsymbol{\rho}_{i,\boldsymbol{\nu},j};\boldsymbol{y}_{t-1}>\right)_j \qquad [3.11]$$

where g_1 is referred to as the link function, $\boldsymbol{\rho}_{i,\boldsymbol{\nu},j}$ is a vector of parameters that summarizes the effect of $\boldsymbol{y}_{t-1}$, which may change in interaction with $i, \boldsymbol{\nu}$ and j, and $< .;. >$ denotes the canonical dot product. The role of function g_1 is to map the value of the linear predictor (right-hand side of [3.11]), which is in $\mathbb{R}^K$, to an admissible value for $(A_{i,\boldsymbol{\nu},\boldsymbol{y}_{t-1}}(j))_j$: it should be non-negative and sum to 1.

Similarly, emission probabilities can be modeled as:

$$g_2\left(E\left[Y_t^c|Z_t^c = j, \boldsymbol{Z}_{t-1} = \boldsymbol{i}, \boldsymbol{y}_{t-1}\right]\right) =< \boldsymbol{\rho}'_{i,j};\boldsymbol{y}_{t-1} >;$$

since in the case of GLMs, it is sufficient to model the conditional expectation of Y_t^c to obtain its conditional distribution.

Note that the above presentation on forward-backward recursions and their computational complexities is not directly impacted by the nature of y_t^c, since these recursions only depend on the conditional independence graph and on the fact that the hidden variables z_t^c take finite values.

3.6. Approximate inference in MHMMs

One of the main conclusions from section 3.5 is that in the families of MHMMs considered in section 3.5, inference with the EM algorithm has time complexity $\mathcal{O}(TK^{2C})$ per iteration in the worst case, T being the chain length, K the number of states per chain and C is the number of chains. This is essentially due to the complexity of computing marginal state conditional probabilities at specified times and chains, given observations such as $\mathbb{P}_{\theta^{(m)}}(Z_t^c = j, \boldsymbol{Z}_{t-1} = \boldsymbol{i} | \boldsymbol{Y}_{0:T} = \boldsymbol{y}_{0:T})$, where $\theta^{(m)}$ is the EM estimate at iteration m, j is a possible state for chain c at time t, $\boldsymbol{i}$ is a whole possible joint configuration of states at time $t-1$ and $\boldsymbol{y}_{0:T}$ is the whole set of observations. The recursive formulas to compute these probabilities are formally the same in monochain HMMs, with the only difference being the state space being much larger in MHMMs, which leads to infeasible computations in practice.

As a consequence of such time complexity, the practical use of the EM algorithm for statistical inference is restricted to small values of C ($C = 2$ is already time-demanding). To circumvent this limitation, different approximations have been developed for reducing the complexity of the E step, which are presented here. Two main families of approximations emerge: Monte Carlo and variational approximations, although less frequently used strategies also exist, such as loopy belief propagation (LBP). Monte Carlo methods are dedicated to approximating expectations directly using simulations and do not require to approximate the model itself. The relevance of Monte Carlo methods appears when considering probabilities as conditional expectations of functions of states. For instance, $\mathbb{P}_{\theta^{(m)}}(Z_t^c = j, \boldsymbol{Z}_{t-1} = \boldsymbol{i} | \boldsymbol{y}_{0:T}) = \mathbb{E}_{\theta^{(m)}}[\mathbb{1}_{\{Z_t^c = j, \boldsymbol{Z}_{t-1} = \boldsymbol{i}\}} | \boldsymbol{y}_{0:T}]$. The variational approach leads to the variational EM (VEM) when used to approximate quantities involved in the E step. It consists of selecting a family of distributions for $(Z_t)_t$ for which inference is easier than for the original model and to select in this family the best approximation of the posterior distribution $p_{\theta^{(m)}}(\boldsymbol{z}_{0:T} \mid \boldsymbol{y}_{0:T})$ (Wainwright and Jordan 2008). Optimality is usually defined as minimizing the Kullback–Leibler divergence. Mean field is a special case of variational approximations in which the approximate model assumes that all hidden variables are independent, given the observed data. As an alternative to Monte Carlo and VEM algorithms, LBP is an algorithm for approximate computation of marginals in complex graphical models. It is derived from the exact belief propagation (BP) procedure, which achieves the sums and products involved in the marginals computation with polynomial complexity. BP is a two-pass procedure, which matches the forward-backward algorithm for HMMs for example. However, BP is restricted to a specific family of graphical models; outside that family (typically in the case of MHMMs, see Figure 3.7) exact BP is not possible and looping, that is, iterative application of BP, is an ad-hoc solution, even though inference may not be exact anymore. LBP and variational approximations are presented together hereafter because a link has been established between these two ways to approximate probabilities in graphical models (see Yedidia et al. 2005; Peyrard et al. 2019).

Among all the structures of MHMMs defined in this chapter, literature on approximate inference can only be found for CHMMs (Figure 3.3) and FHMMs (Figure 3.5). We present here a short review of what has been proposed in the literature using LBP, variational or Monte Carlo approaches. For CHMM inference, we also present works based on ad hoc simplifications of the original model.

3.6.1. *State of the art of approximate inference for CHMMs*

In the case of CHMMs, the EM algorithm is based on quantities of the form $\mathbb{P}_{\theta^{(m)}}(Z_t^c = i, \boldsymbol{Z}_{t-1} = \boldsymbol{j} | \boldsymbol{y}_{0:T})$, that is, posterior probability of one hidden variable and its parents. As a consequence of the results presented in section 3.5, the E step of EM has complexity exponential in the number of chains, C. This explains why approximations are required outside the context of a small number of chains C.

3.6.1.1. *Monte Carlo approximations*

The use of Monte Carlo approximations in EM algorithms, leading to MCEM algorithms (Wei and Tanner 1990), is relevant in cases where conditional expectations involved in the E step have no closed form, but the random variables involved can be simulated either straightforwardly from the required conditional distributions or by efficient approximate schemes, such as Gibbs sampling. In the context of CHMMs, the difficulty comes essentially from the high dimension of the state space of $\boldsymbol{Z}_t$ (i.e. the large number of combinations of states among all chains, each of which is involved in probability computations). As a result, the added value of using MCEM rather than brute force marginalization is not clear for parameter estimation in CHMMs, since naive Monte Carlo approximations in very large dimensions converge very slowly in the general case (Brown 2022). Thus, the Monte Carlo approximations of the probabilities required in the EM algorithm may be crude. MCEM was applied by Kwon and Murphy (2000) on a CHMM where chains are coupled pairwise. However, the authors did not justify their choice nor provide comparisons to other alternatives.

3.6.1.2. *Variational approximations and LBP*

The approach in Heskes et al. (2004) relies on a variational approximation: the authors address the general problem of approximate E step computation in DAGs with hidden variables. It has been shown that the E step can be rewritten as the minimization of the Kullback–Leibler divergence between a distribution to optimize and the conditional distribution of the model. Computations rely on Bethe or Kikuchi approximations of the Kullback–Leibler divergence, which are variational approaches less constrained than mean field approximations in terms of independence assumptions. The minimization of the divergence is an alternative to LBP, which has the potential shortcoming of not converging toward the true marginal conditional expectations. The original algorithm proposed in Heskes et al. (2004) proceeds with two loops, with the inner loop dedicated to solving a convex problem. Their approach is illustrated on parameter estimation in some CHMM where chains are coupled pairwise.

In Wainwright and Jordan (2008), the author also propose to use a less drastic approximation than the mean field one using the family of variational distributions of C independent heterogeneous Markov chains.

LBP has been exploited in Craley et al. (2019), where the authors developed a model for seizure detection in EEG data, that is based on a CHMM augmented by a regime-switching Markov model. For parameter estimation, they used EM with LBP in the E step.

3.6.1.3. *Other approaches*

Other approaches have been proposed to decrease the original complexity of exact inference in CHMMs. These share a common principle with variational approaches: to replace the true posterior distribution of the C hidden chains given the observations by a simpler one. But as opposed to the variational approach, there is no attempt here to justify the choice by an optimization argument.

In Brand (1997) and Brand et al. (1997), the distribution $\mathbb{P}_{\theta^{(m)}}(Z_t^c = z_t^c | \boldsymbol{Z}_{t-1} = \boldsymbol{z}_{t-1})$ is assumed to be the product over all chains c' of $\mathbb{P}(Z_t^c = z_t^c | Z_{t-1}^{c'} = z_{t-1}^{c'})$ with a renormalization. In addition, in order to reduce the number of sums in the forward-backward algorithm, sums are limited to some states with maximal probabilities identified thanks to the Viterbi algorithm. The same hypothesis on the form of $\mathbb{P}_{\theta^{(m)}}(Z_t^c = z_t^c | \boldsymbol{Z}_{t-1} = \boldsymbol{z}_{t-1})$ is used in Montazeri Ghahjaverestan et al. (2016) where an additional hypothesis is made: the C hidden states at time t are independent given all observations $\boldsymbol{y}_{0:T}$. A more general version of this latter additional approximation is defined in Kwon and Murphy (2000): the C chains are grouped into K (non-necessarily disjoint) clusters and $\mathbb{P}_{\theta^{(m)}}(\boldsymbol{Z}_t = \boldsymbol{z}_t | \boldsymbol{Y}_{0:T} = \boldsymbol{y}_{0:T})$ is approximated by the product $\prod_{k=1}^{K} p(\boldsymbol{z}_t^{C_k} | \boldsymbol{y}_{0:T})$, where C_k is the set of chains in cluster k. Note that in practice they develop the case where there is one chain per cluster. A different approach is proposed in Zhong and Ghosh (2002), where an original approximation is used for $\mathbb{P}_{\theta^{(m)}}(Z_t^c = z_t^c | \boldsymbol{Z}_{t-1} = \boldsymbol{z}_{t-1})$, as a linear combination of the $(p_{\theta^{(m)}}(z_t^c | z_{t-1}^{c'}))_{c'=1,\dots,C}$.

3.6.1.4. *A numerical comparison from literature*

We have presented LBP, variational methods and Monte Carlo methods as possible approximations to make EM tractable. They are used in the E step since all methods compute approximations of the marginals of the model of interest. In Peyrard et al. (2019), these approaches are compared on the problem of computing the marginals of all the hidden variables of a CHMM model of pest propagation: exact inference, LBP, mean field and Gibbs sampling. Both hidden and observed variables are binary, and the number of hidden chains varies from $C = 9$ to $C = 40,000$. Exact inference was only possible for $C = 9$ due to space complexity. As expected, LBP and mean field are much more competitive than Gibbs sampling in terms of running time. LBP marginals are more accurate than mean field ones (when $C > 9$, the Gibbs sampling marginals were the reference). The comparison study was limited to marginal inference though. When plugged into the E step of an EM for CHMM, mean field can still lead to a VEM that provides parameters estimates of good quality, as has been observed for other graphical models. It means that a precise approximation of the marginals in the E step may not be necessary.

3.6.2. *State of the art of approximate inference for FHMMs*

In the case of FHMMs, the EM algorithm is based on quantities of the form $\mathbb{P}_{\theta^{(m)}}(\boldsymbol{Z}_t = \boldsymbol{i}|\boldsymbol{Y}_{0:T} = \boldsymbol{y}_{0:T})$ and $\mathbb{P}_{\theta^{(m)}}(Z_{c,t} = i, Z_{c,t-1} = j|\boldsymbol{Y}_{0:T} = \boldsymbol{y}_{0:T})$. The forward recursion to obtain $\mathbb{P}_{\theta^{(m)}}(\boldsymbol{Z}_t = \boldsymbol{i}|\boldsymbol{Y}_{0:T} = \boldsymbol{y}_{0:T})$ consists of computing $\alpha_t(\boldsymbol{z}_t) = \mathbb{P}_{\theta^{(m)}}(\boldsymbol{z}_t, \boldsymbol{y}_0, \ldots, \boldsymbol{y}_t)$ for each t; its overall complexity is $\mathcal{O}(TCK^{C+1})$. Moreover, there are TK^{2C} possible values for $\mathbb{P}_{\theta^{(m)}}(Z_{c,t} = i, Z_{c,t-1} = j|\boldsymbol{Y}_{0:T} = \boldsymbol{y}_{0:T})$: as a result, the E step of EM has complexity at least $\mathcal{O}(TK^{2C})$. We present here the main approaches that have been proposed to circumvent this complexity. They are mainly variational but we can mention the approach by Kolter and Jaakkola (2012) for additive FHMMs (i.e. the shared observation is the sum of the observations from each chain) that makes the hypothesis that only one chain can change state at each time step.

3.6.2.1. *Variational approximations*

The FHMM model was first introduced by Ghahramani and Jordan (1997). The authors pointed out the difficulty for exact inference and they proposed two variational approximations of the conditional distribution of the hidden variables in the E step. The first one corresponds to a mean field approximation: independence is assumed between all the $Z_{c,t}$. With the second approximation, referred to as structured variational, each hidden chain is assumed independent to the others conditionally to the observations but within the chain, the Markov property is preserved.

In two more recent works, the principles of variational approximation of the joint conditional distribution have been combined with deep learning. In these works, the family of tractable distributions (among which the variational distribution is chosen by maximizing the Kullback–Leibler divergence) is parameterized by a neural network. In Ng et al. (2016), an FHMM with binary hidden states is considered. They propose an alternative to the structured variational approximation that enables us to deal with very long sequences. The conditional distribution is approximated by one of C independent chains and each chain is modeled by bivariate Gaussian copulas parameterized using feed-forward recognition neural networks. The original FHMM parameters and the variational parameters are jointly optimized using stochastic gradient descent instead of EM. It would seem that the approach is an adaption of VEM for large datasets, although this is not stated explicitly. In Lange and Mario (2018), only an approximation of $\mathbb{P}_{\theta^{(m)}}(\boldsymbol{Z}_t = \boldsymbol{i}|\boldsymbol{Y}_t = \boldsymbol{y}_t)$ is used. The variational simplifying assumption is that at a single time point each $Z_{c,t}$ is independent of the others given the observations. They deal with the case where these variables are binary, and they model them as Bernoulli variables whose parameter is a recurrent deep neural network. It seems that their approach is precisely a VEM algorithm, although as in Ng et al. (2016), this is not stated explicitly.

3.6.2.2. *Comparison of performances between variational and Monte Carlo approximations*

In Ghahramani and Jordan (1997), the authors compared the two variational approximations to a E step performed using Gibbs sampling. In their FHMM model, the observation $\boldsymbol{Y}_t$ is a Gaussian random vector whose mean is a linear function of the hidden variables $\{Z_{c,t}\}_{1 \leq c \leq C}$, and each $Z_{c,t}$ is a vector in $\{0, 1\}^K$ with exactly one non-zero component. The different approximate EMs are compared in terms of the (approximated) log likelihood. The conclusions are similar to those for CHMMs: a Monte Carlo method can provide accurate results but requires much more running time than a variational method. Among the variational methods, as expected the structured one led to an improved accuracy compared to the mean field one, while still efficient computationally.

3.6.2.3. *Conclusion*

This review of literature shows that there is not a unique way to deal with complexity in MHMM inference. However, since in such models the state space to explore is large and there are not always conditional independencies that can be exploited, MCMC approaches may not be adapted. Deterministic approximations or simplifications of the original models, with additional conditional independence assumptions are more suited. From a practical point of view, implementing one of the approximation approach presented in this section may not be straightforward from the literature we cited. Concrete algorithms are often barely described. We recommend some readings on the foundations of graphical models, LBP and variational methods before applying these principles for approximate inference in MHMM: Wainwright and Jordan (2008), Koller and Friedman (2009), Yedidia et al. (2005) and Peyrard et al. (2019).

3.7. Discussion and conclusion

With this chapter, we proposed a common framework to present in a unified way several notions of MHMMs, and we illustrated the variety of applications that can be modeled with this framework. We also provided the key elements for their exact or approximate inference.

When presenting MHMM parameter estimation with EM, we discussed only the difficulties arising in the E step, corresponding to difficulties in marginals evaluation in the MHMM. This is because the M step is an optimization step and the computational complexity of the evaluation of the function to optimize is that of the E step. However, there may be other difficulties attached to the realization of the M step. They are not the same depending on whether the MHMM is defined with canonical parameterization (i.e. all the initial, transition and emission probabilities defining the model), or if the distributions are parameterized. (Note that, on the contrary, this distinction has no impact on the E step.) The advantage of the canonical representation of the MHMM is that there exists an explicit expression of the updated canonical parameters. However, it may be difficult to obtain estimates of good quality when the number of chains, C, and the state space of hidden variables, Ω_Z, are large. For a parameterized model, there is not always an explicit updating formula, and in this case the M step relies on numerical optimization tools. However, a parametric MHMM is certainly a better choice in terms of interpretability.

Beyond computational difficulties, the quality of model inference is also conditioned by the amount of available data, strongly dependent on the domain of application. For example, measurements on population in metapopulation dynamics are costly and limited to unities or tens, while sensor data (like the ones available for the Deer toy example) collected at high resolution provide chains of lengths tens or hundreds of thousands time points. Thus, the choice of an MHMM structure results simultaneously of prior knowledge regarding the phenomenon of interest and a compromise between the bias due to a too simple model and the variance generated by the model complexity with regard to sample size.

In this chapter, we did not address the questions of model validation and model selection in MHMMs, since these are still open questions. We describe here some elements of perspective. In the conclusion of Chapter 1, we have mentioned the approach of Guédon (1999) for model validation in ED-HMMs, which relies on the distributions of some statistics of the observations. These distributions have never been considered for MHMMs but their computation is a perspective, using either marginal distributions in each separate chain, or joint configurations of discrete observations. Regarding model selection in MHMMs, that is, choice of the number K of hidden states and choice of the structure of the global transition, the difficulty comes from the exponential number of candidate models when exploring all possibilities for conditional independence relationships. A promising way of addressing both selection of the number of states and the conditional independence relationships in MHMMs is Bayesian statistical modeling. This offers the possibilities, first, to handle non-bounded a priori numbers of parameters using Dirichlet processes and, second, to replace the separate estimation of several models with the estimation of posterior distributions of random variables in nested models, using spike-and-slab priors. These specific random variables represent switches between families of models. Both approaches avoid computational complexity overheads due to combinatorics when selecting several meta-parameters. The first method was used in Johnson and Willsky (2013) in hidden semi-Markov chains to select K and could be transposed to MHMMs. The second method was used in Hernández-Lobato et al. (2015) for sparse regression modeling. The transposition of this principle to MHMMs would be to associate a binary random variable with each type of arc in the probabilistic graphical model and to remove arcs whose probabilities are less than a threshold. This approach has not yet been implemented, but it appears as a promising lead to handle graphical model selection.

Finally, a natural extension is to introduce the semi-Markov property instead of the Markov property. We will see in Chapter 4 that the formalization of this idea is not straightforward and we will propose a sound solution that enables us to extend the various MHMMs classes presented in this chapter.

3.8. Notations

We gather here the main notations for MHMM used in this chapter.

Definition	Notation	Domain
Calendar time	t	$\mathbb{N}$
Duration of the time series	$T+1$	$\mathbb{N}$
Number of chains	C	$\mathbb{N}^*$
State space of hidden variable for a chain	$\Omega_Z = \{1, \ldots, K\}$	finite
Cardinal of the hidden variable state space	$K = \|\Omega_Z\|$	$\mathbb{N}^*$
Hidden state of chain c at time t:		
Random variable	Z_t^c	Ω_Z
Realization	z_t^c	Ω_Z
Hidden states of all chains at time t:		
Random variable	$\boldsymbol{Z}_t = (Z_t^1, \ldots, Z_t^C)$	$(\Omega_Z)^C$
Realization	$\boldsymbol{z}_t = (z_t^1, \ldots, z_t^C)$	$(\Omega_Z)^C$
Sequence of hidden states of chain c:		
Random variable	$Z_{0:T}^c = (Z_0^c, \ldots, Z_T^c)$	$(\Omega_Z)^{T+1}$
Realization	$z_{0:T}^c = (z_0^c, \ldots, z_T^c)$	$(\Omega_Z)^{T+1}$
Hidden states, all times, all chains:		
Random variable	$\boldsymbol{Z}_{0:T}$	$(\Omega_Z)^{C(T+1)}$
Realization	$\boldsymbol{z}_{0:T}$	$(\Omega_Z)^{C(T+1)}$
Number of components of observation	O	$\mathbb{N}^*$
State space of one component	Ω_Y	general
Component o of observation at time t:		
Random variable	Y_t^o	Ω_Y
Realization	y_t^o	Ω_Y
All components of observation at time t:		
Random variable	$\boldsymbol{Y}_t = (Y_t^1, \ldots, Y_t^O)$	$(\Omega_Y)^O$
Realization	$\boldsymbol{y}_t = (y_t^1, \ldots, y_t^O)$	$(\Omega_Y)^O$
Sequence of observations for component o:		
Random variable	$Y_{0:T}^o = (Y_0^o, \ldots, Y_T^o)$	$(\Omega_Y)^{T+1}$
Realization	$y_{0:T}^o = (y_0^o, \ldots, y_T^o)$	$(\Omega_Y)^{T+1}$
Observations, all times, all component:		
Random variable	$\boldsymbol{Y}_{0:T}$	$(\Omega_Y)^{O(T+1)}$
Realization	$\boldsymbol{y}_{0:T}$	$(\Omega_Y)^{O(T+1)}$

Table 3.2. *Main notations for hidden and observed variables of an MHMM*

3.9. References

Anandkumar, A., Hsu, D., Kakade, S. (2012). A method of moments for mixture models and hidden Markov models. In *Proceedings of the 25th Annual Conference on Learning Theory*. PMLR, Edinburgh.

Bartolucci, F., Farcomeni, A., Pennoni, F. (2014). Latent Markov models: A review of a general framework for the analysis of longitudinal data with covariates. *TEST*, 23, 433–465.

Bessac, J., Ailliot, P., Cattiaux, J., Monbet, V. (2016). Comparison of hidden and observed regime-switching autoregressive models for (u, v)-components of wind fields in the northeastern Atlantic. *Advances in Statistical Climatology, Meteorology and Oceanography*, 2(1), 1–16.

Brand, M. (1997). Coupled hidden Markov models for modeling interacting processes. Technical Report 405, MIT Media Laboratory, Cambridge.

Brand, M., Oliver, N. and Pentland, A. (1997) Coupled hidden Markov models for complex action recognition. In *Proceedings of IEEE Computer Society Conference on Computer Vision and Pattern Recognition*, San Juan, 17–19 June, 994-999.

Brown, A. (2022). Some convergence results for Metropolis-Hastings algorithms. PhD Thesis, University of Minnesota.

Cappé, O., Buchoux, V., Moulines, E. (1998). Quasi-Newton method for maximum likelihood estimation of hidden Markov models. In *Proceedings of the 1998 IEEE International Conference on Acoustics, Speech and Signal Processing, ICASSP '98*, Seattle, WA, 2265–2268.

Craley, J., Johnson, E., Jouny, C., Venkataraman, A. (2019). Automated noninvasive seizure detection and localization using switching Markov models and convolutional neural networks. In *Medical Image Computing and Computer Assisted Intervention – MICCAI 2019*. Springer, Cham.

De Castro, Y., Gassiat, E., Le Corff, S. (2017). Consistent estimation of the filtering and marginal smoothing distributions in nonparametric hidden Markov models. *IEEE Transactions on Information Theory*, 63(8), 4758-4777.

Franc, A. (2004). Metapopulation dynamics as a contact process on a graph. *Ecological Complexity*, 1, 49–63.

Ghahramani, Z. and Jordan, M. (1997). Factorial hidden Markov models. *Machine Learning*, 29(2–3), 245–273.

Guédon, Y. (1999). Computational methods for discrete hidden semi-Markov chains. *Applied Stochastic Models in Business and Industry*, 15, 195–224.

Hamilton, J. (1989). A new approach to the economic analysis of nonstationary time series and the business cycle. *Econometrica*, 57, 357–384.

Hernández-Lobato, J., Hernández-Lobato, D., Suárez, A. (2015). Expectation propagation in linear regression models with spike-and-slab priors. *Machine Learning*, 99, 437–487.

Heskes, T., Zoeter, O., Wiegerinck, W. (2004). Approximate expectation maximization. *Advances in Neural Information Processing Systems*, 16, 353–360.

Johnson, M. and Willsky, A. (2013). Bayesian nonparametric hidden semi-Markov models. *Journal of Machine Learning Research*, 14(20), 673–701.

Koller, D. and Friedman, N. (2009). *Probabilistic Graphical Models: Principles and Techniques*. MIT Press, Cambridge, MA.

Kolter, J. and Jaakkola, T. (2012). Approximate inference in additive factorial HMMs with application to energy disaggregation. In *Proceedings of the 15th International Conference on Artificial Intelligence and Statistics – AISTATS*. PMLR, Maastricht University.

Kwon, J. and Murphy, K. (2000). Modeling freeway traffic with coupled HMMs. Technical Report, University of California, Berkeley.

Lange, H. and Mario, B. (2018). Variational BOLT: Approximate learning in factorial hidden Markov models with application to energy disaggregation. In *Thirty-Second Conference on Artifical Intelligence – AAAI-18*. AAAI, Palo Alto, CA.

Le Coz, S., Cheptou, P.-O., Peyrard, N. (2019). A spatial Markovian framework for estimating regional and local dynamics of annual plants with dormancy. *Theoretical Population Biology*, 127, 120–132.

Levins, R. (1969). Some demographic and genetic consequences of environmental heterogeneity for biological control. *Bulletin of the Entomological Society of America*, 15(3), 237–240.

McClintock, B.T., Langrock, R., Gimenez, O., Cam, E., Borchers, D.L., Glennie, R., Patterson, T.A. (2020). Uncovering ecological state dynamics with hidden Markov models. *Ecology Letters*, 23, 1878–1903.

McCulloch, C., Searle, S., Neuhaus, J. (2008). *Generalized, Linear, and Mixed Models*, 2nd edition. Wiley, Hoboken, NJ.

McLachlan, G. and Krishnan, T. (2008). *The EM Algorithm and Extensions*, 2nd edition. Wiley, Hoboken, NJ.

Melnyk, I. and Banerjee, A. (2017). A spectral algorithm for inference in hidden semi-Markov models. *Journal of Machine Learning Research*, 35, 1–39.

Monbet, V. and Ailliot, P. (2017). Sparse vector Markov switching autoregressive models. Application to multivariate time series of temperature. *Computational Statistics and Data Analysis*, 108, 40–51.

Montazeri Ghahjaverestan, N., Masoudi, S., Shamsollahi, M.B., Beuchée, A., Pladys, P., Ge, D., Hernández, A.I. (2016). Coupled hidden Markov model-based method for apnea bradycardia detection. *IEEE Journal of Biomedical and Health Informatics*, 20(2), 527–538.

Murphy, K.P. (2012). *Machine Learning: A Probabilistic Perspective*. MIT Press, Cambridge, MA.

Ng, Y.C., Chilinski, P.M., Silva, R. (2016). Scaling factorial hidden Markov models: Stochastic variational inference without messages. In *30th Conference on Neural Information Processing Systems (NIPS 2016)*, Barcelona.

Pearl, J. (1988). *Probabilistic Reasoning in Intelligent Systems, Networks of Plausible Inference*. Morgan Kaufmann, Palo Alto.

Peyrard, N., Cros, M.-J., de Givry, S., Franc, A., Robin, S., Sabbadin, R., Schiex, T., Vignes, M. (2019). Exact or approximate inference in graphical models: Why the choice is dictated by the treewidth, and how variable elimination can be exploited. *Australian and New Zealand Journal of Statistics*, 61(2), 89–133.

Ronce, O. and Olivieri, I. (2004). Life history evolution in metapopulations. In *Ecology, Genetics and Evolution of Metapopulations*, Hanski, I. and Gaggiotti, O.E. (eds). Academic Press, Burlington.

Touloupou, P., Finkenstädt, B., Spencer, S.E.F. (2020). Scalable Bayesian inference for coupled hidden Markov and semi-Markov models. *Journal of Computational and Graphical Statistics*, 29(2), 238–249.

Wainwright, M. and Jordan, M. (eds) (2008). Graphical models, exponential families, and variational inference. In *Foundations and Trends in Machine Learning*, volume 1. Now Publishers, Boston, Delft.

Wang, X., Lebarbier, E., Aubert, J., Robin, S. (2019). Variational inference for coupled hidden Markov models applied to the joint detection of copy number variations. *The International Journal of Biostatistics*, 15(1), 20180023.

Wei, G.C.G. and Tanner, M.A. (1990). A Monte Carlo implementation of the EM algorithm and the poor man's data augmentation algorithms. *Journal of the American Statistical Association*, 85(411), 699–704.

Yedidia, J., Freeman, W., Weiss, Y. (2005). Constructing free energy approximations and generalized belief propagation algorithms. *IEEE Transactions on Information Theory*, 51(7), 2282–2312.

Zhong, S. and Ghosh, J. (2002). HMMs and coupled HMMs for multi-channel EEG classification. In *Proceedings of the 2002 International Joint Conference on Neural Networks (IJCNN '02)*, 12–17 May, Honolulu.

4

Multichain HSMM

The concept of multichain (H)SMM has not been already rigorously formalized, even if a few models have been proposed in literature. In this chapter, after a review of existing multichain (H)SMMs, we propose a sound formalization of two classes of models that extend standard and general semi-Markov models to the multichain setting. Then, we consider the hidden framework and build various classes of multichain-HSMMs (MHSMMs) that generalize some MHMM structures defined in Chapter 3.

4.1. Multichain HSMM in literature

The concept of Multichain HSMM (MHSMM) is not something already rigorously formalized. By extension to the definition of multichain HMM (MHMM) in Chapter 3, the vision we adopt here is that of a stochastic model representing the joint distribution of a set of several hidden and observed processes that evolve through time and that satisfy some semi-Markov property. Having this context in mind, the MHSMMs that have been studied in literature mainly generalize two classes of MHMMs: the coupled HMM (CHMM) (Figure 3.7(d), Chapter 3) that assumes direct dependences between hidden chains and the factorial HMM (FHMM) (Figure 3.5 , Chapter 3) in which hidden chains are marginally independent and emit together a unique observation chain.

The first attempt to extend CHMM to the semi-Markov case is found in Natarajan and Nevatia (2007), with an application to sign language recognition. In the proposed model, a chain is influenced by the others only at its jump times, namely, when the chain ends a state and the next state depends on the current state of the other

Chapter written by Jean-Baptiste DURAND, Nathalie PEYRARD, Sandra PLANCADE and Régis SABBADIN.

chains. However, the distributions defining the model are not presented in a rigorous way, and the assumptions made are not clearly stated. Estimation is obtained with an approximation of the EM algorithm that works with a forward-backward algorithm at the level of the chain. In a more recent work, Touloupou et al. (2020) propose a similar model to describe epidemics dynamics. The parameterization of the transition probabilities is specific to this application and Bayesian inference is performed with a Markov chain Monte Carlo (MCMC) method. These works highlight an interest to extend CHMM to the semi-Markov case, but a general and flexible definition is still missing.

In Kim et al. (2012), the authors define a semi-Markov that extends the FHMM to infer the power load of each appliance knowing the aggregate power consumption over T time periods. Parameter estimation is performed with a Monte Carlo expectation-maximization (MCEM) algorithm using Gibbs sampling in the E step. In this case, the observation is the sum of the state of each chain. In a totally different context, Nicol et al. (2022) proposed an FHSMM for modeling bird migratory networks. Each hidden semi-Markov chain represents the sequence of a bird position in the network, and the observation is a noisy version of the bird count at each node. Two methods are proposed for estimation: MCEM Wei and Tanner (1990) and ABC Csilléry et al. (2010).

4.2. Formalization of an explicit duration coupled semi-Markov model with interaction at jump events

For the sake of simplicity, we first consider a specific coupled SMM with interactions only at its jump times, which corresponds to the hidden layer of the MHSMM in Natarajan and Nevatia (2007). Based on this model, we enhance the difficulties to define multichain SMMs and propose a non-ambiguous formalization as well as the associated graphical representation.

The (H)SMM notations of Chapter 1 are extended to the multichain setting by adding an exponent for chain c. Thus, Z_t^c denotes the state of chain c at time t; $Z_{t+1,t+d}^c = i$ means that chain c is in state i at times $t+1,\ldots,t+d$ without specification of the value of Z_t^c and Z_{t+d+1}^c; $Z_{[t+1,t+d}^c = i$ means that chain c enters state i at time $t+1$ and exits state i at some time $t' \geq t+d$; $Z_{t+1,t+d]}^c = i$ means that chain c enters state i at some time $t' \leq t+1$ and exits state i at time $t+d$; $Z_{[t+1,t+d]}^c = i$ means that chain c enters state i at $t+1$ and exits at $t+d$; $Z_{[t}^c = i$ (resp. $Z_{t]}^c = i$) means that chain c enters (respectively, exits) state i at t. The number of chains is denoted by C and the number of states is denoted by K.

4.2.1. Definition based on literal hypotheses

First of all, we attempt to define the model in a literal way, similarly to Natarajan and Nevatia (2007). The state duration is assumed to depend on the current state only, and the coupling only occurs at transitions, which corresponds to the following set of assumptions:

– (*H-ind*): the chain states at time t are mutually independent given the states of all chains at $t' < t$.

– (*H-trans*): when a chain jumps at time $t+1$, the new state depends on the value of all chains at time t.

– (*H-ED*): the sojourn duration only depends on the current state of the chain.

Note that under this assumption, the model can be seen as a generalization of the explicit duration model ED-HMM (Chapter 1, section 1.2.4 and Figure 1.4).

However, this literal definition may raise ambiguity regarding conditional independence assumptions that are central in the definition of multichain processes. Notably, if one wants to express assumption (*H-ED*) in terms of conditional distribution, at least two formulations could naturally be considered.

$$
\begin{aligned}
&\text{(i)}\ \mathbb{P}\left(Z^c_{[t+1:t+l]} = j | Z^c_{[t+1} = j, (Z^{c'}_{t'} = z^{c'}_{t'})_{c'=1,\ldots,C,c'\neq c,t'\leq t+1}\right)\\
&= \mathbb{P}\left(Z^c_{[t+1:t+l]} = j | Z^c_{[t+1} = j\right)\\
&\text{(ii)}\ \mathbb{P}\left(Z^c_{[t+1:t+l]} = j | Z^c_{[t+1} = j, (Z^{c'}_{t'} = z^{c'}_{t'})_{c'=1,\ldots,C,c'\neq c,t'\leq t+l}\right)\\
&= \mathbb{P}\left(Z^c_{[t+1:t+l]} = j | Z^c_{[t+1} = j\right).
\end{aligned} \qquad [4.1]
$$

The difference lies in the conditioning event: in the first formulation, the conditioning with respect to the other chains is related to all times prior to the jump of chain c at $t+1$: $(Z^{c'}_{t'})_{c'=1,\ldots,C,c'\neq c,t'\leq t+1}$. In the second formulation, this conditioning is related to all times prior to the jump of chain c and inside the time window $[t+1, t+l]$ of the sojourn of chain c.

Moreover, beyond the potential ambiguity in the hypothesis formulation, there is no guarantee that this set of assumptions completely defines the distribution of the model.

4.2.2. Generative definition using a time indexed representation

As an alternative to literal hypotheses, we consider a non-ambiguous formalism based on a generative definition of the model or its equivalent graphical representation of conditional dependences. The classic graphical representation of a (monochain) semi-Markov model is based on the Markov property of the couple (state, duration). But this representation is not appropriate for an extension to coupled chains, as the transitions do not occur simultaneously on all chains. Therefore, a time indexed graphical representation is necessary and may be achieved using auxiliary chains, namely, the time elapsed and the remaining time in the current state, chosen so that the joint chain satisfies Markov property.

4.2.2.1. *Preliminaries: time indexed representation of a monochain SMM*

In this section, we consider the notations for a (monochain) HSMM defined in Chapter 1. Similarly to Limnios and Oprişan (2001), the auxiliary chains corresponding to the elapsed and remaining time in the current state are defined as follows.

DEFINITION 4.1.– *Let $(Z_t)_{t\geq 0}$ be a semi-Markov chain. At each time t, let E_t be the time elapsed in the current state:*

$$E_t = \min\{d \in \mathbb{N}^*, Z_{t-d} \neq Z_t\}$$

and R_t be the remaining time in the current state:

$$R_t = \min\{d \in \mathbb{N}, Z_{t+d+1} \neq Z_t\}.$$

As a consequence of these definitions:

$$\begin{cases} Z_{[t} = i & \Leftrightarrow (Z_t, E_t) = (i, 1) \\ Z_{t]} = i & \Leftrightarrow (Z_t, R_t) = (i, 0) \\ Z_{[t-e+1:t+r]} = i & \Leftrightarrow (Z_t, E_t, R_t) = (i, e, r) \\ & \Leftrightarrow \{(Z_{t'}, E_{t'}, R_{t'}) = (i, t' - t + e, r + t - t'), \forall t' \\ & \quad = t - e + 1, \ldots, t + r\} \end{cases} \qquad [4.2]$$

Under specific classes of HSMMs, the auxiliary chains jointly satisfy the Markov property.

PROPOSITION 4.1.– *Let $(Z_t)_{t\geq 0}$ be a semi-Markov model, then:*

1) The triplet $(Z_t, E_t, R_t)_{t\geq 0}$ is Markovian.

2) If $(Z_t)_{t\geq 0}$ is a standard HSMM (Chapter 1, section 1.2.3 and Figure 1.2), the couple $(Z_t, E_t)_{t\geq 0}$ is Markovian.

3) If $(Z_t)_{t\geq 1}$ is an ED-HMM (Chapter 1, section 1.2.4 and Figure 1.4), the couple $(Z_t, R_t)_t$ and the couple $(Z_t, E_t)_t$ are Markovian.

The proof is in section 4.7. Appendix: proof of proposition 1.

4.2.2.2. *Generative definition using remaining time in current state*

– *Initialization*: we consider the simplifying assumption that all chains enter a new state at time $t = 0$. For each chain $c = 1, \ldots, C$:

$$\begin{cases} Z_0^c \sim (\pi_i^c)_{i=1,\ldots,K}, \\ 1 + R_0^c \sim (h_{Z_0^c}^c(d))_{d=1,\ldots,D} \end{cases}$$

with $\{\pi_i^c\}_{i=1,\ldots,K} \in [0,1]^K$ the initial state distribution satisfying $\sum_{i=1}^{K} \pi_i^c = 1$, and $\{(h_i^c(d))_{d=1,\ldots,D}\}_{i=1,\ldots,K} \in [0,1]^{K\times D}$ the sojourn duration distributions for chain c satisfying for all i, $\sum_{d=1}^{D} h_i^c(d) = 1$. The shift of 1 appearing in the distribution of R_0^c above enables consistency with the monochain framework.

– *Recurrence*: for each $t \geq 1$:

- for all $c = 1, \ldots, C$ such that $R_{t-1}^c > 0$:

$$\begin{cases} Z_t^c = Z_{t-1}^c \\ R_t^c = R_{t-1}^c - 1 \end{cases}$$

- for all $c = 1, \ldots, C$ such that $R_{t-1}^c = 0$:

$$\begin{cases} Z_t^c \sim \mathbb{P}(Z_t^c = j | \{Z_{t-1}^{c'} = i^{c'}\}_{c'=1,\ldots,C}) := P^c_{(i_{c'})_{c'=1,\ldots,C},j} \\ 1 + R_t^c \sim (h_{Z_t^c}^c(d))_{d=1,\ldots,D} \end{cases}$$

with $\{P^c_{(i_{c'})_{c'=1,\ldots,C},j}, j = 1, \ldots, K, i_{c'} = 1, \ldots, K\} \in [0,1]^{C^K}$ the transition probabilities such that for every c and $(i_{c'})_{c'=1,\ldots,C}$, $\sum_{j=1}^{K} P^c_{(i_{c'})_{c'=1,\ldots,C},j} = 1$ and $P^c_{(i_{c'})_{c'=1,\ldots,C},i_c} = 0$.

The resulting model $(Z_t^c)_{t\geq 0, c=1,\ldots,C}$ satisfies assumptions that can be literally described as (H-ind), (H-trans) and (H-ED) equation [4.1]. Moreover, for all $t \geq 0, c = 1, \ldots, C$, R_t^c +1 corresponds to the remaining time in the current state for chain c. Indeed, from time t, $R_{t'}^c$ decreases of 1 at each step, and the state transition of chain c occurs when $R_{t'}^c = 0$, so after R_t^c steps.

4.2.3. *Graphical representation*

According to the generative definition of section 4.2.2.2, for each $t \geq 0$, $c = 1, \ldots, C$:

$$\mathbb{P}(Z_t^c, R_t^c | (Z_{t-1}^{c'}, R_{t-1}^{c'})_{c'=1,\ldots,C}) = \mathbb{P}(R_t^c | Z_t^c, R_{t-1}^c) \mathbb{P}(Z_t^c | (Z_{t-1}^{c'})_{c'=1,\ldots,C}, R_{t-1}^c)$$

which leads to the representation of Figure 4.1.

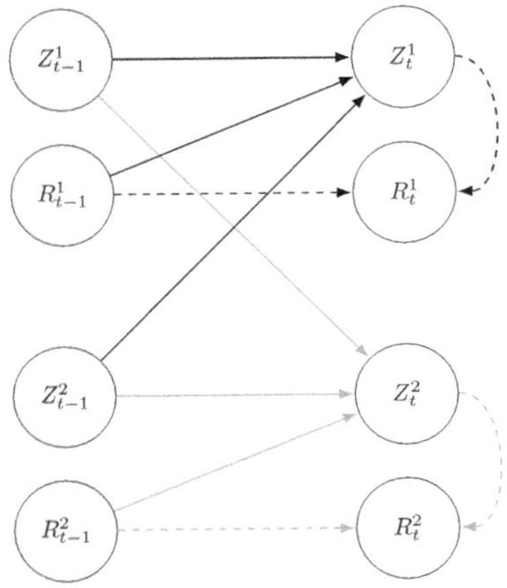

Figure 4.1. *Graphical representation of the coupled explicit duration semi-Markov model with interaction only at jump events, for two chains and two time steps.* Z_t^c *denotes the current state of the semi-Markov chain and* R_t^c *the remaining time in the current state. The solid (resp. dashed) arcs represent the conditional probability of* Z_t^c *(resp.* R_t^c*). The black arcs point to chain 1 and the gray arcs to chain 2*

4.3. Definition of coupled SMM classes based on a time-indexed representation

4.3.1. *Limitations of the* (Z, E, R) *representation*

As previously mentioned, one of the difficulties with a graphical and generative definition of coupled semi-Markov models is the non-synchronicity between the jumps of the different chains. Therefore, representations based on jump indexes are not appropriate, which is enlightened in section 4.2 for a specific model. The first attempt to generalize the model of section 4.2 could be to define the class of coupled process $\{(Z_t^c)_{c=1,\dots,C}\}_{t\geq 0}$ such that the joint chain $\{(Z_t^c, E_t^c, R_t^c)_{c=1,\dots,C}\}_{t\geq 0}$ is Markov with $E_t^c = \min\{d \in \mathbb{N}^*, Z_{t-d}^c \neq Z_t^c\}$ and $R_t^c = \min\{d \in \mathbb{N}, Z_{t+d+1}^c \neq Z_t^c\}$, the auxiliary chains of the elapsed time and the remaining time in the current state. But this definition imposes that the influence of other chains on a given chain only occurs at the jump times of this chain which is not sufficient to represent more general coupling patterns that may naturally arise in applications.

Consider the toy model Deer (Chapter 1) where animals go through successive phases of behavior and assume that each animal tends to imitate its peers. More precisely, let an animal c enter a phase of behavior, say "immobility", at time t, and assume that the probability that the animal stops this behavior is impacted at each time by the other animal behavior (if other animals start to walk or run, the animal c may decide to move and then shorten its immobility phase). In this example, the coupling between a given chain and the others occurs out of the jump times of this chain, since the animal trajectory behavior can be impacted by the other animal trajectories at any time.

The formalization based on a generative definition using the (Z, E, R) representation is not adapted to model such a coupling. Indeed, as a state starts at time t for chain c, by definition the remaining time R_t^c corresponds to the sojourn duration of chain c in the current state. With a Markov generative formalism such as section 4.2.2, R_t^c should be sampled from a distribution depending only on the past $(Z_{t-1}^{c'}, E_{t-1}^{c'}, R_{t-1}^{c'})_{c'}$. But under the coupling by imitation described above, the distribution of R_t^c would depend on the value $Z_{t'}^{c'}$ of the other chains $c' \neq c$ for posterior $t' > t$. Therefore, the use of the auxiliary variable R_t^c corresponding to the remaining time in the current state, which requires us to sample the sojourn duration as soon as the chain enters a new state does not allow a Markov generative representation. In an alternative way, a variable $\tilde{R}_t^c$ could be generated from the marginal distribution of R_t^c given the past $\mathbb{P}(\tilde{R}_t^c | (Z_{t'}^{c'})_{t' \leq t, c'}) = \int \mathbb{P}(R_t^c | (Z_{t'}^c)_{t' \leq t}, (Z_{t'}^{c'})_{t' \in \mathbb{N}^*, c' \neq c}) d(Z_{t'}^{c'})_{t' \geq t, c' \neq c}$, but this distribution would not offer a simplifying decomposition and would be inextricable.

4.3.2. *Hazard rate representation*

A simple way to circumvent the limitation described in the previous section while preserving a step-by-step (Markov) generative definition of multichain SMM is to sample the probability to exit from a state at each time instead of sampling the sojourn duration at the time step when the chain enters a new state. In the survival analysis domain, this corresponds to the concept of *hazard rate*. The hazard rate represents a characterization of the distribution of a random time, equivalent to the probability distribution function. For a discrete random time $X \in \mathbb{N}^*$, the hazard rate is the function of x equal to the probability that X is equal to x given that X is larger than x:

$$\lambda_X(x) = \mathbb{P}(X = x | X \geq x) = \frac{\mathbb{P}[X = x]}{\mathbb{P}(X \geq x)} = \frac{f_X(x)}{1 - \sum\limits_{x'=1}^{x-1} f_X(x')}$$

if $x \geq 2$, and $\lambda_X(1) = f_X(1)$ with f_X the probability distribution function of X. If X has a finite support $X \in \{1, \ldots, D\}$ a.s., for $x > D$, $\mathbb{P}(X = x) = \mathbb{P}(X \geq x) = 0$ and by convention, $\lambda_X(x) = 1$.

A function λ is a hazard rate function if either:

– there exists $D \in \mathbb{N}^*$ such that $\lambda(x) \in [0, 1)$, $\forall x = 1, \ldots, D - 1$ and $\lambda(x) = 1$, $\forall x \geq D$ (finite support random time);

– $\lambda(x) \in [0,1)\ \forall x \in \mathbb{N}^*$ and $\sum_{x=1}^{+\infty} \lambda(x) = +\infty$.

These two distinct properties can be unified in the following formulation: there exists $D \in \{\mathbb{N}^* \cup +\infty\}$ such that:

$$\begin{cases} \lambda(x) \in [0,1),\ \forall x \in \mathbb{N}^* \text{ with } x < D \\ \lambda(x) = 1,\ \forall x \in \mathbb{N}^* \text{ with } x \geq D \\ \sum_{x=1}^{+\infty} \lambda(x) = +\infty. \end{cases} \qquad [4.3]$$

This characterization of a function of hazard rate is the counterpart of the classic conditions on probability distribution function that impose values between 0 and 1 and sum of the function over its definition domain equal to 1.

4.3.3. *Definition and formalization of a class of coupled standard SMMs*

We define the class of coupled SMMs $\{(Z_t^c)_{c=1,\dots,C}\}_{t\geq 0}$ such that the chain $\{(Z_t^c, E_t^c, V_t^c)_{c=1,\dots,C}\}_{t\geq 0}$ is Markov with $E_t^c = \min\{d \in \mathbb{N}^*, Z_{t-d}^c \neq Z_t^c\}$ the elapsed time in current state and $V_t^c = \mathbb{1}_{Z_t^c \neq Z_{t-1}^c}$ the indicator of chain c jumping at time t. This class extends the standard SMM (Chapter 1, section 1.2.3 and Figure 1.2 to the multichain framework.

DEFINITION 4.2.– *The class of coupled standard SMMs is defined as follows.*

– *Initialization: we consider the simplifying assumption that all chains enter a new state at time $t = 0$. For each $c = 1, \dots, C$, let:*

$$\begin{cases} Z_0^c \sim (\pi_i^c)_{i=1,\dots,K} \\ E_0^c = 1 \\ V_0^c = 1 \text{ (by convention)} \end{cases}$$

– *Recurrence: for each $t \geq 1$, for each $c = 1, \dots, C$, the indicator of changing of state is:*

- *$V_t^c \sim Ber(\lambda_t^c)$ with:*

$$\lambda_t^c = \lambda^c(E_{t-1}^c, Z_{t-1}^c, (Z_{t-1}^{c'}, E_{t-1}^{c'})_{c'\neq c}) \in [0,1].$$

Heuristically, the function $\lambda^c(\cdot, Z_{t-1}^c, (Z_{t-1}^{c'}, E_{t-1}^{c'})_{c'\neq c})$ corresponds to the hazard rate of the sojourn duration, and the state of other chains $c' \neq c$ plays the role of "time-dependent covariates" (see section 4.3.5) where Ber denotes the Bernouilli distribution.

- *If $V_t^c = 0$, then:*

$$\begin{cases} Z_t^c = Z_{t-1}^c \\ E_t^c = E_{t-1}^c + 1 \end{cases}$$

and if $V_t^c = 1$, *then:*

$$\begin{cases} Z_t^c \sim \left(P^c((Z_{t-1}^{c'}, E_{t-1}^{c'})_{c'=1,\dots,C}, j)\right)_{j=1,\dots,K} \\ E_t^c = 1. \end{cases}$$

In the definition above, the semi-Markov property lies in the fact that the distribution of $\{Z_t^c\}_{c=1,\dots,C}$ given the past only depends on the states and durations at the previous time $\{(Z_{t-1}^c, E_{t-1}^c)\}_{c=1,\dots,C}$.

Figure 4.3 displays the graphical representation of the complete model as well as sub-models including the coupled SMM of section 4.2.

The parameters must satisfy the following conditions for all $c = 1, \dots, C$ and for a $D \in \mathbb{N}^* \cup \{+\infty\}$.

– (*cond-sojourn*): for all $i_c = 1, \dots, K, (i_{c'}, e_{c'})_{c' \neq c} \in (\{1, \dots, K\} \times \mathbb{N}^*)^{C-1}$, the function $\lambda^c(\cdot, i_c, (i_{c'}, e_{c'})_{c' \neq c})$ satisfies [4.3].

This condition ensures that independent standard semi-Markov chains are included as a particular case in the class of coupled SMMs. Indeed, let $(\{Z_t^c\}_{t \geq 0})_{c=1,\dots,C}$ be independent semi-Markov chains, then $\lambda^c(\cdot, Z_{t-1}^c, (Z_{t-1}^{c'}, E_{t-1}^{c'})_{c' \neq c}) = \lambda^c(\cdot, Z_{t-1}^c)$ is the hazard rate of the sojourn duration in state Z_{t-1}^c for chain c equal to:

$$\lambda^c(d, i) = \frac{h_i^c(d)}{1 - \sum\limits_{d'=1}^{d-1} h_i^c(d')}, \forall d = 1, \dots, D, i = 1, \dots, K \quad \text{and} \quad \lambda^c(1, i) = h_i^c(1)$$

with $(h_i^c(d))_{d=1,\dots,D}$ the distribution of the sojourn duration in state i.

– (*cond-trans*): for all $c' = 1, \dots, C, j = 1, \dots, K, i_{c'} = 1, \dots, K, e_{c'} = 1, \dots, D$, $P^c\left((i_{c'}, e_{c'})_{c'=1,\dots,C}, j\right) \in [0, 1]$:

$$\sum_{j=1}^{K} P^c\left((i_{c'}, e_{c'})_{c'=1,\dots,C}, j\right) = 1$$

and $P^c\left((i_{c'}, e_{c'})_{c'=1,\dots,C}, i_c\right) = 0$.

– (*cond-init*): for all $c = 1, \dots, C, j = 1, \dots, \pi_j^c \in [0, 1]$ and:

$$\sum_{j=1}^{K} \pi_j^c = 1.$$

EXAMPLE.– Consider a model as described in section 4.3.1 with two animals going through successive phases of behaviors with two different behaviors (1 = immobility, 2 = movement) and a coupling in which each animal tends to imitate the behavior of the other. Consider a one-shifted discrete Weibull distribution with shape parameter $\beta = 2$ as a "basis" for sojourn duration, with probability distribution function:

$$d \mapsto \exp\left(-\left(\frac{d-1}{\alpha}\right)^2\right) - \exp\left(-\left(\frac{d}{\alpha}\right)^2\right), \quad d \in \mathbb{N}^*.$$

The corresponding hazard rate function $d \mapsto 1 - \exp(-(2d-1)/\alpha^2) := \overline{\lambda}(d|\alpha)$ is decreasing in α for a given $d \in \mathbb{N}^*$. More precisely, the hazard rate function associated with animal 1 (i.e. to chain 1) involved in definition 4.2 is:

$$\lambda^1(e^1_{t-1}, z^1_{t-1}, e^2_{t-1}, z^2_{t-1}) = \overline{\lambda}(e^1_{t-1}|\alpha_{z^1_{t-1}, z^2_{t-1}}) = 1 - \exp\left(-\frac{2e^1_{t-1} - 1}{\alpha^2_{z^1_{t-1}, z^2_{t-1}}}\right)$$

a distribution with four parameters $(\alpha_{11}, \alpha_{12}, \alpha_{21}, \alpha_{22})$. Change of activity of an animal at time t is assumed to depend on the time already spent in the current state as well as the activity of the other animal at time $t-1$. Then, the setting $\alpha_{11} \leq \alpha_{12}$ and $\alpha_{22} \leq \alpha_{21}$ models an imitation behavior. Indeed, assume that $Z^1_{[t-d:t-1} = 1$, that is, animal 1 has been immobile for a duration d at $t-1$. Then the probability for animal 1 to start moving at t is $h(d|\alpha_{11})$ if animal 2 is immobile and $h(d|\alpha_{12})$ otherwise. Therefore, $\alpha_{11} \leq \alpha_{12}$ implies that $h(d|\alpha_{12}) \leq h(d|\alpha_{11})$, so the probability to move for animal 1 is higher if animal 2 is already moving.

REMARK 4.1.– A more parcimonious representation is theoretically possible but would be untractable. Indeed, theoretically, the variable V could be omitted while preserving the Markovianity of the chain $\{(E^c_t, Z^c_t)_{c=1,\ldots,C}\}_{t\geq 0}$, but it would be at the price of a much more complex expression of the distributions, where the sub-models presented in Figure 4.3 would be difficult to interpret.

REMARK 4.2.– Absorbing states, that is, states from which the chain never exits, are an issue both for theoretical analysis and numerical studies. Indeed, the theory of consistent parameter estimation classically relies on estimation from a single sequence whose length tends to infinity, which requires recurrence of the chain so that each possible transition occurs infinitely many times. Moreover, some algorithms (e.g. some expectation maximization variants) require simulations of long chains with a sufficient number of transitions, which may be hindered by the existence of absorbing states. In a monochain SMM, the definition of the hazard rate [4.3] imposes that the chain exits a state almost surely. If one wants to model an absorbing state, it has to be defined explicitly, without defining the sojourn duration distribution. Nevertheless, without additional assumptions, in the multichain context, a coupled SMM can remain infinitely in a state even if this property has not been explicitely settled, as illustrated by the example in Figure 4.2. Indeed, even if the condition *(cond-sojourn)* imposes that $\lambda^c(\cdot, i_c, (i_{c'}, e_{c'})_{c'\neq c})$ which governs the distribution of V^c_t satisfies the characterization [4.3] of the hazard rate that avoids absorbing state, the value of $i_c, (i_{c'}, e_{c'})_{c'\neq c}$ may change with the time point t. The following proposition gives a sufficient condition to guarantee almost surely finite sojourn durations.

PROPOSITION 4.2.– *Assume that there exists a hazard rate function* $\overline{\lambda}^c$ *satisfying* [4.3] *such that:*

$$\lambda^c(e_c, i_c, (i_{c'}, e_{c'})_{c' \neq c}) \geq \overline{\lambda}^c(e_c), \ \forall (e_c, i_c, (i_{c'}, e_{c'})_{c' \neq c})$$

then the marginal sojourn durations of chain c are finite almost surely, that is, for every c and t_0*:*

$$\mathbb{P}(V_t^c = 0, \forall t = t_0, \ldots, +\infty, V_{t_0-1}^c = 1) = 0.$$

PROOF OF PROPOSITION 2.–

Heuristically, let $\overline{X}_t^c$ be a random time with hazard rate $\overline{\lambda}^c$, and let $\overline{V}_t^c$ be the corresponding jump process: $\overline{V}_t^c = 1$ if $t = \overline{X}_t^c - 1$ and 0 otherwise. As the probability that chain c exits a state i after a elapsed time t' is stochastically larger than the probability that $\overline{V}_{t'}^c$ jumps, then the sojourn duration in state i is stochastically smaller than $\overline{X}_t^c$, and so is finite a.s.

More precisely, let $T \in \mathbb{N}^*$ and $c \in \{1, \ldots, C\}$. First of all, we prove that:

$$\begin{aligned} P_T &= \mathbb{P}(V_t^c = 0, \forall t = t_0, \ldots, T, V_{t_0-1}^c = 1) \\ &\leq \mathbb{P}\left(\overline{X}^c \geq T - t_0 + 1\right) \mathbb{P}\left(V_{t_0}^c = 0, V_{t_0-1}^c = 1\right) \end{aligned} \quad [4.4]$$

where $\overline{X}_t^c$ is a random time with hazard rate $\overline{\lambda}^c$. Indeed

$$P_T = \prod_{t=t_0+1}^{T} \mathbb{P}\left(V_t^c = 0 | V_{t-1}^c = \cdots = V_{t_0}^c = 0, V_{t_0-1}^c = 1\right) \mathbb{P}\left(V_{t_0}^c = 0, V_{t_0-1}^c = 1\right)$$

and

$$\begin{aligned} &\mathbb{P}\left(V_t^c = 0 | V_{t-1}^c = \cdots = V_{t_0}^c = 0, V_{t_0-1}^c = 1\right) \\ &= \sum_{(e_{t-1}^{c'}, i_{t-1}^{c'})_{c'=1,\ldots,C}} \mathbb{P}\left(V_t^c = 0 | \left((E_{t-1}^{c'}, Z_{t-1}^{c'}) = (e_{t-1}^{c'}, i_{t-1}^{c'})\right)_{c'=1,\ldots,C}, \right. \\ &\left. V_{t-1}^c = \cdots = V_{t_0}^c = 0, V_{t_0-1}^c = 1\right) \\ &\times \mathbb{P}\left(\left((E_{t-1}^{c'}, Z_{t-1}^{c'}) = (e_{t-1}^{c'}, i_{t-1}^{c'})\right)_{c'=1,\ldots,C} | V_{t-1}^c = \cdots = V_{t_0}^c = 0, V_{t_0-1}^c = 1\right) \end{aligned}$$

Consider two chains with geometric basic distribution for sojourn durations, exponentially decreasing with the sojourn duration of the other chain, namely:

$$\lambda^1(e^1_{t-1}, z^1_{t-1}, e^2_{t-1}, z^2_{t-1}) = p\exp(-e^2_{t-1})$$

$$\lambda^2(e^2_{t-1}, z^2_{t-1}, e^1_{t-1}, z^1_{t-1}) = p\exp(-e^1_{t-1})$$

with (z^c_t, e^c_t) the current state and elapsed time in the state at time t for chain c, and $p \in (0, 1)$. These distributions satisfy condition (*cond-sojourn*) for $D = +\infty$. More specifically, when $(z^1_{t-1}, e^2_{t-1}, z^2_{t-1})$ is fixed, λ_1 is constant and thus is equal to the hazard rate of an exponential distribution and symmetrically for λ_2.

To prove that the sojourn duration is not finite a.s, it is sufficient to show that the probability $\mathbb{P}(\forall t \geq 1, V^1_t = 0 \text{ and } V^2_t = 0)$ to remain indefinitely in the first state for both chains in non-zero. First of all,

$$\mathbb{P}(\forall t \geq 1, V^1_t = 0 \text{ and } V^2_t = 0)$$

$$= \mathbb{P}(V^1_1 = 0, V^2_1 = 0) \times \prod_{t=2}^{+\infty} \mathbb{P}(V^1_t = 0, V^2_t$$

$$= 0 | V^1_{t-1} = 0, V^2_{t-1} = 0, V^1_1 = 0, V^2_1 = 0)$$

and

$$\{V^1_t = 0, V^2_t = 0, \ldots, V^1_1 = 0, V^2_1 = 0\} \Leftrightarrow$$
$$\{V^1_t = 0, V^2_t = 0, \ldots, V^1_1 = 0, V^2_1 = 0, E^1_{t'} = t', E^2_{t'} = t', \forall t'$$
$$\leq t, Z^1_t = \cdots = Z^1_0, Z^2_t = \cdots = Z^2_0\}$$

Moreover,

$$\mathbb{P}(V^1_1 = 0, V^2_1 = 0) = \mathbb{P}(V^1_1 = 0)\mathbb{P}(V^2_1 = 0) = (1 - pe^{-1})^2,$$

$$\mathbb{P}(V^1_2 = 0, V^2_2 = 0 | V^1_1 = 0, V^2_1 = 0) = (1 - pe^{-2})^2.$$

Therefore,

$$\mathbb{P}(\forall t \geq 1, V^1_t = 0 \text{ and } V^2_t = 0) = (1 - pe^{-1})^2 \times (1 - pe^{-2})^2 \times \cdots = \left[\prod_{t=1}^{+\infty}(1 - pe^{-t})\right]^2.$$

Besides, for every $t \geq t_0 = -\log(\log 2/(2p))$ $\log(1 - pe^{-t}) > -2pe^{-t}$ so:

$$\sum_{t=1}^{+\infty} \log(1 - pe^{-t}) > \sum_{t=1}^{t_0-1} \log(1 - pe^{-t}) + \sum_{t=t_0}^{+\infty} -2pe^{-t} > -\infty$$

and finally

$$\prod_{t=1}^{+\infty}(1 - pe^{-t}) > 0 \quad \square$$

Figure 4.2. *Example of a coupled SMM with non-finite sojourn durations*

Now, V_t^c only depends on the current state and the elapsed time of the other chains at $t-1$, so:

$$\begin{aligned}
&\mathbb{P}\left(V_t^c = 0 | \left((E_{t-1}^{c'}, Z_{t-1}^{c'}) = (e_{t-1}^{c'}, i_{t-1}^{c'})\right)_{c'=1,\dots,C}, V_{t-1}^c = \dots\right.\\
&\left.= V_{t_0}^c = 0, V_{t_0-1}^c = 1\right)\\
&= \mathbb{P}\left(V_t^c = 0 | \left((E_{t-1}^{c'}, Z_{t-1}^{c'}) = (e_{t-1}^{c'}, i_{t-1}^{c'})\right)_{c'=1,\dots,C}\right)\\
&= 1 - \lambda^c(e_{t-1}^c, i_{t-1}^c, (e_{t-1}^{c'}, i_{t-1}^{c'})_{c'\neq c})
\end{aligned}$$

Therefore,

$$\begin{aligned}
&\mathbb{P}(V_t^c = 0 | V_{t-1}^c = \dots = V_{t_0}^c = 0, V_{t_0-1}^c = 1)\\
&= \sum_{(e_{t-1}^{c'}, i_{t-1}^{c'})_{c'=1,\dots,C}} (1 - \lambda^c(e_{t-1}^c, i_{t-1}^c, (e_{t-1}^{c'}, i_{t-1}^{c'})_{c'\neq c}))\\
&\times \mathbb{P}\left(\left((E_{t-1}^{c'}, Z_{t-1}^{c'}) = (e_{t-1}^{c'}, i_{t-1}^{c'})\right)_{c'=1,\dots,C}\right.\\
&\left.\times | V_{t-1}^c = \dots = V_{t_0}^c = 0, V_{t_0-1}^c = 1\right).
\end{aligned}$$

By definition of E_{t-1}^c the elapsed time in current state at $t-1$, $\mathbb{P}\left(\left((E_{t-1}^{c'}, Z_{t-1}^{c'}) = (e_{t-1}^{c'}, i_{t-1}^{c'})\right)_{c'=1,\dots,C} | V_{t-1}^c = \dots = V_{t_0}^c = 0, V_{t_0-1}^c = 1\right)$ is equal to 0 if $e_{t-1}^c \neq t - t_0$. Therefore,

$$\begin{aligned}
&\mathbb{P}(V_t^c = 0 | V_{t-1}^c = \dots = V_{t_0}^c = 0, V_{t_0-1}^c = 1)\\
&\leq \sum_{(e_{t-1}^{c'}, i_{t-1}^{c'})_{c'=1,\dots,C}} (1 - \overline{\lambda}^c(t - t_0))\\
&\times \mathbb{P}\left(\left((E_{t-1}^{c'}, Z_{t-1}^{c'}) = (e_{t-1}^{c'}, i_{t-1}^{c'})\right)_{c'=1,\dots,C}\right.\\
&\left.\times | V_{t-1}^c = \dots = V_{t_0}^c = 0, V_{t_0-1}^c = 1\right)\\
&= (1 - \overline{\lambda}^c(t - t_0)).
\end{aligned}$$

Then,

$$
\begin{aligned}
P_T &\leq \prod_{t=t_0+1}^{T} (1-\overline{\lambda}^c(t-t_0))\mathbb{P}(V^c_{t_0}=0, V^c_{t_0-1}=1) \\
&= \prod_{t=1}^{T-t_0} (1-\overline{\lambda}^c(t))\mathbb{P}[V^c_{t_0}=0, V^c_{t_0-1}=1) \\
&= \mathbb{P}(\overline{X}^c \geq T-t_0+1]\mathbb{P}[V^c_{t_0}=0, V^c_{t_0-1}=1).
\end{aligned}
$$

The last equality is a classic result in survival analysis, indeed for all $T' \in \mathbb{N}^*$:

$$
\begin{aligned}
\prod_{t=1}^{T'-1} (1-\overline{\lambda}^c(t)) &= \prod_{t=1}^{T'-1} \frac{\mathbb{P}(\overline{X}^c \geq t) - \mathbb{P}(\overline{X}^c = t)}{\mathbb{P}(\overline{X}^c \geq t)} \\
&= \prod_{t=1}^{T'-1} \frac{\mathbb{P}(\overline{X}^c \geq t+1)}{\mathbb{P}(\overline{X}^c \geq t)} = \frac{\mathbb{P}(\overline{X}^c \geq T')}{\mathbb{P}(\overline{X}^c \geq 1)}.
\end{aligned}
$$

Thus, equation [4.4] is proved.

Consequently, since $\overline{X}^c_t$ is a real-valued (implying, finite) random variable, $\lim_{T\to+\infty} \mathbb{P}[\overline{X}^c \geq T-t_0+1] = 0$. Finally, $\lim_{T\to+\infty} P_T = 0$.

Since

$$
\begin{aligned}
&\mathbb{P}[V^c_t = 0, \forall t = t_0, \ldots, +\infty, V^c_{t_0-1}=1] \\
&= \prod_{t=t_0+1}^{+\infty} \mathbb{P}[V^c_t = 0 | V^c_{t-1} = \cdots = V^c_{t_0} = 0, V^c_{t_0-1} = 1]\mathbb{P}[V^c_{t_0}=0, V^c_{t_0-1}=1] \\
&= \lim_{T\to+\infty} \prod_{t=t_0+1}^{T} \mathbb{P}[V^c_t = 0 | V^c_{t-1} = \cdots = V^c_{t_0} = 0, V^c_{t_0-1} = 1]\mathbb{P}[V^c_{t_0}=0, V^c_{t_0-1}=1] \\
&= \lim_{T\to+\infty} P_T,
\end{aligned}
$$

we have

$$
\forall c, \mathbb{P}(V^c_t = 0, \forall t = t_0, \ldots, +\infty, V^c_{t_0-1} = 1) = 0.
$$

□

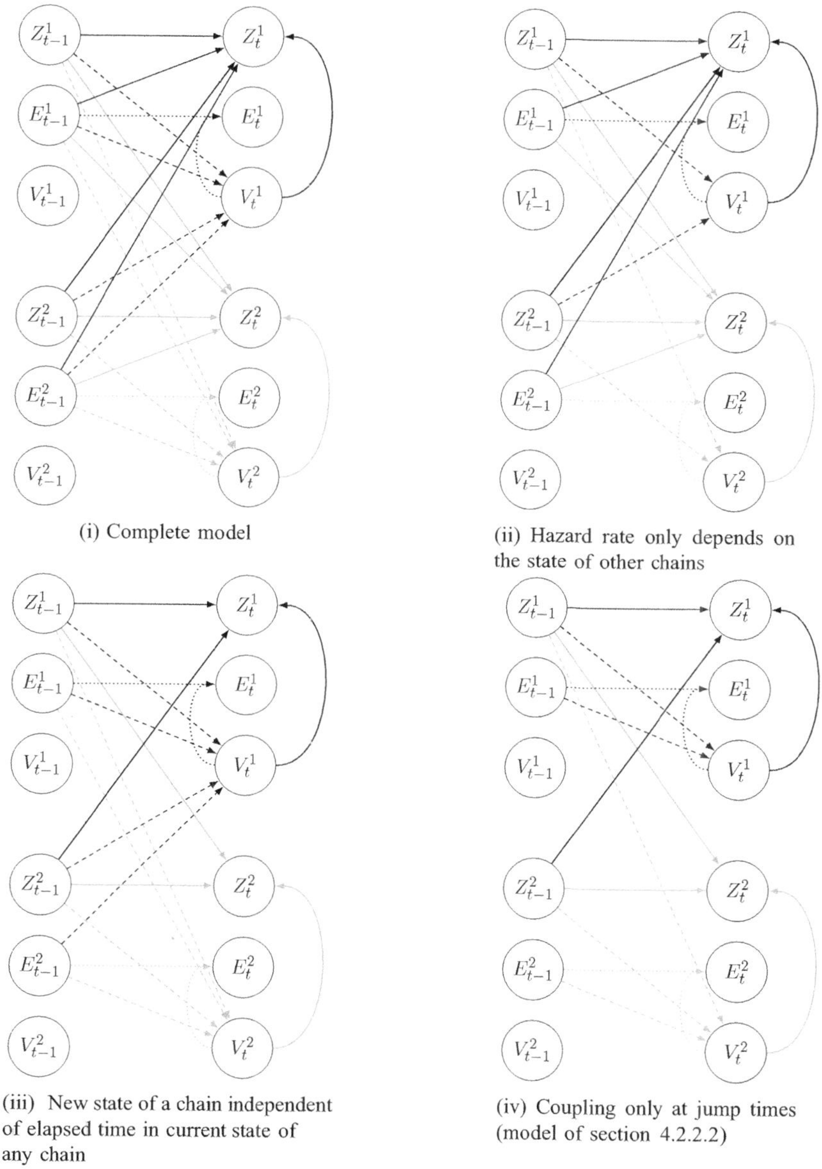

(i) Complete model

(ii) Hazard rate only depends on the state of other chains

(iii) New state of a chain independent of elapsed time in current state of any chain

(iv) Coupling only at jump times (model of section 4.2.2.2)

Figure 4.3. *Graphical representation of coupled SMMs. Z_t^c denotes the value of the chain c at time t, E_t^c the time elapsed in the current state, and V_t^c the indicator for chain c to jump at t. The solid arcs represent the conditional probability of Z_t^c, the dashed arcs represent the conditional probability of V_t^c and the dotted arrow represent the conditional probability of E_t^c. The black arcs point to chain 1 and the gray arcs to chain 2*

4.3.4. *Extension to general semi-Markov property*

The class of models defined in section 4.3.3 does not include general monochain SMMs (definition 1.1 in Chapter 1) where the sojourn duration may depend on the sojourn duration of the previous state. A class of general coupled SMMs could be defined by introducing an additional auxiliary variable equal to the couple of the previous state and its duration: $W_t^c = (Z_{t-E_t^c}^c, E_{t-E_t^c}^c)$.

– *Initialization*: for each chain $c = 1, \dots, C$:

$$\begin{cases} Z_0^c \sim (\pi_i^c)_{i=1,\dots,K}, \ \forall c \\ E_0^c = 1 \ \forall c \\ W_0^c = (\text{NA}, \text{NA}). \end{cases}$$

The NA for W_t^c is an arbitrary value to indicate that the chain c is still in its first state at time t, therefore the previous state is undefined.

– *Recurrence*: for each $t \geq 1$, $c = 1, \dots, C$:

- $V_t^c \sim \text{Ber}(\lambda_t^c)$ with

$$\lambda_t^c = \lambda^c(E_{t-1}^c, Z_{t-1}^c, (Z_{t-1}^{c'}, E_{t-1}^{c'})_{c' \neq c}, W_t^c)$$

and the convention $\sum_1^0 \cdot = 0$, and with $\lambda^c(\cdot, z_{t-1}^c, (e_{t-1}^{c'}, z_{t-1}^{c'})_{c' \neq c}, w_t^c)$ satisfying assumption [4.3] for every $(Z_{t-1}^c, (z_{t-1}^{c'}, e_{t-1}^{c'})_{c' \neq c}, w_t^c) \in \{1, \dots, K\} \times (\{1, \dots, K\} \times \{1, \dots, D\})^C \times \{1, \dots, K, \text{NA}\} \times \{1, \dots, D, \text{NA}\}$.

- If $V_t^c = 0$, then:

$$\begin{cases} Z_t^c = Z_{t-1}^c \\ E_t^c = E_{t-1}^c + 1 \\ W_t^c = W_{t-1}^c \end{cases}$$

and if $V_t^c = 1$, then:

$$\begin{cases} Z_t^c \sim \left(P^c((Z_{t-1}^{c'}, E_{t-1}^{c'})_{c'=1,\dots,C}, W_{t-1}^c, i)\right)_{i=1,\dots,K} \\ E_t^c = 1 \\ W_t^c = (Z_{t-1}^c, E_{t-1}^c). \end{cases}$$

4.3.5. *Other uses of time-indexed representation*

The time-indexed representation of SMM based on the hazard rate is not a new idea. Notably, Pertsinidou and Limnios (2015) proposed an efficient Viterbi algorithm for ED-HSMM using hazard rate. Moreover, the hazard rate was used in approximating HSMM by HMM (Langrock and Zucchini 2011). In this approach, the authors duplicate the values of the hazard rate of the HSMM sojourn duration distributions to be approximated, up to some upper bound, and model the distribution tail as a geometric tail.

Notably, this time indexed representation enables us to account for a modulation of sojourn duration by effect of a time-dependent covariate $(\overline{x}_t)_{t\geq 0}$, either in monochain or multichain semi-Markov setting. In practice, the effect of the covariate is included by replacing the expression of λ_t^c in definition 4.2 by:

$$\lambda_t = \lambda(E_{t-1}, Z_{t-1}, \overline{x}_t)$$

in the monochain setting and:

$$\lambda_t^c = \lambda^c(E_{t-1}^c, Z_{t-1}^c, \overline{x}_t, (Z_{t-1}^{c'}, E_{t-1}^{c'})_{c'\neq c})$$

in the multichain setting. The parameterization of the covariate effect can be adapted from classic survival analysis models (e.g. Cox multiplicative hazard rate model).

Moreover, in Chapter 6 (section 6.3), the concept of hazard rate is involved in the definitions of several continuous-time Markovian and semi-Markovian processes; in that context, the hazard rate does not depend on covariates but on different variables involved in the process definition.

4.4. Extension of some MHMM classes to semi-Markov framework

In the former section, classes of coupled SMM were defined without hiding layer. In this section, multichain HSMM classes are defined by extensions of some MHMM classes defined in Chapter 3. We consider the FHMM shown in Figure 3.5 of Chapter 3, where hidden chains are marginally independent and jointly emit a unique observation chain, as well as the four elementary structures displayed in Chapter 3 (Figure 3.7) of the class 1to1-HMM-CI, in which each hidden chain is associated with a unique observation chain, and conditional independence of all observed and hidden chains given the past is assumed at each time. Similarly to section 4.3.3, classes of models are first characterized by generative definitions, then the associated graphical representations are displayed.

4.4.1. *Generative definition of MHSMM classes*

Consider first the extension of the structure in Figure 3.5 and of structures (a) and (c) in Figure 3.7 (Chapter 3). For these MHMM classes, hidden chains are mutually independent and the marginal distribution of each hidden chain is that of a Markov chain. Therefore, a generative definition of these classes would amount to first generate independent Markov chains that form the hidden layers, then generate observations conditionally to hidden states. Therefore, the generalization to the semi-Markov framework is straightforward by replacing each hidden Markov chain by a semi-Markov chain, without additional difficulty.

On the contrary, in structures (b) and (d) in Figure 3.7 in Chapter 3, the hidden chains are not marginally Markov and cannot be generated independently of the hidden states or observations of the other chains. Indeed, the new state of a hidden chain is impacted by the observations of the other chains at the previous time step (structure (b)) or by the states of the other hidden chains at the previous time step (structure (d)). Therefore, extension of these classes from MHMM to MHSMM makes use of the definition of coupled SMMs (definition 4.2) of the previous section. More precisely, to define the MHSMM class that extends structure (d), a coupled SMM is first generated following definition 4.2, then observations are independently generated conditionally to hidden states. This class will be referred to as coupled HSMMs (CHSMMs). The extension of structure (b) to the semi-Markov framework requires a modification of definition 4.2. The parameter λ_t^c that governs changing of state must depend on the observation of all chains at time $t - 1$: $\lambda_t^c = \lambda^c(E_{t-1}^c, Z_{t-1}^c, (Y_{t-1}^{c'})_{c' \neq c})$ with $Y_t^{c'}$ the observation of chain c' at time t. Then observations $(Y_t^c)_{c=1,\dots,C}$ are independently generated conditionally to hidden states $(Z_t^c)_{c=1,\dots,C}$.

4.4.2. *Graphical representation of MHSMM classes*

The graphical representation of each of the five extensions described in section 4.4.1 is given in Figure 4.4. Structures (a)–(d) are the extensions of Figure 3.7(a)–(d) (Chapter 3) and structure (e) is the extension of the FHMM to the semi-Markov case (FHSMM). For structures (a), (c) and (e), the hidden layer is composed of independent semi-Markov chains; Arcs to the observations are added, whose origins are identical to the MHMM counterpart.

For structure (d) (CHSMM), the graph of the hidden layer could be either the complete model or a simpler one as displayed in Figure 4.3, and we opted for the complete one. Then arcs are added from the current hidden state of each chain to the current observed state of the same chain. The graphical representation of structure (b) is similar to (d), except regarding the origin of the variables that influence the current state of the hidden chain: observed variables instead of hidden states at previous time step.

Note that all graphical representations in Figure 4.4 make use of the indicator V of changing of state, which is equivalent in the generative definition 4.2 to make use of the hazard rate instead of a probability distribution function (section 4.2.2.2). This representation is necessary for structures (b) and (d), but not for (a), (c) and (e). Nevertheless, the generative definition based on hazard rate is more general and covers all structures.

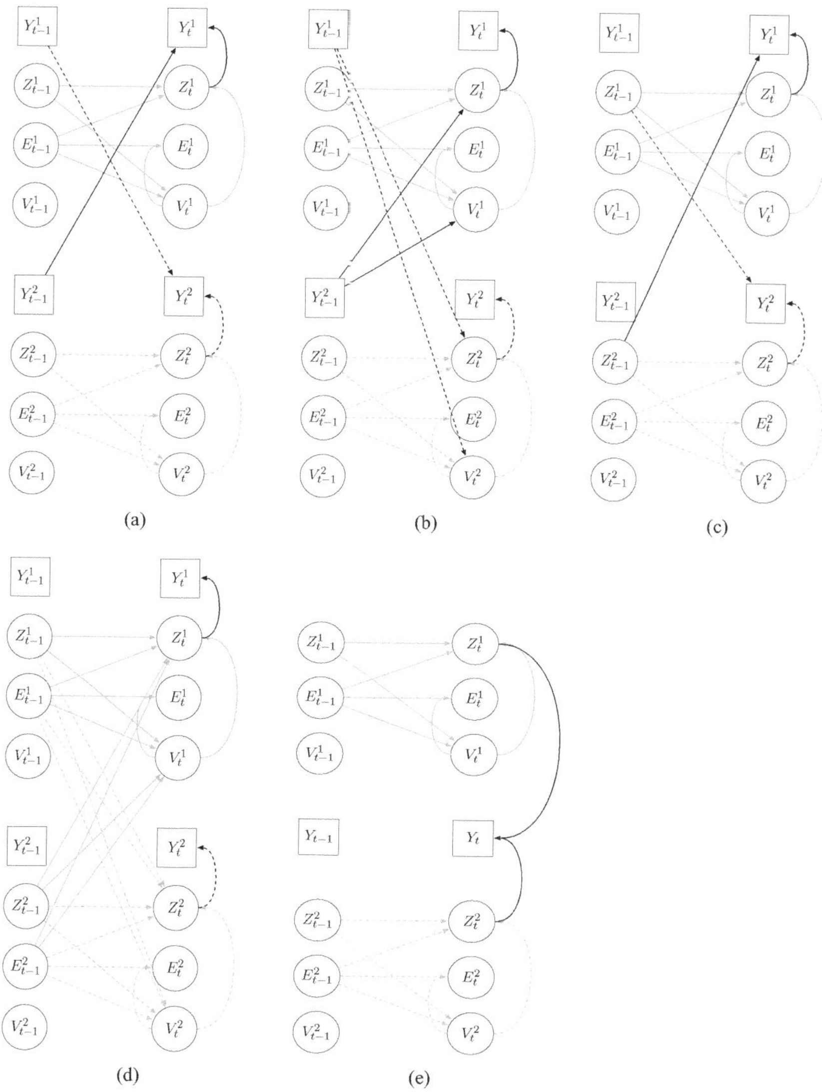

Figure 4.4. *Graphical representation of various structures of multichain HSMMs. Graphics (a,b,c,d) (resp. (e)) are extensions of multichain HMM structures of Figure 3.7 Chapter 3 (resp. of Factorial HMM) in the semi-Markov frameworks. The gray arcs represents the structure of the underlying multichain semi-Markov model and the black arcs represents the conditional probability of the observations Y. The solid arcs point to chain 1 and the dashed arcs to chain 2, except for Factorial HSMM in Figure (e) where only one chain is observed*

4.4.3. *About inference*

A challenge is to develop efficient inference algorithms for these models. Let us consider the five examples in Figure 4.4. For structures (a) and (b), the C hidden chains are independent semi-Markov chains conditionally to the observations. Therefore, it is possible to derive an EM algorithm with complexity linear in C, in the same way as for the corresponding MHMM structures (Chapter 3, Figure 3.7). For structures (c), (d) and (e), the complexity of EM would be exponential in C and only approximate inference would be conceivable. For structure (e) corresponding to the FHSMM, approximate inference has been tackled in Nicol et al. (2022) with MCEM and ABC algorithms (see details in Chapter 3, section 3.6.2.2). The proposed algorithms rely on a FHSMM definition based on the sojourn duration distribution without introducing the time index representation, since hidden chains are marginally independent. But the time indexed representation is necessary for structures (c) and (d) and would lead to an exact EM with three hidden variables by chain. Therefore, the corresponding forward/backward algorithms that must be run in the E step would generate an additional computational complexity.

4.5. Discussion and conclusion

The goal of this chapter was to develop the first sound and rigorous definition of classes of multichain HSMMs that extends a large part of the existing MHMM classes and covers a variety of practical situations. We showed that some classes could be easily extended from the Markov to the semi-Markov framework, while others (and notably the popular coupled HMM) require specific developments. A generative definition based on hazard rate instead of probability distribution function enables us to account for flexible interactions between dynamics of observed and hidden chains.

Adaptation of these general classes into models for practical situations still raises challenges in terms of inference, but also in terms of parameterization. Indeed, the dimension of the functions (hazard rates and probability distribution functions) involved in the multichain distribution increases with the model richness. Therefore, choices will be necessary to keep the model tractable both in terms of statistical variance and in terms of inference algorithms.

4.6. Notations

We gather here the main notations for MHMM used in this chapter.

Definition	Notation	Domain
Number of chains	C	$\mathbb{N}^*$
Cardinal of variable state space	K	$\mathbb{N}$
Maximum sojourn duration	D	$\mathbb{N}^* \bigcup +\infty$
State of chain c at time t	Z_t^c	$\{1, \ldots, K\} \subset \mathbb{N}^*$
Elapsed time in current state for chain c at time t	E_t^c	$\mathbb{N}^*$
Remaining time in current state for chain c at time t	R_t^c	$\mathbb{N}$
Jump indicator of chain c at time t	V_t^c	$= \mathbb{1}_{Z_t^c \neq Z_{t-1}^c} \in \{0, 1\}$
Previous state and its duration for chain c at time t	W_t^c	$= (Z_{t-E_t^c}^c, E_{t-E_t^c}^c) \in \{1, \ldots, K\} \times \mathbb{N}^*$
Observation of chain c at time t	Y_t^c	$\mathbb{R}$

Table 4.1. *Main notations for hidden and observed variables of an MHSMM*

Definition	Notation	Domain
Initial state distribution of chain c	$(\pi_i^c)_{i=1,\ldots,K}$	$[0,1]^K$
Probability distribution of sojourn duration in state i for chain c	$(h_i(d))_{d=1,\ldots,D}$	$[0,1]^D$
Transition probability of chain c given current states of all chains	$P^c_{(i_{c'})_{c'=1,\ldots,C}, i} = \mathbb{P}(Z_t^c = i \mid (Z_{t-1}^{c'} = i_{c'})_{c'=1,\ldots,C}, Z_t^c \neq Z_{t-1}^c)$	$[0,1]$

Table 4.2. *Parameters for explicit duration coupled SMM with interaction at jump times*

Definition	Notation	Domain
Initial state distribution of chain c	$(\pi_i^c)_{i=1,\ldots,K}$	$[0,1]^K$
Parameter of the jump indicator of chain c given current states and elapsed times $(i_{c'}, e_{c'})_{c'=1,\ldots,C}$ of all chains	$\lambda^c(e_c, i_c, (i_{c'}, e_{c'})_{c' \neq c})$	$[0,1]$
Transition probability of chain c given current states and elapsed times of all chains	$P^c_{(i_{c'}, e_{c'})_{c'=1,\ldots,C}, i} = \mathbb{P}(Z_t^c = i \mid (Z_{[t-e_{c'}:t-1}^{c'} = i_{c'})_{c'=1,\ldots,C}, Z_t^c \neq Z_{t-1}^c)$	$[0,1]$

Table 4.3. *Parameters for standard coupled SMM*

Definition	Notation	Domain
Initial state distribution of chain c	$(\pi_i^c)_{i=1,\ldots,K}$	$[0,1]^K$
Parameter of the jump indicator of chain c given current states and elapsed times $(i_{c'}, e_{c'})_{c'=1,\ldots,C}$ of all chains, and the couple w_c of previous state and duration of chain c	$\lambda^c(e_c, i_c, (i_{c'}, e_{c'})_{c' \neq c}, w_c)$	$[0,1]$
Transition probability of chain c given current states and elapsed times $(i_{c'}, e_{c'})_{c'=1,\ldots,C}$ of all chains, and the couple w_c of previous state and duration of chain c	$P^c_{(i_{c'},e_{c'})_{c'=1,\ldots,C}, w_c, i} = \mathbb{P}\big(Z_t^c = i \mid (Z^{c'}_{[t-e_{c'}:t-1} = i_{c'})_{c'=1,\ldots,C}, Z_t^c \neq Z_{t-1}^c, Z^c_{[t-e_c-w_{c,2}, t-e_c-1]]} = w_{c,1}\big)$	$[0,1]$

Table 4.4. *Parameters for general coupled SMM*

4.7. Appendix: proof of proposition 1

The definition of standard HSMM and ED-HMM in Chapter 1, sections 1.2.3 and 1.2.4 , makes use of the distribution of the couple $(J_{n-1}, X_n)_{n\geq 1}$ of successive states and sojourn durations. They can be reformulated in terms of calendar time as follows. For every time $t \in \mathbb{N}$, sojourn durations $d, l \in \mathbb{N}^*$ and states $i, j \in \{1, \ldots, K\}$, the semi-Markov property leads to:

$$
\begin{aligned}
&\mathbb{P}(Z_{[t+1:t+l]} = j | Z_{[t-d+1:t]} = i, (Z_{t'})_{t' \leq t}) \\
&= \mathbb{P}(Z_{[t+1:t+l]} = j | Z_{[t-d+1:t]} = i) \\
&= \mathbb{P}(Z_{[t+1:t+l]} = j | Z_{[t+1} = j, Z_{[t-d+1:t]} = i) \\
&\times \mathbb{P}(Z_{[t+1} = j | Z_{[t-d+1:t]} = i).
\end{aligned}
$$

Under the standard HSMM:

$$
\begin{aligned}
&\mathbb{P}(Z_{[t+1:t+l]} = j | Z_{[t-d+1:t]} = i, (Z_{t'})_{t' \leq t}) \\
&= \mathbb{P}(Z_{[t+1:t+l]} = j | Z_{[t+1} = j) \times \mathbb{P}(Z_{[t+1} = j | Z_{[t-d+1:t]} = i)
\end{aligned}
$$

and under the ED-HMM:

$$
\begin{aligned}
&\mathbb{P}(Z_{[t+1:t+l]} = j | Z_{[t-d+1:t]} = i, (Z_{t'})_{t' \leq t}) \\
&= \mathbb{P}(Z_{[t+1:t+l]} = j | Z_{[t+1} = j) \times \mathbb{P}(Z_{[t+1} = j | Z_{t]} = i).
\end{aligned}
$$

PROOF OF PROPOSITION 1-1.– Let:

$$\begin{aligned} P_0 &= \mathbb{P}\left((Z_{t+1}, E_{t+1}, R_{t+1}) = (i', e', r')|(Z_t, E_t, R_t)\right. \\ &= \left.(i, e, r), (Z_{t'}, E_{t'}, R_{t'})_{t' \leq t-1}\right). \end{aligned}$$

If $r > 0$, then (i', e', r') is determined:

$$\begin{aligned} P_0 &= \mathbb{1}_{i'=i, r'=r-1, e'=e+1} \\ &= \mathbb{P}\left((Z_{t+1}, E_{t+1}, R_{t+1}) = (i', e', r')|(Z_t, E_t, R_t) = (i, e, r)\right). \end{aligned}$$

According to [4.2], if $r = 0$, there is a transition at $t + 1$, so:

$$P_0 = \mathbb{1}_{i' \neq i, e'=1} \mathbb{P}\left(Z_{[t+1:t+1+r']} = i'|Z_{[t-e+1:t]} = i, (Z_{t'}, E_{t'}, R_{t'})_{t' \leq t-1}\right).$$

Moreover,

$$P_0 = \mathbb{1}_{i' \neq i, e'=1} \mathbb{P}\left(Z_{[t:t+1+r']} = i'|Z_{[t-e+1:t+r]} = i, (Z_{t'}, E_{t'}, R_{t'})_{t' \leq t-e}\right)$$

and with the semi-Markov property:

$$\begin{aligned} P_0 &= \mathbb{1}_{i' \neq i, e'=1} \mathbb{P}\left(Z_{[t:t+1+r']} = i'|Z_{[t-e+1:t+r]} = i\right) \\ &= \mathbb{P}\left((Z_{t+1}, E_{t+1}, R_{t-1}) = (i', e', r')|(Z_t, E_t, R_t) = (i, e, r)\right). \end{aligned}$$

□

PROOF OF PROPOSITION 1-2.–

First of all, $\{(Z_{t'}, E_{t'})_{t' \leq t}\} = \{(Z_{t'})_{t' \leq t}\}$ since $E_{t'}$ is completely determined by $(Z_{t''})_{t'' \leq t'}$. Let us prove the Markov property.

$$\begin{aligned} P_1 &= \mathbb{P}\left((Z_{t+1}, E_{t+1}) = (i', e')|(Z_t, E_t) = (i, e), (Z_{t'}, E_{t'})_{t' \leq t-1}\right) \\ &= \mathbb{P}\left((Z_{t+1}, E_{t+1}) = (i', e')|Z_{[t-e+1:t} = i, (Z_{t'})_{t' \leq t-e}\right) \\ &= \mathbb{1}_{\{i' \neq i, e'=1\}} \mathbb{P}\left(Z_{[t+1} = i'|Z_{[t-e+1:t} = i, (Z_{t'})_{t' \leq t-e}\right) \\ &\quad + \mathbb{1}_{\{i'=i, e'=e+1\}} \mathbb{P}\left(Z_{[t-e+1:t+1} = i|Z_{[t-e+1:t} = i, (Z_{t'})_{t' \leq t-e}\right) \\ &= \mathbb{1}_{\{i' \neq i, e'=1\}} P_{1a} + \mathbb{1}_{\{i'=i, e'=e+1\}} P_{1b}. \end{aligned}$$

We will demonstrate that the dependence with $(Z_{t'})_{t'\leq t-e}$ can be removed from P_{1a} and P_{1b}. Let $i \neq i'$, then:

$$\begin{aligned}P_{1a} &= \mathbb{P}\left(Z_{[t+1} = i', Z_{[t-e+1:t]} = i | Z_{[t-e+1:t} = i, (Z_{t'})_{t'\leq t-e}\right) \\ &= \mathbb{P}\left(Z_{[t+1} = i' | Z_{[t-e+1:t]} = i, (Z_{t'})_{t'\leq t-e}\right) \\ &\quad \times \mathbb{P}\left(Z_{[t-e+1:t]} = i | Z_{[t-e+1:t} = i, (Z_{t'})_{t'\leq t-e}\right) \\ &= \mathbb{P}\left(Z_{[t+1} = i' | Z_{[t-e+1:t]} = i\right) \times P_{1c}.\end{aligned}$$

The last equality results only from the semi-Markov property, indeed:

$$\begin{aligned}&\mathbb{P}\left(Z_{[t+1} = i' | Z_{[t-e+1:t]} = i, (Z_{t'})_{t'\leq t-e}\right) \\ &= \sum_{r\geq 0} \mathbb{P}\left(Z_{[t+1} = i', R_{t+1} = r | Z_{[t-e+1:t]} = i, (Z_{t'})_{t'\leq t-e}\right) \\ &= \sum_{r\geq 0} \mathbb{P}\left(Z_{[t+1,t+1+r]} = i' | Z_{[t-e+1:t]} = i, (Z_{t'})_{t'\leq t-e}\right) \\ &= \sum_{r\geq 0} \mathbb{P}\left(Z_{[t+1,t+1+r]} = i' | Z_{[t-e+1:t]} = i\right) \text{ by semi-Markov property} \\ &= \sum_{r\geq 0} \mathbb{P}\left(Z_{[t+1} = i', R_{t+1} = r | Z_{[t-e+1:t]} = i\right) \\ &= \mathbb{P}\left(Z_{[t+1} = i' | Z_{[t-e+1:t]} = i\right).\end{aligned}$$

Moreover,

$$\begin{aligned}P_{1c} &= \mathbb{P}\left(Z_{[t-e+1:t]} = i | Z_{[t-e+1:t} = i, Z_{[t-e+1} = i, (Z_{t'})_{t'\leq t-e}\right) \\ &= \frac{\mathbb{P}\left(Z_{[t-e+1:t]} = i, Z_{[t-e+1:t} = i | Z_{[t-e+1} = i, (Z_{t'})_{t'\leq t-e}\right)}{\mathbb{P}\left(Z_{[t-e+1:t} = i | Z_{[t-e+1} = i, (Z_{t'})_{t'\leq t-e}\right)}.\end{aligned}$$

According to the assumption of the Standard HSMM:

$$P_{1c} = \frac{\mathbb{P}\left(Z_{[t-e+1:t]} = i | Z_{[t-e+1} = i\right)}{\sum_{u\geq 0} \mathbb{P}\left(Z_{[t-e+1:t+u]} = i | Z_{[t-e+1} = i\right)}.$$

Finally, following the same decomposition backwards:

$$\begin{aligned}P_{1c} &= \frac{\mathbb{P}\left(Z_{[t-e+1:t]} = i, Z_{[t-e+1:t} = i | Z_{[t-e+1} = i,\right)}{\mathbb{P}\left(Z_{[t-e+1:t} = i | Z_{[t-e+1} = i\right)} \\ &= \mathbb{P}\left(Z_{[t-e+1:t]} = i | Z_{[t-e+1:t} = i\right).\end{aligned}$$

Thus,

$$P_{1a} = \mathbb{P}\left(Z_{[t+1} = i', Z_{[t-e+1:t]} = i | Z_{[t-e+1:t} = i\right).$$

Moreover, with a computing similar to P_{1c}:

$$\begin{aligned}
P_{1b} &= \mathbb{P}\left(Z_{[t-e+1:t+1} = i | Z_{[t-e+1:t} = i, Z_{[t-e+1} = i, (Z_{t'})_{t' \leq t-e}\right) \\
&= \mathbb{P}\left(Z_{[t-e+1:t+1} = i | Z_{[t-e+1} = i, (Z_{t'})_{t' \leq t-e}\right) \\
&\quad \times \mathbb{P}\left(Z_{[t-e+1:t} = i | Z_{[t-e+1} = i, (Z_{t'})_{t' \leq t-e}\right) \\
&= \mathbb{P}\left(Z_{[t-e+1:t+1} = i | Z_{[t-e+1} = i\right) \mathbb{P}\left(Z_{[t-e+1:t} = i | Z_{[t-e+1} = i\right) \\
&= \mathbb{P}\left(Z_{[t-e+1:t+1} = i | Z_{[t-e+1:t} = i\right).
\end{aligned}$$

Therefore, P_{1a} and P_{1b} are independent of $(Z_{t'})_{t' \leq t-e}$. Thus:

$$\begin{aligned}
P_1 &= \mathbb{P}\left((Z_{t+1}, E_{t+1}) = (i', e') | (Z_t, E_t) = (i, e), (Z_{t'}, E_{t'})_{t' \leq t-1}\right) \\
&= \mathbb{P}\left((Z_{t+1}, E_{t+1}) = (i', e') | (Z_t, E_t) = (i, e)\right)
\end{aligned}$$

and the couple $(Z_t, E_t)_t$ is Markovian. □

PROOF OF PROPOSITION 1-3.– For every $i, i' = 1, \ldots, K, r, r' \in \mathbb{N}$:

$$\begin{aligned}
&\mathbb{P}\left((Z_{t+1}, R_{t+1}) = (i', r') | (Z_t, R_t) = (i, r), (Z_{t'}, R_{t'})_{t' \leq t-1}\right) \\
&= \mathbb{P}\left(Z_{t+1:t+1+r']} = i' | Z_{t:t+r]} = i, (Z_{t'}, R_{t'})_{t' \leq t-1}\right) \\
&= \mathbb{1}_{\{r=0, i' \neq i\}} \mathbb{P}\left(Z_{[t+1:t+1+r']} = i' | Z_{t]} = i, (Z_{t'}, R_{t'})_{t' \leq t-1}\right) \\
&\quad + \mathbb{1}_{\{r>0, i'=i, r'=r-1\}}.
\end{aligned}$$

Moreover,

$$\{Z_{t]} = i, (Z_{t'}, R_{t'})_{t' \leq t-1}\} = \{Z_{t]} = i, (Z_{t'})_{t' \leq t-1}\}$$

and under the assumption of model ED-HSMM, $\mathbb{P}(Z_{[t+1:t+1+r']} = i' | Z_{t]} = i, (Z_{t'})_{t' \leq t-1})$ is independent of $(Z_{t'})_{t' \leq t-1}$. Therefore,

$$\begin{aligned}
&\mathbb{P}\left((Z_{t+1}, R_{t+1}) = (i', r') | (Z_t, R_t) = (i, r), (Z_{t'}, R_{t'})_{t' \leq t-1}\right) \\
&= \mathbb{P}\left((Z_{t+1}, R_{t+1}) = (i', r') | (Z_t, R_t) = (i, r)\right)
\end{aligned}$$

and $(Z_t, R_t)_{t \geq 0}$ is Markov. Note that this property does not necessarily hold under the less restrictive model Standard HSMM. □

4.8. References

Csilléry, K., Blum, M.G.B., Gaggiotti, O.E., François, O. (2010). Approximate Bayesian Computation (ABC) in practice. *Trends in Ecology and Evolution*, 25(7), 410–418.

Kim, H., Marwah, M., Arlitt, M., Lyon, G., Han, J. (2012). Unsupervised disaggregation of low frequency power measurements. In *SIAM Conference on Data Mining*, Philadelphia, PA.

Langrock, R. and Zucchini, W. (2011). Hidden Markov models with arbitrary state dwell-time distributions. *Computational Statistics & Data Analysis*, 55(1), 715–724.

Limnios, N. and Oprişan, G. (2001). *Semi-Markov Processes and Reliability*. Birkhäuser, Boston, MA.

Natarajan, P. and Nevatia, R. (2007). Coupled hidden semi-Markov models for activity recognition. In *2007 IEEE Workshop on Motion and Video Computing (WMVC'07)*, Austin, TX.

Nicol, S., Cros, M.-J., Peyrard, N., Sabbadin, R., Trépos, R., Fuller, R.A., Woodworth, B.K. (2022). Flywaynet: A hidden semi-Markov model for inferring the structure of migratory bird networks from count data. *Methods in Ecology and Evolution*, 14(1), 265–279.

Pertsinidou, C. and Limnios, N. (2015). Viterbi algorithms for hidden semi-Markov models with application to DNA analysis. *RAIRO – Operations Research*, 49, 511–526.

Touloupou, P., Finkenstädt, B., Spencer, S.E.F. (2020). Scalable Bayesian inference for coupled hidden Markov and semi-Markov models. *Journal of Computational and Graphical Statistics*, 29(2), 238–249.

Wei, G.C.G. and Tanner, M.A. (1990). A Monte Carlo implementation of the EM algorithm and the poor man's data augmentation algorithms. *Journal of the American Statistical Association*, 85(411), 699–704.

5

The Forward-backward Algorithm with Matrix Calculus

The calculation of the probability of observations in a hidden Markov model (HMM) knowing the parameters of the model can be done classically with the forward or the forward-backward algorithm. Knowing that, we formalize the notion of un-normalized heterogeneous Markov distributions (UHMD), as a tensor associated with a joint distribution. We show how to map the joint distribution of a HMM into an UHMD for computing the distribution of the observations after marginalization over the hidden variables, and these algorithms correspond to an elimination and marginalization algorithm, respectively, based on a series of matrix $\times$ vector products on a tensor network. It leads to a complexity in $\mathcal{O}(TK^2)$ for computing the normalizing constant and computing all unary marginals, where T is the duration of the sequence, and K is the number of states of the hidden variable. We show how the sparsity of the transition matrix allows us to decrease the complexity of the calculation. We use this dictionary between a HMM and an UHMD to extend within a common framework the evaluation of the complexity of forward and forward-backward algorithms for a diversity of Markov-based models with hidden variables. For multichain HMM (MHMM), it leads to bound the complexity according to the type of couplings using or not the decomposition of the transition matrix as a Kronecker product of univariate matrices. We use the generative model for an explicit duration hidden Markov model (ED-HMM) to rewrite it as an UHMD, to show that the transition matrix is sparse and show how the sparsity can control the upper bound of the complexity of the forward-backward algorithm. We present ways to possibly extend this approach toward some cases of multichains HSMM. Finally, we show how such an approach can be adapted to recover the Viterbi algorithm for finding the most likely state.

Chapter written by Alain FRANC.

5.1. Introduction

In his seminal paper on HMMs, (Rabiner 1989, HMM,), Rabiner raised three issues:

– How to compute the probabilities of the observations, the parameters of the model being known?

– How to restore the most likely states of the hidden variables, the parameters and observations being known?

– How to estimate the parameters of the model, the observations being known?

A fourth issues has emerged since, associating *smoothing* and *filtering*, which is to calculate at any time t the marginal distribution of each hidden variable, the observations and parameters of the model being known up to t for filtering, and up to the end of the sequence in smoothing. All these issues are classically solved by running *forward* or *forward-backward* algorithms (Rabiner 1989; Bishop 2006; Murphy 2012). The first issue is solved by forward algorithm, the second by Viterbi algorithm which is akin to the forward algorithm and the third one by EM algorithm, whose E step is a forward-backward algorithm; the smoothing is solved by a forward-backward algorithm and filtering by forward algorithm.

Since then, the family of Markov-based models with hidden variables has grown into two directions: the development of hidden semi-Markov models (HSMMs), presented in Chapter 1 (see references therein), and MHMM and multichain HSMM (MHSMM) that are discussed in detail in Chapters 3 and 4 of this book. The issues raised by Rabiner and the smoothing are relevant inference topics for each of these models. The forward, forward-backward and EM algorithm have flourished with these different guises, and are presented in Chapters 1 and 3 of this book.

They are presented for each model one after the other. Here, we propose a common framework and, within this framework, a standardized procedure that provides a way to write in a systematic way a forward and a forward-backward algorithm and an evaluation of their complexity expressed as the number of products they induce. We do not address explicitly the EM algorithm, even if its key step, the E step, is solved by a forward-backward algorithm. However, the approach presented here will lead to an estimate of the complexity of the EM algorithm. Therefore, we define the notion of an UHMD. For a HMM or any of its extensions, we map the joint distribution of the hidden and observed variables, where the observations are fixed, into an UHMD. It is then easy to see that the forward algorithm is an elimination algorithm on the UHMD, and the forward-backward algorithm is a marginalization algorithm on the UHMD. These algorithms are simply a recursive use of a matrix $\times$ vector product, one way for elimination and two ways for marginalization. Knowing the complexity of the matrix $\times$ vector product, it is simple to evaluate the complexity of the forward and forward-backward algorithms.

This chapter is organized as follows. UHMD and elimination/marginalization algorithms are presented in section 5.2, with some complements in section 5.3, which will be useful when addressing issues on multichain models. The presentation of the standard procedure is explained with the example of HMM in section 5.4, and developed on MHMMs in section 5.5, on HSMMs in section 5.6 and for some types of MHSMMs in section 5.7. Finally, we show in Appendix 5.11 how the Viterbi algorithm can be read as an elimination algorithm on an UHMD for restoring most likely hidden states, for each of these models (see as well Peyrard et al. 2019).

5.2. UHMDs, with elimination and marginalization algorithms

This section is the backbone of all calculations, which will be developed further. We define the notion of UHMD, which is a tensor whose terms generalize a Markov distribution. We then define an elimination and a marginalization algorithm on an UHMD.

5.2.1. *Un-normalized heterogeneous Markov-based distribution*

Let $T \in \mathbb{N}^*$ and denote $0:T = (0,\dots,T) \subset \mathbb{N}$ a succession of time steps. Let $(Z_t)_{t\in\mathbb{N}}$ be a stochastic process indexed on t, with $Z_t \in \mathcal{K}$, where $\mathcal{K}$ is discrete and finite. We will denote $|\mathcal{K}| = K$, and assume that $\mathcal{K} = \{1,\dots,K\} \subset \mathbb{N}$. We first recall the definition of a Markov model (MM) in order to introduce the definition of an UHMD.

DEFINITION 5.1.– *Markov model.*

$(Z_t)_t$ is an MM if:

$$\forall t \in 1:T, \qquad \mathbb{P}(Z_t = z_t \mid Z_0 = z_0,\dots,Z_{t-1} = z_{t-1})$$
$$= \mathbb{P}(Z_t = z_t \mid Z_{t-1} = z_{t-1}). \qquad [5.1]$$

An MM is defined by:

– an initial distribution, $\pi_0 \in \mathbb{R}^K$ with $\pi_0[z] = \mathbb{P}(Z_0 = z)$;

– a transition matrix, $A \in \mathbb{R}^{K\times K}$ with $A[z,z'] = \mathbb{P}(Z_t = z \mid Z_{t-1} = z')$.

Let $\pi_t \in \mathbb{R}^K$ be the probability distribution of the states of Z_t at time t: $\pi_t[z] = \mathbb{P}(Z_t = z)$. We have, recursively, $\pi_t = A\,\pi_{t-1}$. This is the matrix calculus expression of:

$$\mathbb{P}(Z_t = z) = \sum_{z'\in\mathcal{K}} \mathbb{P}(Z_t = z \mid Z_{t-1} = z')\,\mathbb{P}(Z_{t-1} = z'). \qquad [5.2]$$

Let us denote $\boldsymbol{z} = (z_0, \dots, z_\mathrm{T})$, with $\boldsymbol{z} \in \Omega_\mathrm{Z} = \mathcal{K}^{\mathrm{T}+1}$, and $p(\boldsymbol{z}) = \mathbb{P}(\boldsymbol{Z} = \boldsymbol{z})$. The joint probability of a sequence $\boldsymbol{z}$ in an MM is:

$$p(\boldsymbol{z}) = A[z_\mathrm{T}, z_{\mathrm{T}-1}] \dots A[z_t, z_{t-1}] \dots A[z_1, z_0]\, \pi_0[z_0].$$

This is a normalized probability distribution ($\sum_{\boldsymbol{z}\in\Omega_\mathrm{Z}} p(\boldsymbol{z}) = 1$), homogeneous (all transition matrices are equal to A), and the matrix A is stochastic.

A multivariate distribution given by $p(\boldsymbol{z})$ for $\boldsymbol{z} \in \Omega_\mathrm{Z}$ defines a tensor $\boldsymbol{T}$ (a multiway array, here with $T + 1$ entries) whose indices are the discrete states $z_0, \dots, z_T$ and whose coefficients are:

$$\boldsymbol{T}[z_0, \dots, z_\mathrm{T}] = p(\boldsymbol{z}).$$

An UHMD is a tensor $\boldsymbol{T}$ which generalizes the probability distribution $p(\boldsymbol{z})$ of an MM, such that the associated distribution is not normalized, the transitions matrices depend on time t and need not to be stochastic.

DEFINITION 5.2.– *Un-normalized heterogeneous Markov distribution (UHMD).*

An UHMD is a tensor of order $T + 1$, with dimension K on each entry t, if there exists a vector $\pi_0 \in \mathbb{R}_+^\mathrm{K}$ and T matrices $A_1, \dots, A_\mathrm{T} \in \mathbb{R}_+^{\mathrm{K}\times\mathrm{K}}$ such that each component can be decomposed as:

$$\boldsymbol{T}[z_0 \dots z_\mathrm{T}] = A_\mathrm{T}[z_\mathrm{T}, z_{\mathrm{T}-1}] \dots A_1[z_1, z_0]\, \pi_0[z_0]. \qquad [5.3]$$

To emphasize the link with heterogeneous Markov chains, we will sometimes call a *chain* the tuple $\boldsymbol{z} = (z_0, \dots, z_\mathrm{T})$, which is a at the same time multi-index of the tensor $\boldsymbol{T}$ and a succession of states $(z_t)_{0\le t\le T}$. The coefficients of $\boldsymbol{T}$ define an (un-normalized) distribution in Ω_Z. We sometimes will denote $\boldsymbol{T}[\boldsymbol{z}] := \boldsymbol{T}[z_0 \dots z_\mathrm{T}]$. We can assume that $\forall\, t, \forall\, z_t, z_{t-1} \in \mathcal{K},\ \ A_t[z_t, z_{t-1}] \ge 0$, but it is not necessary.

5.2.2. *Elimination algorithm*

DEFINITION 5.3.– *Normalizing constant.*

The normalizing constant W of $\boldsymbol{T}$ is defined by:

$$W = \sum_{\boldsymbol{z}\in\Omega_\mathrm{Z}} \boldsymbol{T}[\boldsymbol{z}].$$

It can be computed with the following algorithm, called the *elimination algorithm*. We recall that $\forall\, t \ge 1,\ \pi_t = A_t\pi_{t-1}$, which can be developed as $\pi_t[z_t] = \sum_{z_{t-1}=1}^{\mathrm{K}} A_t[z_t, z_{t-1}]\pi_{t-1}[z_{t-1}]$. We can write by induction

$\pi_t = A_t A_{t-1} \dots A_1 \pi_0$ until $\pi_{\mathrm{T}} = A_{\mathrm{T}} \dots A_1 \pi_0$. This matrix product can be developed componentwise as:

$$\pi_{\mathrm{T}}[z_{\mathrm{T}}] = \sum_{z_{\mathrm{T}-1}=1}^{\mathrm{K}} \dots \sum_{z_0=1}^{\mathrm{K}} A_{\mathrm{T}}[z_{\mathrm{T}}, z_{\mathrm{T}-1}] \dots A_1[z_1, z_0]\pi_0[z_0].$$

Then

$$W = \sum_{\boldsymbol{z}\in\Omega_{\mathrm{Z}}} \boldsymbol{T}[\boldsymbol{z}] = \sum_{\boldsymbol{z}\in\Omega_{\mathrm{Z}}} A_{\mathrm{T}}[z_{\mathrm{T}}, z_{\mathrm{T}-1}] \dots A_1[z_1, z_0]\pi_0[z_0] = \sum_{z_{\mathrm{T}}} \pi_{\mathrm{T}}[z_{\mathrm{T}}],$$

which leads to the following pseudo-code:

Algorithm 5.1 Elimination algorithm ELIM() in an UHMD

```
1: input T, K
2: input (A_t)_{1≤t≤T} with A_t ∈ ℝ^{K×K}
3: input π_0 ∈ ℝ^K
4: for t ∈ {1, ..., T} do
5:     compute π_t = A_t π_{t-1}      with   π_t[z] = Σ_{z'=1}^{K} A_t[z, z'] π_{t-1}[z']
6: end for
7: compute W = Σ_{z∈K} π_T[z]
8: return W
```

.

PROPOSITION 5.1.– *The complexity of the elimination algorithm for an UHMD is in* $\mathcal{O}(TK^2)$.

PROOF.– The product $A_t\,\pi_{t-1}$ in line 5 of Algorithm 5.1 requires K^2 multiplications ($A_t \in \mathbb{R}^{\mathrm{K}\times\mathrm{K}}$, $\pi_{t-1} \in \mathbb{R}^{\mathrm{K}}$). The number of such computations in the loop over t starting at line 4 is T. So, the complexity of the algorithm is in $\mathcal{O}(T\,K^2)$. □

This algorithm will be used for writing the forward algorithm of several models in a unified framework in the remaining chapter. Therefore, we present it as a function.

Algorithm 5.2 Function ELIM$(T, K, (A_t)_{1\le t\le \mathrm{T}}, \pi_0)$ in an UHMD

```
for t ∈ {1, ..., T} do
    compute π_t = A_t π_{t-1}      with   π_t[z] = Σ_{z'=1}^{K} A_t[z, z'] π_{t-1}[z']
end for
compute W = Σ_{z∈K} π_T[z]
return W
```

REMARK 5.1.– TK^2 is also the number of terms in the set of matrices $(A_t)_t$, that is, necessary to specify the UHMD: each matrix has K^2 terms and there are T of them. We observe that, for all models studied in this chapter, the complexity of the elimination algorithm is precisely the number of terms in the set of matrices, which specify the UHMD. Indeed, each matrix A_t is involved once and only once in a matrix $\times$ vector product of complexity K^2. This is true as well up to a constant for the marginalization algorithm presented below, which is a double elimination: one forward and one backward.

5.2.3. *Marginalization algorithm*

DEFINITION 5.4.– *Unary marginal of an UHMD.*

Let $\boldsymbol{T}$ be an UHMD. We call a unary marginal of $\boldsymbol{T}$ at time t for state z, and denote $\mathfrak{m}(t, z)$*, the quantity:*

$$\mathfrak{m}(t, z) = \sum_{\boldsymbol{z} \in \mathcal{K}^{\mathrm{T}+1} \,:\, z_t = z} \boldsymbol{T}[\boldsymbol{z}].$$

In the case of $\boldsymbol{T}$ being a probability distribution (i.e. A_t stochastic and $W(\boldsymbol{T}) = 1$), it is equal to $\mathbb{P}(Z_t = z)$. Marginalization algorithm is about computing simultaneously all unary marginals of $\boldsymbol{T}$, with t running over $0 : T$ and z over $\mathcal{K}$. It is a double elimination algorithm, one forward and one backward over the sequence $\boldsymbol{z}$.

Let us recall it (see e.g. Murphy 2012, Section 17.4.3). We have:

$$\begin{aligned} \mathfrak{m}(t, z) &= \sum_{\boldsymbol{z} \in \mathcal{K}^{\mathrm{T}+1} \,:\, z_t = z} A_{\mathrm{T}}[z_{\mathrm{T}}, z_{\mathrm{T}-1}] \dots A_1[z_1, z_0]\, \pi_0[z_0] \\ &= \underbrace{\left(\sum_{z_{\mathrm{T}}, \dots, z_{t+1}} A_{\mathrm{T}}[z_{\mathrm{T}}, z_{\mathrm{T}-1}] \dots A_{t+1}[z_{t+1}, z] \right)}_{\beta_t(z)} \times \\ &\quad \underbrace{\left(\sum_{z_{t-1}, \dots, z_0} A_t[z, z_{t-1}] \dots A_1[z_1, z_0]\, \pi_0[z_0] \right)}_{\alpha_t(z)}. \end{aligned} \qquad [5.4]$$

which can be summarized by:

$$\mathfrak{m}(t, z) = \alpha_t(z)\, \beta_t(z). \qquad [5.5]$$

Both terms $\alpha_t(z)$ and $\beta_t(z)$ can be computed by recurrence on t and all z with an elimination algorithm, which enables us to compute efficiently all unary marginals with a procedure akin to message passing.

REMARK ABOUT NOTATIONS.– We notice that what we call α_t here corresponds to what we called π_t in the context of MMs and an UHMD: $\alpha_t = \pi_t$. However, we, keep both notations π_t or α_t according to the context: both are standard notations. We will use systematically α_t in the context of HMM.

So $\alpha_t = A_t \alpha_{t-1}$.

Similarly, from:

$$\beta_t(z) = \sum_{z_{\mathrm{T}},\dots,z_{t+1}} A_{\mathrm{T}}[z_{\mathrm{T}}, z_{\mathrm{T}-1}] \dots A_{t+1}[z_{t+1}, z],$$

we have

$$\begin{aligned}
\beta_t(z) &= \sum_{z_{t+1}} \left(\sum_{z_{\mathrm{T}},\dots,z_{t+2}} A_{\mathrm{T}}[z_{\mathrm{T}}, z_{\mathrm{T}-1}] \dots A_{t+2}[z_{t+2}, z_{t+1}] \right) A_{t+1}[z_{t+1}, z] \\
&= \sum_{z_{t+1}} \beta_{t+1}(z_{t+1})\, A_{t+1}[z_{t+1}, z] \qquad [5.6] \\
&= \sum_{z_{t+1}} A^{\mathrm{T}}_{t+1}[z, z_{t+1}]\, \beta_{t+1}(z_{t+1})
\end{aligned}$$

(A^{T} is the transpose of A). If β_t is the vector in $\mathbb{R}^{\mathrm{K}}$ of coordinates $\beta_t[z]$ for $z \in \mathcal{K}$, we have:

$$\beta_t = A^{\mathrm{T}}_{t+1}\, \beta_{t+1}. \qquad [5.7]$$

The initialization is made with $\beta_{\mathrm{T}} = \mathbf{1}$.

From now on, we consider vector α_t (respectively, β_t) whose component $\alpha_t[z]$ (respectively, $\beta_t[z]$) is precisely $\alpha_t(z)$ (respectively, $\beta_t(z)$). Then, $\mathfrak{m}(t, z) = \alpha_t[z]\, \beta_t[z]$, and we define:

$$\mathfrak{m}(t) = \alpha_t \odot \beta_t, \qquad [5.8]$$

where $\odot$ denotes the Hadamard product (i.e. componentwise product). Here is a visualization of the marginalization algorithm as a series of matrix $\times$ vector products for $T = 4$:

$$\begin{array}{ccccccccccc}
\alpha_0 & \xrightarrow{\times A_1} & \alpha_1 & \xrightarrow{\times A_2} & \alpha_2 & \xrightarrow{\times A_3} & \alpha_3 & \xrightarrow{\times A_4} & \alpha_4 & & \\
\odot & \xleftarrow[\times A_4^{\mathrm{T}}]{} & \odot & \xleftarrow[\times A_3^{\mathrm{T}}]{} & \odot & \xleftarrow[\times A_2^{\mathrm{T}}]{} & \odot & \xleftarrow[\times A_1^{\mathrm{T}}]{} & \odot & & \odot \\
\beta_4 & & \beta_3 & & \beta_2 & & \beta_1 & & \beta_0 & & \\
= & & = & & = & & = & & = & & \\
\mathfrak{m}(0) & & \mathfrak{m}(1) & & \mathfrak{m}(2) & & \mathfrak{m}(3) & & \mathfrak{m}(4) & &
\end{array}$$

where the symbols $\odot$ mean at any time t that $\mathfrak{m}(t) = \alpha_t \odot \beta_{T-t}$ (visualized columnwise) and, for example, $\alpha_0 \xrightarrow{\times A_1} \alpha_1$ means that $\alpha_1 = A_1 \alpha_0$.

Here is the pseudo-code of the marginalization algorithm for an UHMD:

Algorithm 5.3 Marginalization algorithm MARGIN() in an UHMD

1: **input** T, K
2: **input** $(A_t)_{1 \le t \le T}$ with $A_t \in \mathbb{R}^{K \times K}$
3: **input** $\pi_0 \in \mathbb{R}^K$
4: set $\alpha_0 = \pi_0$
5: **for** $t \in \{1, \dots, T\}$ **do**
6:     **compute** $\alpha_t = A_t \, \alpha_{t-1}$
7: **end for**
8: set $\beta_T = \mathbf{1}$
9: **for** $t \in \{T-1, \dots, 0\}$ **do**
10:     **compute** $\beta_t = A_{t+1}^T \, \beta_{t+1}$
11: **end for**
12: **for** $t \in \{0, \dots, T\}$ **do**
13:     **compute** $\mathfrak{m}_t = \alpha_t \odot \beta_{T-t}$
14:     **return** $\mathfrak{m}_t$
15: **end for**

PROPOSITION 5.2.– *The complexity of the marginalization algorithm is in* $\mathcal{O}(TK^2)$.

PROOF.– Line 6 of Algorithm 5.3 requires K^2 multiplications. The loop in line 5 repeats it T times. Hence, the calculations of all α_t requires TK^2 multiplications. The same calculation can be made for computing all β_t from line 10 in the loop starting at line 9, so $2TK^2$ multiplications altogether. The Hadamard product in line 13 costs K multiplications (one for each z), and it is done $T+1$ times in loop 12. The calculation of all $\mathfrak{m}_t$ knowing the α_t and β_t costs $(T+1)K$ multiplications. The overall cost is $2TK^2 + (T+1)K \simeq 2TK^2$. Hence, the complexity of the *marginalization algorithm* is in $\mathcal{O}(T\,K^2)$. □

This algorithm can be written as a function as well, with the vectors $\mathfrak{m}(t)$ as outputs, for use in a unified framework for writing the forward-backward algorithm of the models studied in this chapter.

5.3. Complements on the complexity of elimination and marginalization algorithms for an UHMD

Here, we present some complements which will be useful for multichain models.

5.3.1. *Multichain UHMD*

DEFINITION 5.5.– *Multichain Markov model.*

An MM is called a multichain Markov model *if its state space* $\mathscr{K}$ *is a product of spaces, that is,* $\mathscr{K} = \mathcal{K}_1 \times \ldots \times \mathcal{K}_{\mathrm{C}}$.

Then,

$$z_t = (z_t^1, \ldots, z_t^{\mathrm{C}}), \quad \text{with} \quad z_t^c \in \mathcal{K}_c, \quad 1 \leq c \leq C.$$

NOTATIONS.– The probability of a transition $\boldsymbol{z}' \to \boldsymbol{z}$ at time t is written by $\mathbb{P}(\boldsymbol{Z}_t = \boldsymbol{z} \mid \boldsymbol{Z}_{t-1} = \boldsymbol{z}')$. It is written sometimes, when there are no ambiguities, $p(\boldsymbol{z}_t \mid \boldsymbol{z}_{t-1})$. The coefficients of the transition matrix are written $A[\boldsymbol{z}, \boldsymbol{z}']$. The row indices $\boldsymbol{z}$ and column indices $\boldsymbol{z}'$ are tuples, that is, the indices are multiplexed. Finally, we denote $\boldsymbol{z}_{0:\mathrm{T}} = (\boldsymbol{z}_0, \ldots, \boldsymbol{z}_{\mathrm{T}})$. For the sake of simplicity, and without loss of generality, one assumes that for any c, $\mathcal{K}_c := \mathcal{K}$.

The notion of multichain can be extended to multichains UHMD as follows:

DEFINITION 5.6.– *Multichains UHMD (MUHMD).*

An MUHMD is an UHMD $\boldsymbol{T}$ *such that each element is identified by a set* $\boldsymbol{z}_{0:\mathrm{T}}$ *of multi-indices, with:*

$$\boldsymbol{z}_{0:\mathrm{T}} = ((z_0^1, \ldots, z_0^{\mathrm{C}}), \ldots, (z_{\mathrm{T}}^1, \ldots, z_{\mathrm{T}}^{\mathrm{C}})) = (\boldsymbol{z}_0, \ldots, \boldsymbol{z}_T).$$

An element indexed by $\boldsymbol{z}_{0:\mathrm{T}}$ *can be decomposed as:*

$$\boldsymbol{T}[\boldsymbol{z}_0, \ldots, \boldsymbol{z}_T] = A_{\mathrm{T}}[\boldsymbol{z}_{\mathrm{T}}, \boldsymbol{z}_{\mathrm{T}-1}] \ldots A_1[\boldsymbol{z}_1, \boldsymbol{z}_0]\pi_0[\boldsymbol{z}_0]. \tag{5.9}$$

Each matrix A_t *is a TT-matrix (see Oseledets and Dolgov 2012, for the notion of TT-matrix). The chain c is the tuple* $\boldsymbol{z}^c = (z_0^c, \ldots, z_{\mathrm{T}}^c)$.

Multiplexing/demultiplexing: the multi-indices $\boldsymbol{z}_{0:\mathrm{T}}$ of T are said to be multiplexed. It is possible to map $\boldsymbol{z}_{0:\mathrm{T}}$ onto a list of integer indices by demultiplexing each $\boldsymbol{z}_t$ for $1 \leq t \leq T$. Then, the multi-index $\boldsymbol{z}_t = (z_t^1, \ldots, z_t^{\mathrm{C}}) \in \mathcal{K}^{\mathrm{C}}$ can be demultiplexed as an integer index $z_t' \in \{1, \ldots, K^{\mathrm{C}}\} \subset \mathbb{N}$. For example, if $C = 2$, $(z_t^1, z_t^2) \in \mathcal{K}^2$ is mapped onto $z_t' \in \{1, \ldots, K^2\}$ by $z_t' = K(z_t^1 - 1) + z_t^2 \in \{1, \ldots, K^2\}$. This can be generalized to $C > 2$ by $z_t' = 1 + \sum_{c=1}^{\mathrm{C}} K^{c-1}(z_t^c - 1)$ (c is an index in z_t^c and $c-1$ is an exponent in K^{c-1}).

By multiplexing/demultiplexing indices, an MUHMD with multi-indices can be rewritten as an UHMD with integer indices, with state space $\mathscr{K}$, with $|\mathscr{K}| = K^{\mathrm{C}}$. Then, the elimination and marginalization algorithms can be run, yielding a complexity in $\mathcal{O}(TK^{2C})$, since $K^{2C} = (K^C)^2$. Hence, we have the following proposition:

PROPOSITION 5.3.– *The complexity of the elimination or marginalization algorithms on an MUHMD is in $\mathcal{O}(TK^{2C})$ if the multi-indices are tuples of C elements each in $\mathcal{K}$ with $|\mathcal{K}| = K$.*

This exponential growth with C of the complexity makes it very easy to hit the time wall when the algorithm is run. However, there exists situations where the complexity of the algorithms can be significantly decreased. Let us mention two of them which will be presented:

– when the transition matrix is sparse (most of its coefficients are zero), presented in section 5.3.2;

– when, in the multichain case, the MUHMD can be decomposed as a product, which means that the chains of the MUHMD are independent, presented in section 5.3.3.

5.3.2. *Sparsity*

The expression $\mathcal{O}(TK^2)$ of the complexity of elimination or marginalization algorithm comes from the term matrix $\times$ vector product in $\alpha_t = A_t\alpha_{t-1}$ (line 6, and the same for β_t line 10, in Algorithm 5.3). If $A_t \in \mathbb{R}^{K\times K}$ and $\alpha_t \in \mathbb{R}^K$, the product $A_t\alpha_t$ requires K^2 multiplications. This is a plain consequence of the general setting of a matrix $\times$ vector product $y = Ax$ developed as $y_i = \sum_{j=1}^{K} a_{ij}x_j$. The complexity (number of multiplications) is the number of terms $a_{ij}x_j$, that is, of terms a_{ij}. Having a term a_{ij} equal to zero decreases by one the complexity. Let us denote by $\#A$ the number of non-zero terms in A. Then, the complexity of the calculation of $y = Ax$ is in $\mathcal{O}(\#A)$, where $\#A$ can be lower than K^2. Classically, a sparse matrix is a $K \times K$ matrix such that the number of non-zero terms is in $\mathcal{O}(K)$. For example, a permutation matrix is a sparse matrix. The identity matrix is a sparse matrix. Here, we adopt the convention that a sparse matrix is a $K \times K$ matrix such that the number of non-zero terms is in $\mathcal{O}(K^\gamma)$ with $1 \leq \gamma < 2$. If A_t is such a sparse matrix, the complexity of the calculation of $\alpha_t = A_t\alpha_{t-1}$ is in $\mathcal{O}(K^\gamma)$.

For each pair (i, j) such that $a_{ij} = 1$ or $a_{ij} = -1$, no product is involved in $a_{ij}x_j = \pm x_j$, and the complexity of the calculation of $y = Ax$ is decreased by one. So, we extend the definition of $\#A$ as the number of terms a_{ij}, which do not belong to $\{-1, 0, 1\}$. For example, if A is the identity matrix $\mathbb{I}$, $\#\mathbb{I} = 0$, and the calculation of $y = \mathbb{I}x$ requires no multiplication: the complexity is 0 (indeed, $y = x$).

If the UHMD is homogeneous, that is, all matrices A_t are equal to the same matrix A, it suffices to check the sparsity of A. But for a general UHMD, the sparsity of A_t

must be checked and the quantity $\#A_t$ evaluated at each time step t. We have the following proposition:

PROPOSITION 5.4.– *In an UHMD, in the case of distinct matrices A_t for $1 \leq t \leq T$, the complexity of elimination or marginalization algorithms are in $\mathcal{O}(\sum_{t=1}^{\mathrm{T}} \#A_t)$, where $\#A_t$ is the number of terms in A_t which are not in $\{-1, 0, 1\}$.*

We call *sparsity* the value of $1 - \frac{\#A}{K^2}$. The higher the sparsity, the sparser the matrix.

5.3.3. *Independence between chains in an UHMD*

Independence is presented for two chains for sake of clarity of notations, knowing that extension to C chains is straightforward. A bivariate UHMD is an MUHMD with $C = 2$, and is a tensor $\boldsymbol{T}$ such that:

$$\boldsymbol{T}[\boldsymbol{z}_0, \ldots, \boldsymbol{z}_{\mathrm{T}}] = A_{\mathrm{T}}[(z^1_{\mathrm{T}}, z^2_{\mathrm{T}}), (z^1_{\mathrm{T}-1}, z^2_{\mathrm{T}-1})] \ldots A_1[(z^1_1, z^2_1), (z^1_0, z^2_0)]\, \pi_0[z^1_0, z^2_0], \quad [5.10]$$

with $\boldsymbol{z}_0 = (z^1_0, z^2_0), \ldots, \boldsymbol{z}_{\mathrm{T}} = (z^1_{\mathrm{T}}, z^2_{\mathrm{T}})$.

DEFINITION 5.7.– *Multichains UHMD with independent chains. The chains $\boldsymbol{z}^1 = (z^1_0, \ldots, z^1_{\mathrm{T}})$ and $\boldsymbol{z}^2 = (z^2_0, \ldots, z^2_{\mathrm{T}})$ are called* independent *in MUHMD $\boldsymbol{T}$ if:*

– there are two sets $(A^1_t)_t$ and $(A^2_t)_t$ of $K \times K$ matrices such that, for any time t with $1 \leq t \leq T$:

$$A_t[(z^1_t, z^2_t), (z^1_{t-1}, z^2_{t-1})] = A^1_t[z^1_t, z^1_{t-1}] \times A^2_t[z^2_t, z^2_{t-1}]; \quad [5.11]$$

– there are two distributions (π^1_0, π^2_0) such that $\pi_0[z^1_0, z^2_0] = \pi^1_0[z_1]\pi^2_0[z^2_0]$.

This can be written more compactly $A_t = A^1_t \otimes A^2_t$ and $\pi_0 = \pi^1_0 \otimes \pi^2_0$, where $\otimes$ denotes the Kronecker product between matrices or vectors. We then have, by reordering the products:

$$\begin{aligned}
&\boldsymbol{T}[\boldsymbol{z}_0, \ldots, \boldsymbol{z}_{\mathrm{T}}] \\
&= \left(A^1_{\mathrm{T}}[z^1_{\mathrm{T}}, z^1_{\mathrm{T}-1}] \ldots A^1_1[z^1_1, z^1_0]\, \pi^1_0[z^1_0]\right) \times \left(A^2_{\mathrm{T}}[z^1_{\mathrm{T}}, z^1_{\mathrm{T}-1}] \ldots A^2_1[z^1_1, z^1_0]\, \pi^2_0[z^1_0]\right) \\
&= \boldsymbol{T}^1[z^1_0, \ldots, z^1_{\mathrm{T}}]\, \boldsymbol{T}^2[z^2_0, \ldots, z^2_{\mathrm{T}}]
\end{aligned}$$

where

$$\boldsymbol{T}^c[z^c_0, \ldots, z^c_{\mathrm{T}}] = A^c_{\mathrm{T}}[z^c_{\mathrm{T}}, z_{\mathrm{T}-1}] \ldots A^c_1[z^c_1, z^c_0]\, \pi_0[z^c_0], \qquad c = 1, 2.$$

Then,

PROPOSITION 5.5.– *Let $\boldsymbol{T}$ be a bivariate UHMD with chains $\boldsymbol{z}^1 = (z_0^1, \ldots, z_{\mathrm{T}}^1)$ and $\boldsymbol{z}^2 = (z_0^2, \ldots, z_{\mathrm{T}}^2)$ being independent. Then, the complexity of the elimination and marginalization algorithms on $\boldsymbol{T}$ is in $\mathcal{O}(TK^2)$ instead of $\mathcal{O}(TK^4)$ if they were not independent.*

PROOF.– Indeed:

$$\begin{aligned} W(\boldsymbol{T}) &= \sum_{\boldsymbol{z}_{0:\mathrm{T}}} \boldsymbol{T}[\boldsymbol{z}_{0:\mathrm{T}}] && \boldsymbol{z}_{0:\mathrm{T}} \in \mathcal{K}^{2(\mathrm{T}+1)} \\ &= \sum_{\boldsymbol{z}^1, \boldsymbol{z}^2} \boldsymbol{T}^1[\boldsymbol{z}^1]\boldsymbol{T}^2[\boldsymbol{z}^2] && \boldsymbol{z}^1, \boldsymbol{z}^2 \in \mathcal{K}^{\mathrm{T}+1} \\ &= \left(\sum_{\boldsymbol{z}^1} \boldsymbol{T}^1[\boldsymbol{z}^1]\right)\left(\sum_{\boldsymbol{z}^2} \boldsymbol{T}^2[\boldsymbol{z}^2]\right) \\ &= W(\boldsymbol{T}^1)\, W(\boldsymbol{T}^2). \end{aligned}$$

Both $\boldsymbol{T}^1$ and $\boldsymbol{T}^2$ are monochain UHMD. So, the complexity of the computations of $W(\boldsymbol{T}^1)$ and $W(\boldsymbol{T}^2)$ are each in $\mathcal{O}(TK^2)$. Then, the calculation of both costs $2TK^2$ products, and the computation of $W(\boldsymbol{T})$ is in $\mathcal{O}(2TK^2)$. For C independent chains, it is in $\mathcal{O}(TCK^2)$. The same can be shown for the marginalization algorithm, which is a double elimination algorithm. □

5.4. Hidden Markov model

A HMM is defined in Chapter 1 (Section 1.2.5). We recall here the main notations:

– the hidden variables are denoted Z_t; their state space is $\mathcal{K}$, with $|\mathcal{K}| = K$;

– the observed variables are denoted Y_t; their state space is $\mathcal{S}$ with $|\mathcal{S}| = S$;

– the initial state of the hidden variable is π_0, with $\pi_0[z] = \mathbb{P}(Z_0 = z)$;

– the transition matrix for hidden variables is denoted A, with:

$$A[z, z'] = P(Z_t = z \mid Z_{t-1} = z');$$

– the emission matrix is denoted B, with:

$$B[y, z] = \mathbb{P}(Y_t = y \mid Z_t = z).$$

The parameters of the model are $\theta = (\pi_0, B, A)$. If $\boldsymbol{z} = (z_0, \ldots, z_{\mathrm{T}}) \in \mathcal{K}^{\mathrm{T}+1}$ and $\boldsymbol{y} = (y_0, \ldots, y_{\mathrm{T}}) \in \mathcal{S}^{\mathrm{T}+1}$, the joint distribution is:

$$\begin{aligned} p(\boldsymbol{z}, \boldsymbol{y} \mid \theta) &= B[y_{\mathrm{T}}, z_{\mathrm{T}}]A[z_{\mathrm{T}}, z_{\mathrm{T}-1}] \cdots \\ &\quad \ldots B[y_t, z_t]A[z_t, z_{t-1}] \ldots B[y_1, z_1]A[z_1, z_0]B[y_0, z_0]\pi_0[z_0]. \end{aligned} \quad [5.12]$$

We show here how elimination (respectively, marginalization) algorithms on an appropriate UHMD, which will be given below, are the mirror of the forward (respectively, forward-backward) algorithm on a HMM. This leads to the answers to the three questions raised in Rabiner (1989) recalled in the introduction and smoothing, that is, design the forward and forward-backward algorithm, estimate their complexity and project the result as an estimate of the complexity of EM algorithm.As far as naming things is concerned, the elimination algorithm is referred to as the *forward* algorithm in the context of HMMs, and marginalization as the *forward-backward* algorithm in same context. We will use one or other of the names, depending on the context. We start with the forward algorithm.

5.4.1. *Computing the probability of the observations*

We are interested in computing $p(\boldsymbol{y} \mid \theta)$, which is the probability of the observations knowing the parameters of the model, also called *likelihood*, and solved with the forward algorithm. We have:

$$p(\boldsymbol{y}) = \sum_{\boldsymbol{z} \in \Omega_Z} p(\boldsymbol{z}, \boldsymbol{y}), \qquad [5.13]$$

(we drop the conditioning by the parameters of the models, because they are fixed throughout this section). We consider that the observations are fixed.

As $\boldsymbol{y}$ is fixed, and following equation [5.12], one can define for each time t a matrix $A_t \in \mathbb{R}^{\mathrm{K} \times \mathrm{K}}$ by:

$$A_t[z_t, z_{t-1}] = B[y_t, z_t]A[z_t, z_{t-1}]. \qquad [5.14]$$

Then

$$p(\boldsymbol{z}, \boldsymbol{y}) = A_{\mathrm{T}}[z_{\mathrm{T}}, z_{\mathrm{T}-1}] \dots A_t[z_t, z_{t-1}] \dots A_1[z_1, z_0]\pi_0'[z_0], \qquad [5.15]$$

with $\pi_0'[z_0] = B[y_0, z_0]\pi_0[z_0]$. This defines an UHMD $\mathcal{A}$ whose coefficients $\mathcal{A}[\boldsymbol{z}]$ for $\boldsymbol{z} \in \Omega_Z$ is the RHS of equation [5.15]. $\mathcal{A}$ is called the UHMD associated with the HMM defined by $\theta = (A, B, \pi_0)$. The following observation is simple, and is the key for using the UHMD approach.

PROPOSITION 5.6.– *$p(\boldsymbol{y})$ is the normalizing constant of $\mathcal{A}$.*

PROOF.– Indeed:

$$\begin{aligned} p(\boldsymbol{y}) &= \sum_{\boldsymbol{z} \in \mathcal{K}^{\mathrm{T}+1}} A_{\mathrm{T}}[z_{\mathrm{T}}, z_{\mathrm{T}-1}] \dots A_t[z_t, z_{t-1}] \dots A_1[z_1, z_0]\pi_0'[z_0] \\ &= \sum_{\boldsymbol{z} \in \mathcal{K}^{\mathrm{T}+1}} \mathcal{A}[\boldsymbol{z}] \qquad [5.16] \\ &= W(\mathcal{A}). \end{aligned}$$

□

As $p(\boldsymbol{y})$ is the normalizing constant of the UHMD $\mathcal{A}$, it can be computed with the elimination algorithm (Algorithm 5.2, section 5.2.1) as follows:

Algorithm 5.4 Computation of $p(\boldsymbol{y})$ for a HMM with algorithm ELIM()

1: **input** $A \in \mathbb{R}^{K\times K}, B \in \mathbb{R}^{S\times K}, \pi_0 \in \mathbb{R}^{K}, \boldsymbol{y} \in \mathcal{S}^{T+1}, T \in \mathbb{N}$
2: # *preparation: building UHMD* $\mathcal{A}$
3: **compute** $\pi_0' \in \mathbb{R}^K$ such that $\pi_0'[z_0] = B[y_0, z_0]\pi_0[z_0]$
4: **for** $t \in \{1, \ldots, T\}$ **do**
5: **compute** $A_t \in \mathbb{R}^{K\times K}$ s.t. $A_t[z_t, z_{t-1}] = B[y_t, z_t]A[z_t, z_{t-1}]$
6: **end for**
7: # *elimination with Algorithm* 5.2
8: **compute** $W = \text{ELIM}(T, K, (A_t)_t, \pi_0')$
9: **return** W

REMARK 5.2.– There are two steps in this algorithm that have been identified as a comment: the preparation phase that builds the matrices $(A_t)_t$ of the UHMD knowing the parameters of the HMM and the observations, and the elimination phase, which is simply Algorithm 5.2 run on the UHMD $\mathcal{A}$ (recall that $T, K, (A_t)_t, \pi_0'$ define $\mathcal{A}$). This two-phases structure for building a solution will be used for all the models studied in this chapter.

PROPOSITION 5.7.– *The complexity of the calculation of $p(\boldsymbol{y})$ knowing the parameters θ is in $\mathcal{O}(TK^2)$.*

PROOF.– This is a direct consequence of the use of Algorithm 5.2 in line 8 of Algorithm 5.4. The preparation costs SK^2 multiplications: for any pair (z, z'), one computes $B[y, z]A[z, z']$ with S different values of y. In general, $S \ll T$. □

5.4.2. *Smoothing and EM algorithm*

Smoothing is the calculation of $\mathbb{P}(Z_t = z \mid \boldsymbol{y}, \theta)$ for any $z \in \mathcal{K}$ and any t with $0 \leq t \leq T$. It can be calculated by (dropping θ which is fixed):

$$\mathbb{P}(Z_t = z \mid \boldsymbol{y}) = \sum_{\boldsymbol{z}\in\Omega_Z\,:\,z_t=z} p(\boldsymbol{z} \mid \boldsymbol{y}) = \frac{\sum_{\boldsymbol{z}\in\Omega_Z\,:\,z_t=z} p(\boldsymbol{z}, \boldsymbol{y})}{p(\boldsymbol{y})}.$$

As the joint law $p(\boldsymbol{z}, \boldsymbol{y})$ can be written as an UHMD (see equation [5.15], $\mathfrak{m}(t, z) = \sum_{\boldsymbol{z}\,:\,z_t=z} p(\boldsymbol{z}, \boldsymbol{y})$ can be computed with the marginalization algorithm (Algorithm 5.3). This allows us to compute all marginals $\mathfrak{m}(t, z)$ for $0 \leq t \leq T$ and $z \in \mathcal{K}$. Selecting a t and summing up over z_t leads to the computation of $p(\boldsymbol{y})$. This leads to the following proposition.

PROPOSITION 5.8.– *The calculation of* $\mathbb{P}(Z_t = z \mid \boldsymbol{y}, \theta)$ *for any* $z \in \mathcal{K}$ *and any* t *with* $0 \leq t \leq T$ (smoothing) *can be done with the application of the marginalization algorithm on the UHMD* $\mathcal{A}$ *associated with the HMM* (A, B, π_0). *Its complexity is in* $\mathcal{O}(TK^2)$.

The marginalization algorithm can also be used to compute binary marginals as in the E step in EM algorithm for parameter inference given the observations. These calculations are not detailed here (see Chapter 1, Appendix 1.10 for the presentation and calculation of EM algorithm for a HMM). Their complexity is in $\mathcal{O}(TK^2)$ from the general result on an UHMD and the size of transition matrices A_t.

The elimination algorithm is known as the *forward* algorithm in the context of HMMs. However, often, the algorithm used for computing $p(\boldsymbol{y} \mid \theta)$ is the *forward-backward* algorithm. The latter is the marginalization algorithm. One reason for this is that it allows us to compute all marginals, and once all marginals are known for a given t, it is very simple to recover the quantity $p(\boldsymbol{y} \mid \theta)$ by summing up over the values of z_t. So, it is possible to have more with a very small effort by using the forward-backward algorithm.

5.4.3. *Presentation of the general approach*

The approach developed in this section for different issues in inference of a HMM can be transported to address the same issues in different Markov-based models with hidden variables, such as multichain models or HSMMs. The approach is summarized here, step-by-step, and will be applied in following sections.

Inputs: we are given:

– a transition matrix A in the hidden layer;

– an emission matrix B;

– an initial state π_0 for the hidden layer;

– a sequence of observations $\boldsymbol{y}$.

Steps: the steps of the general approach are as follows:

1) To write the joint distribution $p(\boldsymbol{z}, \boldsymbol{y})$ as a factorization involving A, B, π_0. It has the form, in general:

$$p(\boldsymbol{z}, \boldsymbol{y}) = \left(\prod_{t=T}^{1} B[y_t, z_t] A[z_t, z_{t-1}] \right) B[y_0, z_0] \pi_0[z_0]$$

but it can be more complicated due to the structure of dependencies of observations.

2) To compute all matrices A_t, for $1 \leq t \leq T$. It has the form, in general:

$$A_t[z_t, z_{t-1}] = B[y_t, z_t]A[z_t, z_{t-1}],$$

but it can be more complicated, and the initial distribution $\pi'_0[z_0] = B[y_0, z_0]\pi_0[z_0]$.

3) To define an UHMD $\mathcal{A}$, as a tensor of order $T + 1$, with coefficients:

$$\mathcal{A}[z_0, \ldots, z_T] = A_T[z_T, z_{T-1}] \ldots A_1[z_1, z_0]\, \pi'_0[z_0].$$

Applications for some inferential issues: $\mathcal{A}$ being known from $(T, K, (A_t)_t, \pi'_0)$:

– The computation of the probabilities of the observations given the parameters of the model is done with the elimination algorithm on $\mathcal{A}$.

– The restoration of the hidden state (computation of the most likely state) knowing the observations and the parameters of the model can be done by the Viterbi algorithm, which can be done as an elimination algorithm on $\mathcal{A}$ known as the `max-prod` algorithm (see Peyrard et al. 2019, and section 5.11).

– The smoothing can be done by applying the marginalization algorithm on $\mathcal{A}$.

– The inference of the parameters knowing the observations is classically done with the EM algorithm. This algorithm is an iteration between an E step and a M step. The E step is the most complex, and includes a computation of binary marginals on the distribution defined by an UHMD $\mathcal{A}^{(m)}$ updated at each step. Its complexity can be evaluated by the complexity of the marginalization algorithm on $\mathcal{A}$.

5.4.4. *Sparsity of the transition matrix*

Let us recall that the sparsity of a matrix A is defined as $\Sigma(A) = 1 - \frac{\#A}{K^2}$. The terms in UHMD $\mathcal{A}$ are a product of terms of distinct matrices A_t, and the complexity of the elimination algorithm is in $\mathcal{O}(\sum_{t=1}^{T} \#A_t)$. If we compute it naively, one must evaluate $\#A_t$ at each time step t with $1 \leq t \leq T$. Here, we show that the sparsity of each matrix A_t is at least equal to the sparsity of A.

PROPOSITION 5.9.– *If A, the transition matrix between hidden variables, is sparse, then the sparsity of any matrix A_t is bounded below by the sparsity of A ($\forall t, \Sigma(A_t) \geq \Sigma(A)$), and the complexity of the elimination and marginalization algorithm for $\mathcal{A}$ is in $\mathcal{O}(T\#A)$.*

PROOF.– Indeed, the terms of matrix A_t can be written as (see equation [5.14]):

$$A_t[z_t, z_{t-1}] = B[y_t, z_t]A[z_t, z_{t-1}].$$

It can be seen immediately that, if $A[z_t, z_{t-1}] = 0$, then $A_t[z_t, z_{t-1}] = 0$. And the sparsity of A_t is the same as that of matrix A. It can even be larger if $B[y_t, z_t] = 0$. □

It is possible to go one step further and show that the sparsity of A controls the complexity of the EM algorithm. However, when running the EM algorithm, A is not known: it is inferred by the EM algorithm as an output. So, the sparsity of A means here the sparsity of the initial value of A or $A^{(0)}$. There exist situations where some transitions $z' \to z$ are known to be impossible. Then, the coefficient $A^{(0)}[z, z']$ can be set to 0. We show here that, in such a situation, it will be null for any iteration (m), or $\forall\, m > 0,\ A^{(m)}[z, z'] = 0$.

PROOF.– We follow the calculation of $A^{(m+1)}$ knowing $A^{(m)}$ given in Chapter 1 (Appendix 1.10) to prove that if $A^{(m)}[z_t, z_{t-1}] = 0$, then $A^{(m+1)}[z_t, z_{t-1}] = 0$. Let us note that the α_t and β_t defined in this chapter at each iteration m are the same as those defined in equation [1.20] in this appendix. At each step m, there is a transition matrix $A^{(m)}$ for the MM of the hidden variables. What we denote here $A^{(m)}[z_t, z_{t-1}]$ is denoted as $a^m_{z_{t-1}z_t}$ in this appendix, as in equation [1.21]. $A^{(m+1)}$ is computed in M step with $A^{(m+1)}[z, z']$ being proportional to the sum $\sum_{t=1}^{T} \gamma_t^{(m)}[z, z']$. Each $\gamma_t^{(m)}[z, z']$ is proportional to $A^{(m)}[z, z']$ (see equation [1.24] in the appendix). Then, if $A^{(m)}[z, z'] = 0$, $\gamma_t^{(m)}[z, z'] = 0$, and $A^{(m+1)}[z, z'] = 0$. It is then easy to show by induction that $A^{(m)}[z, z'] = 0$ for any m if $A^{(0)}[z, z'] = 0$. □

Then, we have the following proposition:

PROPOSITION 5.10.– *The complexity of EM algorithm with initial value of A being* $A^{(0)}$ *is bounded by* $\mathcal{O}(T\#A^{(0)})$.

PROOF.– The sparsity of matrix $A^{(0)}$ is propagated to all matrices $A^{(m)}$, and consequently to all matrices $A_t^{(m)}$ in E step of EM algorithm. The complexity of the marginalization algorithm at each iteration m is in $\mathcal{O}(T\#A^{(0)})$. □

This will prove to be useful for MHMM, where the transition matrix A can be sparse. Let us mention that, if it is not sparse, the exponential growth of the number of coefficients to estimate will be discouraging. Only sparse large matrix lead to efficient inference algorithms.

5.5. Multichain hidden Markov models

In this section, we show how the general approach presented in section 5.4.3 can lead to defining the forward and forward-backward algorithm on MHMMs, and evaluating their complexities.

NOTATIONS.– We recall and complete the notations for an MHMM. The individual chains are identified by their label $c \in \{1, \ldots, C\} \subset \mathbb{N}$. The state space of hidden variables for chain c is denoted as $\mathcal{K}_c$. The (multivariate) hidden state at time t is denoted as z_t, with:

$$z_t = (z_t^1, \ldots, z_t^c, \ldots, z_t^C), \qquad z_t^c \in \mathcal{K}_c.$$

For the sake of simplicity of notations, we assume that without loss of generality the spaces $\mathcal{K}_c$ are the same for all chains: $\forall\, c,\ \mathcal{K}_c := \mathcal{K}$. Then, $z_t \in \mathcal{K}^{\mathrm{C}}$. The series of hidden dates (all chains, all times) is denoted as $z_{0:\mathrm{T}}$. There are O observations, each in the same space $\mathcal{S}$ for the sake of simplicity with $|\mathcal{S}| = S$. The same notations can be adopted for observations: an observation y_t at time t is a tuple in $\mathcal{S}^{\mathrm{O}}$. The series of observations for all chains and all times is denoted as $y_{0:\mathrm{T}}$.

5.5.1. *General MHMM*

The most general case is called *general MHMM* and is defined in Chapter 3, section 3.2.1. The definition is recalled here.

DEFINITION 5.8.– *General MHMM.*

– *The hidden layer consists of $C-$dimensional variables: $Z_t \in \mathcal{K}^{\mathrm{C}}$.*

– *The observations consist of $O-$dimensional variables: $Y_t \in \mathcal{S}^o$.*

– *$(Z_t, Y_t)_t$ is Markovian.*

$(z_t, y_t) \in \Omega = \mathcal{K}^{\mathrm{C}} \times \mathcal{S}^{\mathrm{O}}$, with $|\Omega| = K^{\mathrm{C}} S^{\mathrm{O}}$ denoted as N. The transition matrix denoted as $A_{\mathrm{G}} \in \mathbb{R}^{\mathrm{N}\times\mathrm{N}}$ (G as subscript for *general*) is defined by:

$$A_{\mathrm{G}}[(z, y), (z', y')] = \mathbb{P}(Z_t = z, Y_t = y \mid Z_{t-1} = z', Y_{t-1} = y'). \qquad [5.17]$$

It has multiplexed indices, N rows and N columns, and it is also denoted as $p(z, y \mid z', y')$.

Computing the probability of the observations

It is straightforward to apply the general procedure presented in section 5.4.3 to general MHMMs.

The joint law can be written as:

$$p(z_{0:\mathrm{T}}, y_{0:\mathrm{T}}) =$$
$$A_{\mathrm{G}}[(z_{\mathrm{T}}, y_{\mathrm{T}}), (z_{\mathrm{T}-1}, y_{\mathrm{T}-1})] \ \ldots \ A_{\mathrm{G}}[(z_1, y_1), (z_0, y_0)]\, \pi_0^{\mathrm{G}}[z_0, y_0] \qquad [5.18]$$

where $\pi_0^{\mathrm{G}}[z_0, y_0]$ is the probability of the initial state of (z_0, y_0).

If the observations are fixed, it is possible to define at each time t with $1 \le t \le T$ a $K^{\mathrm{C}} \times K^{\mathrm{C}}$ matrix A_t^{G} with multiplexed indices as:

$$A_t^{\mathrm{G}}[z_t\,,\, z_{t-1}] = A_{\mathrm{G}}[(z_t, y_t), (z_{t-1}, y_{t-1})],$$

and a vector $\pi_0' \in \mathbb{R}^{\mathrm{K^C}}$ with multiplexed indices:

$$\pi_0'[z_0] = \pi_0[z_0, y_0].$$

Then, from [5.18], $y_{0:\mathrm{T}}$ being fixed:

$$p(z_{0:\mathrm{T}}, y_{0:\mathrm{T}}) = A_{\mathrm{T}}^{\mathrm{G}}[z_{\mathrm{T}}, z_{\mathrm{T}-1}] \;\ldots\; A_1^{\mathrm{G}}[z_1, z_0]\, \pi_0'[z_0]. \qquad [5.19]$$

This allows us to define an MUHMD with multi-indices $z_{0:\mathrm{T}}$ and coefficients $\mathcal{A}_{\mathrm{G}}[z_{0:\mathrm{T}}]$ defined by:

$$\mathcal{A}_{\mathrm{G}}[z_{0:\mathrm{T}}] = A_{\mathrm{T}}^{\mathrm{G}}[z_{\mathrm{T}}, z_{\mathrm{T}-1}] \;\ldots\; A_1^{\mathrm{G}}[z_1, z_0]\, \pi_0'[z_0].$$

Then, $p(y_{0:\mathrm{T}})$ is the normalizing constant of $\mathcal{A}_{\mathrm{G}}$ and can be computed by Algorithm 5.1. Here is the forward algorithm, adapted for most general cases of multichains HMM:

Algorithm 5.5 Computation of $p(y_{0:\mathrm{T}})$ for a general MHMM with algorithm ELIM()

1: **input** $K, C, S, O, T, N = K^{\mathrm{C}} S^{\mathrm{O}}$
2: **input** $y_{0:\mathrm{T}} = (y_0, \ldots, y_{\mathrm{T}})$
3: **input** $A \in \mathbb{R}^{\mathrm{N\times N}}$ with elements $A[(z, y), (z', y')]$ for $z, z' \in \mathcal{K}^{\mathrm{C}}$ and $y, y' \in \mathcal{S}^o$
4: **input** π_0 with elements $\pi_0[(z, y_{\mathrm{C}})]$ for $z_0 \in K^{\mathrm{C}}$
5: *# preparation*
6: **for** $t \in \{1, \ldots, T\}$ **do**
7:     **compute** $A_t \in \mathbb{R}^{\mathrm{K^C \times K^C}}$ s.t. $A_t[z_t, z_{t-1}] = A[(z_t, y_t), (z_{t-1}, y_{t-1})]$
8: **end for**
9: **compute** $\pi_0' \in \mathbb{R}^{\mathrm{K^C}}$ with $\pi_0'[z] = \pi_0[(z, y_0)]$ for each $z \in \mathbb{R}^{\mathrm{K^C}}$
10: *# elimination*
11: **compute** $W = $ ELIM$(T, K^{\mathrm{C}}, (A_t)_t, \pi_0')$
12: **return** W

PROPOSITION 5.11.– *The complexity of the forward and forward-backward algorithms for a general MHMM is in* $\mathcal{O}(TK^{2\mathrm{C}})$.

PROOF.– This is a direct consequence of the use of the elimination algorithm in line 11 of Algorithm 5.5. There are no multiplications in the preparation phase. □

Similarly, the unary marginals $m(t, z) = \mathbb{P}(Z_t = z \mid Y_{0:\mathrm{T}} = y_{0:\mathrm{T}})$, whose calculation is called "smoothing", can be computed using Algorithm 5.3, yielding an upper bound in $\mathcal{O}(T\,K^{2\mathrm{C}})$. The estimation of the parameters knowing the observations can be done with the EM algorithm, whose complexity is in step E which is a marginalization. Its complexity is in $\mathcal{O}(TK^{2\mathrm{C}})$ as well.

5.5.2. *A hierarchy of models*

Particular cases of the general MHMM have been defined in Chapter 3 (section 3.2) corresponding to different independence assumptions. The forward-backward algorithm and its complexity for these particular cases are studied here using the UHMD approach. Each term of the general matrix A_{G} of a general MHMM can be decomposed as the product of the term of an emission matrix denoted as B and the term of a transition matrix denoted as A:

$$A_{\mathrm{G}}[(\boldsymbol{z},\boldsymbol{y}),(\boldsymbol{z}',\boldsymbol{y}')] = B[\boldsymbol{y},(\boldsymbol{z},\boldsymbol{z}',\boldsymbol{y}')]\,A[\boldsymbol{z},(\boldsymbol{z}',\boldsymbol{y}')]. \qquad [5.20]$$

Conditional independence: MHMM-CI

A factorization assumption of conditional independence of $\boldsymbol{z}_t$ (respectively, $\boldsymbol{y}_t$) relatively to $\boldsymbol{z}_{t-1}$ (respectively, $\boldsymbol{y}_{t-1}$) has been introduced in Chapter 3 (Section 3.2.2), such that each of the matrix A and B is written as:

$$\begin{cases} B[\boldsymbol{y}_t,(\boldsymbol{z}_t,\boldsymbol{z}_{t-1},\boldsymbol{y}_{t-1})] &= \prod_{o=1}^{\mathrm{O}} B^o[y_t^o,(\boldsymbol{z}_t,\boldsymbol{z}_{t-1},\boldsymbol{y}_{t-1})], \\ A[\boldsymbol{z}_t,(\boldsymbol{z}_{t-1},\boldsymbol{y}_{t-1})] &= \prod_{c=1}^{\mathrm{C}} A^c[z_t^c,(\boldsymbol{z}_{t-1},\boldsymbol{y}_{t-1})]. \end{cases} \qquad [5.21]$$

These are equations [3.4] and [3.3] of Chapter 3. This is called an MHMM-CI. It is possible to define an MUHMD $\mathcal{A}^{\mathrm{CI}}$ with coefficients indexed by $\boldsymbol{z}_{0:\mathrm{T}}$ as:

$$\mathcal{A}^{\mathrm{CI}}[\boldsymbol{z}_{0:\mathrm{T}}] = A_{\mathrm{T}}^{\mathrm{CI}}[\boldsymbol{z}_{\mathrm{T}},\boldsymbol{z}_{\mathrm{T}-1}]\ldots A_1^{\mathrm{CI}}[\boldsymbol{z}_1,\boldsymbol{z}_0]\,\pi_0'[\boldsymbol{z}_0],$$

with

$$\begin{aligned} A_t^{\mathrm{CI}}[\boldsymbol{z}_t,\boldsymbol{z}_{t-1}] &= \left(\prod_{o=1}^{\mathrm{O}} B^o[y_t^o,(\boldsymbol{z}_t,\boldsymbol{z}_{t-1},\boldsymbol{y}_{t-1})]\right)\left(\prod_{c=1}^{\mathrm{C}} A^c[z_t^c,(\boldsymbol{z}_{t-1},\boldsymbol{y}_{t-1})]\right) \\ &= \left(\prod_{o=1}^{\mathrm{O}} B_t^o[\boldsymbol{z}_t,\boldsymbol{z}_{t-1}]\right)\left(\prod_{c=1}^{\mathrm{C}} A_t^c[z_t^c,\boldsymbol{z}_{t-1}]\right), \end{aligned} \qquad [5.22]$$

with

$$\begin{cases} A_t^c[z_t^c,\boldsymbol{z}_{t-1}] &= A^c[z_t^c,(\boldsymbol{z}_{t-1},\boldsymbol{y}_{t-1})] \\ B_t^o[\boldsymbol{z}_t,\boldsymbol{z}_{t-1}] &= B^o[y_t^o,(\boldsymbol{z}_t,\boldsymbol{z}_{t-1},\boldsymbol{y}_{t-1})]. \end{cases}$$

Then, we have the following proposition:

PROPOSITION 5.12.– *The complexity of the MHMM-CI model is in* $\mathcal{O}(T(C+O)K^{2\text{C}})$.

PROOF.– We can apply the general approach presented in section 5.4.3 and apply the elimination or marginalization algorithm on $\mathcal{A}^{\text{CI}}$. As $\#A^{\text{CI}} = K^{2\text{C}}$, the complexity of the elimination algorithms is in $\mathcal{O}(TK^{2\text{C}})$. To get the complexity of the forward or forward-backward algorithm of the MHMM-CI, one must add the complexity of the preparation phase, which is implementing the calculation of equation [5.22]. This costs $C+O$ products for each term $A_t^{\text{CI}}[z_t, z_{t-1}]$, and there are $TK^{2\text{C}}$ of them. This leads to a complexity in $\mathcal{O}(T(C+O)K^{2\text{C}})$. □

To the best of our knowledge, there is no known way to use the factorization [5.22] of A_t^{CI} to decrease the complexity of the elimination algorithm on the UHMD in full generality for an MHMM-CI. However, we will see below that, for some types of couplings between chains defined in Chapter 3 (section 3.3), the matrix A_t^{CI} is a Kronecker product, a situation which significantly decreases the complexity of elimination and marginalization algorithms, as it means that hidden chains are independent conditionally to the observations.

5.5.3. *Correspondence between hidden and observed variables: 1to1-MHMM-CI*

Let us consider further assumptions leading to the 1to1-MHMM-CI, presented in Chapter 3 (section 3.2.3):

– $C = O$.

– Each observation is attached through an emission at time t to one hidden variable at time t and one only (a one to one correspondence).

Then, we can label the observation o with the same label c as the one of the hidden chain to which it is attached. Equation [5.22] can be written as, $y_{0:\text{T}}$ being fixed:

$$A_t^{\text{CI}}[z_t, z_{t-1}] = \prod_{c=1}^{\text{C}} B_t^c[z_t, z_{t-1}] A_t^c[z_t^c, z_{t-1}]. \qquad [5.23]$$

This allows us to derive the elimination algorithm, and evaluate its complexity, for the MUHMD with coefficients indexed by $z_{0:\text{T}}$ as follows:

Algorithm 5.6 Elimination or forward algorithm for a 1to1-MHMM-CI

1: **input** T, K, C
2: **input** A_t^c, B_t^c for $1 \leq t \leq T$ and $1 \leq c \leq C$, with $A_t^c, B_t^c \in \mathbb{R}^{K \times K^C}$,
3: **input** $\pi_0 \in \mathbb{R}^{K^C}$, $\quad \pi_0[z] = \mathbb{P}(Z_0 = z)$
4: *# preparation*
5: **for** $t \in \{1, \ldots, T\}$ **do**
6: **compute** $A_t^{CI} \in \mathbb{R}^{K^C \times K^C}$
7: with $A_t^{CI}[z_t, z_{t-1}] = \prod_{c=1}^{C} B_t^c[z_t, z_{t-1}] A_t^c[z_t^c, z_{t-1}]$
8: **end for**
9: *# elimination*
10: **compute** $W = \text{ELIM}(T, K^C, (A_t^{CI})_t, \pi_0)$
11: **return** W

PROPOSITION 5.13.– *The complexity of the forward and forward-backward algorithm for a 1to1-MHMM-CI is in* $\mathcal{O}(TCK^{2C})$.

PROOF.– The complexity of the elimination step is in $\mathcal{O}(TK^C)$ by direct application of the evaluation of the complexity of the elimination algorithm (see Algorithm 5.2).

As there are T time steps in the preparation phase (loop on line 5 of Algorithm 5.6), with $2C - 1$ multiplications for each of the K^{2C} terms in $A_t^{CI}[z_t, z_{t-1}]$ at each time step (line 7), this leads to a cost of $T(2C - 1)K^{2C}$ multiplications.

Altogether, preparation and elimination phases lead to a complexity in $\mathcal{O}(TCK^{2C})$. Let us note that the complexity of the preparation phase is higher than the complexity of the elimination phase.

As the forward-backward algorithm is simply a double elimination algorithm, one forward and one backward, the complexity of the forward-backward algorithm is in $\mathcal{O}(TCK^{2C})$. This allows us to evaluate the complexity of the smoothing and E step in EM algorithm. □

Then, the complexity of the forward algorithm (and the forward-backward algorithm) for the 1to1-MHMM-CI is the same as for the general MHMM up to the extra cost of the preparation phase. However, some situations have been identified in Chapter 3 (section 3.3.1) where the complexity of the EM algorithm is decreased to $\mathcal{O}(TCK^2)$. We will show that these situations correspond for the forward or forward-backward algorithm to the possibility of writing matrix A_t^{CI}, of dimension $K^C \times K^C$, as a Kronecker product of C matrices of dimension $K \times K$.

Four different geometries of coupling

These geometries correspond to situation (a), (b), (c) and (d) in Chapter 3 (section 3.3). We have (see equation [5.23]):

$$A_t^{\mathrm{CI}}[z_t, z_{t-1}] = \prod_{c=1}^{\mathrm{C}} B_t^c[z_t, z_{t-1}] A_t^c[z_t^c, z_{t-1}].$$

Let us consider the graphical representation of dependencies of the MHMM. Let us assume now that at any time t and for any chain c, the arcs emitted from the hidden variable z_{t-1}^c are toward variables in the same chain only, that is, toward z_t^c, y_{t-1}^c or y_t^c only: there are no arcs emitted toward $z_t^{c'}$, $y_t^{c'}$ or $y_{t-1}^{c'}$ with $c' \neq c$. Then, we can write (where $\rightarrow$ means: is simplified to):

$$\begin{cases} B_t^c[z_t, z_{t-1}] \rightarrow B_t^c[z_t^c, z_{t-1}^c] \\ A_t^c[z_t^c, z_{t-1}] \rightarrow A_t^c[z_t^c, z_{t-1}^c] \end{cases}$$

For B_t^c, let us recall that:

$$B_t^c[z_t, z_{t-1}] = \mathbb{P}(Y_t^c = y_t^c \mid \boldsymbol{Z}_t = z_t, \boldsymbol{Z}_{t-1} = z_{t-1}, \boldsymbol{Y}_{t-1} = \boldsymbol{y}_{t-1}).$$

All arcs arriving at Y_t^c and emitted from a hidden variable $Z_t^{c'}$ or $Z_{t-1}^{c'}$ are emitted from the same chain, and $c' = c$. So $B_t^c[z_t, z_{t-1}] \rightarrow B_t^c[z_t^c, z_{t-1}^c]$. The same reasoning applies to A_t^c for arcs arriving at z_t^c, and $A_t^c[z_t^c, z_{t-1}] \rightarrow A_t^c[z_t^c, z_{t-1}^c]$. We then have (where indices of A_t^{CI} are multiplexed):

$$A_t^{\mathrm{CI}}[(z_t^1, \dots, z_t^{\mathrm{C}}), (z_{t-1}^1, \dots, z_{t-1}^{\mathrm{C}})] = \prod_{c=1}^{\mathrm{C}} B_t^c[z_t^c, z_{t-1}^c] A_t^c[z_t^c, z_{t-1}^c],$$

which can be written as:

$$A_t^{\mathrm{CI}}[(z_t^1, \dots, z_t^{\mathrm{C}}), (z_{t-1}^1, \dots, z_{t-1}^{\mathrm{C}})] = \prod_{c=1}^{\mathrm{C}} (B_t^c \odot A_t^c)[z_t^c, z_{t-1}^c],$$

where $\odot$ is the Hadamard product. From this, we see that chains $\boldsymbol{z}^c$ are independent as:

$$A_t^{\mathrm{CI}} = \bigotimes_{c=1}^{\mathrm{C}} (B_t^c \odot A_t^c). \qquad [5.24]$$

The elimination or forward algorithm now is as follows:

Algorithm 5.7 Elimination or forward algorithm for a MHMM with independent chains

---

1: **input** T, K, C
2: **input** A_t^c, B_t^c for $1 \leq t \leq T$ and $1 \leq c \leq C$, with $A_t^c, B_t^c \in \mathbb{R}^{K \times K}$,
3: **input** $\pi_0^c \in \mathbb{R}^K$ for $1 \leq c \leq C$
4: *# loop on all chains*
5: **for** $c \in \{1, \ldots, C\}$ **do**
6:     *# preparation for chain c*
7:     **for** $t \in \{1, \ldots, T\}$ **do**
8:         **compute** $M_t^c = B_t^c \odot A_t^c$
9:     **end for**
10:     *# elimination for chain c*
11:     **compute** $W^c = \text{ELIM}(T, K, (M_t^c)_t, \pi_0^c)$
12: **end for**
13: **compute** $W = \prod_{c=1}^{C} W^c$
14: **return** W

PROPOSITION 5.14.– *The complexity of elimination algorithm for a 1to1-MHMM-CI with C independent chains is in $\mathcal{O}(TCK^2)$.*

PROOF.– Per chain, the elimination algorithm, line 11 of Algorithm 5.7, requires TK^2 products, and the preparation step, line 8, TK^2 products. So, $2TK^2$ products are required for one chain, for preparation and elimination steps. As there are C chains, the complexity of the whole algorithm is in $\mathcal{O}(TCK^2)$. □

This allows us to recover the results of Chapter 3. Indeed, Figure 3.6 of Chapter 3 presents the four cases (a)–(d), where it can be seen that in cases (a) and (b), all arcs starting from a chain $c = 1$ or $c = 2$ point to a variable in the same chain, whereas in case (c) it points to the observation of another chain, and in case (d) to a hidden variable of another chain. This implies that the complexity for cases (a) and (b) is in $\mathcal{O}(TCK^2)$ and for cases (c) and (d) the complexity is in $\mathcal{O}(TCK^{2C})$, as presented in Table 3.1 in Chapter 3 for EM.

5.6. Hidden semi-Markov models

HSMMs are a generalization of HMMs, in which the hidden layer is a semi-Markov model (SMM). All questions addressed by Rabiner (1989) and smoothing can be addressed for HSMM (see Chapter 1 where these inference issues are well developed, and references therein). We start from the observation that a SMM, which is defined by indexing jumps, can be written as an MM in calendar time with relevant auxiliary variables (for the generative model, see Chapter 4, section 4.2). It follows that a HSMM can be mapped on a HMM, and the general approach derived for a HMM in section 5.4.3 can be applied to a HSMM.

The definition of the general HSMM is given in Chapter 1, section 1.2.2. We recall here the main notations, which will be used in this section:

– The hidden layer is an MM $(J_n, X_{n+1})_n$ where $n \in \mathbb{N}^*$ indexes the jumps, J_n is the new state at jump n and X_{n+1} the sojourn duration in this state.

– The multiplexed transition matrix of (J_n, X_{n+1}) is denoted as A_J:

$$A_J[(j,\ell), (i,d)] = \mathbb{P}\{J_n = j, X_{n+1} = \ell \mid J_{n-1} = i, X_n = d\}. \quad [5.25]$$

– The emissions are modeled by an emission matrix B such that:

$$B[y, z] = \mathbb{P}(Y_t = y \mid Z_t = z).$$

– The initial hidden state is given by:

$$\pi_0[i, d] = \mathbb{P}(J_0 = i, X_1 = d).$$

5.6.1. *General SMM as an MM in calendar time*

Let us assume that $X_n \in \mathcal{D}^*$ with $\mathcal{D}^* = \{1, \dots, D\} \subset \mathbb{N}$, with D possibly infinite (in such a case, $\mathcal{D}^* = \mathbb{N}^*$), and that $J_n \in \mathcal{K}$ with $|\mathcal{K}| = K$. The state space of the chain $(J_n, X_{n+1})_n$ is $\mathcal{K} \times \mathcal{D}^*$. If $D = +\infty$, or $\mathcal{D}^* = \mathbb{N}^*$, we may need to consider summations involving an infinite number of terms, which could make proving convergence difficult. To avoid this situation, and keep the state space finite, we assume that $D < +\infty$ for the rest of this chapter. Then, a state is a pair $(i, d) \in \mathcal{K} \times \mathcal{D}^*$. We define as well $\mathcal{D} = \{0, \dots, T-1\} \subset \mathbb{N}$.

We map here a general SMM into an MM in calendar time, using the trick of auxiliary variables R_t and E_t, presented in Chapter 4 (see as well Barbu and Limnios 2008). We have:

$$R_t = \min_d \{d \in \mathbb{N} \mid Z_{t+d+1} \neq Z_t\}, \qquad E_t = \min_d \{d \in \mathbb{N}^* \mid Z_{t-d} \neq Z_t\}.$$

When the chain is in state $Z_t = z$ at time t, R_t is the *remaining time*, that is, the time remaining until the next jump (counted as 0 if the jump occurs at $t+1$), and E_t is the *elapsed time*, that is, the time elapsed since the previous jump, counted as 1 at time t.

The following proposition has been shown in Chapter 4, section 4.2: if $(Z_t)_t$ follows a general SMM, then $(Z_t, R_t, E_t)_t$ follows a MM. We denote A_{ZRE} the multiplexed transition matrix of this MM in calendar time, with:

$$\begin{aligned}&A_{\text{ZRE}}[(z_t, r_t, e_t), (z_{t-1}, r_{t-1}, e_{t-1})]\\&= \mathbb{P}(Z_t = z_t, R_t = r_t, E_t = e_t \mid Z_{t-1} = z_{t-1}, R_{t-1} = r_{t-1}, E_{t-1} = e_{t-1}).\end{aligned}$$

NOTATIONS.– We denote $\mathcal{C} = \mathcal{K} \times \mathcal{D} \times \mathcal{D}^*$. We have $(z_t, r_t, e_t) \in \mathcal{C}$, and $|\mathcal{C}| = KD^2$. Then, the dimensions of A_{ZRE} are $KD^2 \times KD^2$. We denote $\boldsymbol{r} = (r_0, \ldots, r_{\text{T}})$ and $\boldsymbol{e} = (e_0, \ldots, e_{\text{T}})$. We denote $\Omega_{\text{ZRE}} = \mathcal{C}^{\text{T}+1}$. Then $(\boldsymbol{z}, \boldsymbol{r}, \boldsymbol{e}) \in \Omega_{\text{ZRE}}$.

We show that matrix A_{ZRE} is sparse, which will be useful for estimating the complexity of elimination and marginalization algorithms.

PROPOSITION 5.15.– *The matrix A_{ZRE} is sparse, and $\#A_{\text{ZRE}} = K^2D^2$, instead of K^2D^4 for a full matrix of the same dimension.*

PROOF.– The sparsity comes from the impossibility of changing the state J_n between two jumps. A jump n occurs at time t if $r_{t-1} = 0$ and $e_t = 1$. In such a case, we set the correspondence $z_{t-1} = \jmath_{n-1} = i$, $z_t = \jmath_n = j$, $e_{t-1} = x_n = d$ and $r_t = x_{n+1} - 1 = \ell - 1$ between indexing with calendar time and with jumps. Then, the jump n at time t modeled by terms of matrix A_{J} is from $(\jmath_{n-1} = z_{t-1}, x_n = e_{t-1})$ to $(\jmath_n = z_t, x_{n+1} = r_t + 1)$, and:

$$A_{\text{ZRE}}[(z_t, r_t, 1), (z_{t-1}, 0, e_{t-1})] = A_{\text{J}}[(z_t, r_t + 1), (z_{t-1}, e_{t-1})].$$

If t does not correspond to a jump ($r_{t-1} > 0$), the only possible transition is from $(z_{t-1}, r_{t-1}, e_{t-1})$ to $(z_{t-1}, r_{t-1} - 1, e_{t-1} + 1)$: the state Z does not change, and the counters R and E are updated with $r_t = r_{t-1} - 1$ and $e_t = e_{t-1} + 1$. Its probability is one. All other transitions have a null probability.

This fully specifies matrix A_{ZRE}. The terms not in $\{0, 1\}$ are indexed by $(z, r, 1)$ for the rows and $(z', 0, e)$ for the columns, with $z, z' \in \mathcal{K}$, $r \in \mathcal{D}$ and $e \in \mathcal{D}^*$. There are $KD \times KD = K^2D^2$ such terms (ignoring that $z \neq z'$). □

5.6.2. *General HSMM in calendar time*

Rewriting the MM $(J_n, X_{n+1})_n$ indexed by jumps as an MM $(Z_t, R_t, E_t)_t$ in calendar time allows us to map a general HSMM into a general HMM with the following parameters:

– The transition matrix A_{ZRE} of the hidden layer is defined by:

$$A_{\text{ZRE}}[(z_t, r_t, e_t), (z_{t-1}, r_{t-1}, e_{t-1})] = \\ \mathbb{P}(Z_t = z_t, R_t = r_t, E_t = e_t \mid Z_{t-1} = z_{t-1}, R_{t-1} = r_{t-1}, E_{t-1} = e_{t-1}). \quad [5.26]$$

– The initial state of the hidden layer is:

$$\pi_0[z_0, r_0, e_0] = \mathbb{P}(Z_0 = z_0, R_0 = r_0, E_0 = e_0). \quad [5.27]$$

– The emission matrix is:

$$B[y, z] = \mathbb{P}(Y_t = y \mid Z_t = z). \quad [5.28]$$

The joint law between hidden and observed layers $\mathbb{P}(\boldsymbol{Z}, \boldsymbol{R}, \boldsymbol{E}, \boldsymbol{Y})$ is given by:

$$\begin{aligned}&\mathbb{P}(\boldsymbol{Z}=\boldsymbol{z}, \boldsymbol{R}=\boldsymbol{r}, \boldsymbol{E}=\boldsymbol{e}, \boldsymbol{Y}=\boldsymbol{y})\\&= B(y_{\mathrm{T}}, z_{\mathrm{T}})\, A_{\mathrm{ZRE}}[(z_{\mathrm{T}}, r_{\mathrm{T}}, e_{\mathrm{T}}), (z_{\mathrm{T}-1}, r_{\mathrm{T}-1}, e_{\mathrm{T}-1})] \dots\\&\quad \dots B(y_1, z_1)\, A_{\mathrm{ZRE}}[(z_1, r_1, e_1), (z_0, r_0, e_0)]\, B[y_0, z_0]\, \pi_0[z_0, r_0, 0]. \quad [5.29]\end{aligned}$$

To simplify the writing, we will denote, for example:

$$p(\boldsymbol{z}, \boldsymbol{r}, \boldsymbol{e}, \boldsymbol{y}) := \mathbb{P}(\boldsymbol{Z}=\boldsymbol{z}, \boldsymbol{R}=\boldsymbol{r}, \boldsymbol{E}=\boldsymbol{e}, \boldsymbol{Y}=\boldsymbol{y}),$$

and the same for other related distributions. We are now in the framework of HMM, and we can develop the approach presented in section 5.4.3. We present it based on the example of elimination algorithm for computing the probability of the observations.

5.6.3. *Computing the probability of the observations*

We wish to compute the probability of the observations $\boldsymbol{y}$ knowing the parameters of the model $(A_{\mathrm{ZRE}}, B, \pi_0)$, that is, to compute:

$$\sum_{\boldsymbol{z}\in\Omega_{\mathrm{Z}}} p(\boldsymbol{z}, \boldsymbol{y}) = \sum_{(\boldsymbol{z},\boldsymbol{r},\boldsymbol{e})\in\Omega_{\mathrm{ZRE}}} p(\boldsymbol{z}, \boldsymbol{r}, \boldsymbol{e}, \boldsymbol{y}). \quad [5.30]$$

Let us define for any t, $\boldsymbol{y}$ being fixed:

$$\begin{aligned}&A_t^{\mathrm{ZRE}}[(z_t, r_t, e_t), (z_{t-1}, r_{t-1}, e_{t-1})]\\&= B[y_t, z_t]\, A_{\mathrm{ZRE}}[(z_t, r_t, e_t), (z_{t-1}, r_{t-1}, e_{t-1})]. \quad [5.31]\end{aligned}$$

Then

$$\begin{aligned}p(\boldsymbol{z}, \boldsymbol{r}, \boldsymbol{e}, \boldsymbol{y}) &= A_{\mathrm{T}}^{\mathrm{ZRE}}[(z_{\mathrm{T}}, r_{\mathrm{T}}, e_{\mathrm{T}}), (z_{\mathrm{T}-1}, r_{\mathrm{T}-1}, e_{\mathrm{T}-1})] \dots\\&\dots A_1^{\mathrm{ZRE}}[(z_1, r_1, e_1), (z_0, r_0, e_0)]\, \pi_0'[z_0, r_0, e_0], \quad [5.32]\end{aligned}$$

with

$$\pi_0'[z_0, r_0, e_0] = B[y_0, z_0]\, \pi_0[z_0, r_0, e_0].$$

We build a MUHMD $\mathcal{A}_{\mathrm{ZRE}}$ with multiplexed indices, the coefficients of which $\mathcal{A}_{\mathrm{ZRE}}[(\boldsymbol{z}, \boldsymbol{r}, \boldsymbol{e})]$ are given by the RHS of [5.32]. This leads to the following proposition:

PROPOSITION 5.16.– *The complexity of the forward and the forward-backward algorithm for a general HSMM with T time steps, sojourn duration bounded by D and state space of hidden variables with cardinality K is in $\mathcal{O}(T(KD)^2)$.*

PROOF.– The forward algorithm on a general HSMM, with parameters $(A_{\text{ZRE}}, B, \pi_0)$ for the corresponding HMM in calendar time, is the elimination algorithm (Algorithm 5.1) on $\mathcal{A}_{\text{ZRE}}$. As $\#A_{\text{ZRE}} = K^2D^2 = (KD)^2$, the complexity of the forward algorithm of a HSMM is in $\mathcal{O}(T(KD)^2)$. The same for the forward-backward algorithm, which is a double forward algorithm. □

REMARK 5.3.– We will see next that different types of models may lead to different evaluations of the complexity of the forward, forward-backward and EM algorithms. We notice that, for the general HSMM, $\#A_{\text{ZRE}} = (KD)^2$ is the number of terms in $A_\jmath$ which is a $KD \times KD$ matrix, that is, the term in the complexity of forward and forward-backward algorithms for Markov chain $(J_n, X_{n+1})_n$. This leads to the more general hypothesis that, for all models derived from the general HSMM, the complexity of elimination and marginalization algorithms in calendar time (Markov chain $(Z_t, R_t, E_t)_t$) for a HSMM is the same as for the SMM $(J_n, X_{n+1})_n$.

5.6.4. *Particular cases of HSMM*

Particular cases of HSMM are presented in Chapter 1 (section 1.2.2), namely, *standard HSMM* and *ED-HMM* (see Chapter 1, section 1.2.4). We show here how the evaluation of the complexity of the forward and forward-backward algorithm as complexity of the elimination and marginalization algorithm on an associated UHMD can be adapted to each situation. Therefore, we start from the proposition 1 (Chapter 4) briefly recalled here:

– for the General HSMM, $(Z_t, R_t, E_t)_t$ is Markovian;

– for the standard HSMM, $(Z_t, E_t)_t$ is Markovian;

– for the ED-HMM, both $(Z_t, R_t)_t$ and (Z_t, E_t) are Markovian.

This suggests that the complexity of the forward and forward-backward algorithms for standard model and ED-HMM are lower than for the general model. To show it, one can write the respective transition matrix $(A_{\text{ZRE}}, A_{\text{ZE}}, A_{\text{ZR}})$ of each MM in calendar time of the hidden variables, evaluate its sparsity and conclude with the UHMD approach. Let us present this approach for the ED-HMM.

5.6.5. *Explicit duration hidden Markov model*

The transition matrix of an ED-HMM as a Markov model indexed by jumps is expressed as follows:

$$A_\jmath[(j,\ell)\,;\,(i,d)] = \mathbb{P}(X_{n+1} = \ell \mid J_n = j) \times \mathbb{P}(J_n = j \mid J_{n-1} = i). \qquad [5.33]$$

Neither the new state J_n at a jump nor its sojourn duration X_{n+1} depend on the sojourn time X_n in state J_{n-1}. Knowing that, we denote (see Chapter 1):

$$h_j(\ell) = \mathbb{P}(X_{n+1} = \ell \mid J_n = j), \qquad p_{ji} = \mathbb{P}(J_n = j \mid J_{n-1} = i).$$

Then:

$$A_{\text{J}}[(j,\ell)\,;\,(i,d)] = h_j(\ell)\,p_{ji}$$

Proposition 1 in Chapter 4 shows that $(Z_t, R_t)_t$ is Markovian. Here, we build the transition matrix of this Markovian model.

PROPOSITION 5.17.– *The hidden layer of an ED-HMM, as an MM in $(J_n, X_{n+1})_n$, can be mapped into a calendar time MM with variables $(Z_t, R_t)_t$ only.*

PROOF.– Let us denote it A_{ZR}, that is:

$$A_{\text{ZR}}[(z_t, r_t), (z_{t-1}, r_{t-1})] = \mathbb{P}(Z_t = z_t, R_t = r_t \mid Z_{t-1} = z_{t-1}, R_{t-1} = r_{t-1}).$$

A jump occurs at time t when $r_{t-1} = 0$. We then have:

$$A_{\text{ZR}}[(z_t, r_t), (z_{t-1}, 0)] = h_{z_t}(r_t - 1)\,p_{z_t z_{t-1}}.$$

If $r_{t-1} > 0$, the only possible transition is from (z_{t-1}, r_{t-1}) to $(z_{t-1}, r_{t-1} - 1)$, that is, variable Z does not change and the counter R is decreased by one unit.

All other transitions have a null probability, and this specifies the matrix A_{ZR}. □

We develop now the same approach as for the general HSMM. Therefore, some details will be omitted. We define the $KD \times KD$ matrix A_t^{ZR} by:

$$A_t^{\text{ZR}}[(z_t, r_t), (z_{t-1}, r_{t-1})] = \mathcal{B}[y_t, z_t]A_{\text{ZR}}[(z_t, r_t), (z_{t-1}, r_{t-1})].$$

Then, $\boldsymbol{y}$ being fixed, the joint law $p(\boldsymbol{z}, \boldsymbol{r}, \boldsymbol{y})$ is defined as:

$$p(\boldsymbol{z}, \boldsymbol{r}, \boldsymbol{y}) =$$
$$A_{\text{T}}^{\text{ZR}}[(z_{\text{T}}, r_{\text{T}}), (z_{\text{T}-1}, r_{\text{T}-1})] \ldots A_1^{\text{ZR}}[(z_1, r_1), (z_0, r_0)]\, B[y_0, z_0]\pi_0[z_0]. \qquad [5.34]$$

We define an MUHMD $\mathcal{A}_{\text{ZR}}$ the coefficients of which $\mathcal{A}^{\text{ZR}}[(\boldsymbol{z}, \boldsymbol{r})]$ are multiplexed and given by the RHS of [5.34]. This leads to design an algorithm for the forward (respectively, the forward-backward) algorithm on an ED-HMM: these are the elimination (respectively, marginalization) algorithm on $\mathcal{A}_{\text{ZR}}$, that is, Algorithm 5.1 (respectively, Algorithm 5.3).

This leads to the following proposition:

PROPOSITION 5.18.– *The complexity of the forward and forward-backward algorithm for an ED-HMM with T time steps, D as maximum sojourn duration in any state and K as cardinality of the state space of the hidden layer, is in $\mathcal{O}(TK^2D)$.*

PROOF.– Following the general approach of section 5.4.3 leads us to evaluate the complexity of these algorithm as $\mathcal{O}(T\#A_{\text{ZR}})$. A_{ZR} is sparse. The only terms that are not equal to either 0 (an impossible transition) or 1 (a compulsory transition) are those related to a jump. These terms are $A_{\text{ZR}}[(z_t, r_t), (z_{t-1}, 0)]$. There are $K(K-1)D$ of them ($K(K-1)$ and not K^2 because $z_t \neq z_{t-1}$). However, we approximate $\#A_{\text{ZR}} \approx K^2D$. □

5.7. Multichain HSMM

A MHSMM is a set of interacting HSMMs. The family of MHSMMs includes different types of interactions between chains. To study this, we can map the semi-Markovian hidden layer into a Markovian chain with auxiliary variables R_t and E_t. A state at time t is:

$$(\boldsymbol{z}_t, \boldsymbol{r}_t, \boldsymbol{e}_t) = ((z_t^1, \ldots, z_t^C), (r_t^1, \ldots, r_t^C), (e_t^1, \ldots, r_e^C)).$$

As a consequence of the definition of R_t and E_t, a jump occurs at time t on chain c if, and only if, $r_{t-1}^c = 0$. If not, the state z_t^c remains unchanged. A consequence of this choice, which is not the only possibility, is that once a sojourn duration X_{n+1}^c has been selected for a chain c at jump n, it will not change along its course. In Chapter 4, other types of multichain SMM have been defined, with another approach, the instantaneous risk, where the sojourn duration in one state for one chain can be affected along the course of its realization by the state of other chains. In other words, here, for a chain c, the interaction with other chains will occur at a jump only, where the choice of the new state and its sojourn duration will depend on the states of the other chains as well. There is an interaction with other chains at a jump, and independence from other chains between jumps.

We denote by J-MHSMM (J for "at jumps") the family of MHSMM for which interactions occur at jumps only. This family is rich, because it combines the different cases of HSMM presented in Chapter 1 and the different cases of MHMM presented in Chapter 3. Here, for presenting an example of the application of the UHMD approach, we select the 1to1-J-MHMM-CI model of Chapter 3 with a hidden semi-Markov layer. We call it 1to1-J-MHSMM-CI.

5.7.1. *1to1-J-MHSMM-CI*

DEFINITION 5.9.– *1to1-J-MHSMM-CI.*

A 1to1-J-MHSMM-CI is composed of:

– a hidden layer composed of an MM with C chains, with states $(\boldsymbol{z}_t, \boldsymbol{r}_t, \boldsymbol{e}_t)_t$;

– an observations layer, composed of C chains as well $(\boldsymbol{y}_t)_t$;

– a distribution of the initial state $\pi_0[\boldsymbol{z}_0, \boldsymbol{r}_0, \boldsymbol{e}_0] = \mathbb{P}\{\boldsymbol{Z}_0 = \boldsymbol{z}_0, \boldsymbol{R}_0 = \boldsymbol{r}_0, \boldsymbol{E}_0 = \boldsymbol{e}_0\}$*;*

and the dynamics:

– $(\boldsymbol{Z}_t, \boldsymbol{R}_t, \boldsymbol{E}_t)_t$ *is an MM with a factorized transition matrix:*

$$A^{\mathrm{CI}}[(\boldsymbol{z}_t, \boldsymbol{r}_t, \boldsymbol{e}_t), (\boldsymbol{z}_{t-1}, \boldsymbol{r}_{t-1}, \boldsymbol{e}_{t-1})] = \prod_{c=1}^{C} A^c_{\mathrm{CI}}[(z^c_t, r^c_t, e^c_t), (\boldsymbol{z}_{t-1}, r^c_{t-1}, e^c_{t-1})] \quad [5.35]$$

because of conditional independence, detailed below;

– emission from the hidden variables Z_t *to observations* Y_t *are modeled by a set of matrices* $(B^c)_c$ *with:*

$$B[\boldsymbol{y}_t, (\boldsymbol{z}_t, \boldsymbol{z}_{t-1})] = \prod_{c=1}^{C} B^c[y^c_t, (\boldsymbol{z}_t, \boldsymbol{z}_{t-1})]$$

with

$$B^c[y^c_t, (\boldsymbol{z}_t, \boldsymbol{z}_{t-1})] = \mathbb{P}(Y^c_t = y^c_t \mid \boldsymbol{Z}^c_t = \boldsymbol{z}, \boldsymbol{Z}_{t-1} = \boldsymbol{z}_{t-1}).$$

Dynamics of the hidden Markov layer: at time t, for any chain c: (i) If the transition is not a jump, that is, if $r^c_{t-1} > 0$, then the state z^c_{t-1} does not change and (r^c_{t-1}, e^c_{t-1}) are updated as counters:

$$z^c_t = z^c_{t-1}, \qquad r^c_t = r^c_{t-1} - 1, \qquad e^c_t = e^c_{t-1} + 1.$$

(ii) If t corresponds to a jump, that is, $r^c_{t-1} = 0$, the new state z^c_t and the sojourn duration r^c_t in this new state are updated according to:

$$\begin{aligned} &A^c[(z^c_t, r^c_t, 1), (\boldsymbol{z}_{t-1}, 0, e^c_{t-1})] = \\ &\mathbb{P}\{Z^c_t = z^c_t, R^c_t = r^c_t, E^c_t = 1 \mid \boldsymbol{Z}_{t-1} = \boldsymbol{z}_{t-1}, R^c_{t-1} = 0, E^c_{t-1} = e^c_{t-1}\} \\ &\text{and} \quad z^c_t \neq z^c_{t-1}. \end{aligned}$$

They depend not only on (z^c_{t-1}, e^c_{t-1}), but on all states at time $t-1$ of all chains $c' \neq c$, that is, on $\boldsymbol{z}_{t-1}$. They do not depend however on sojourn durations $E^{c'}_{t-1}$ and $R^{c'}_{t-1}$ with $c' \neq c$ of other chains. To simplify the calculations in what follows, we make now the assumption that for any chain c, and any time t, the observation y^c_t depends on z^c_t only. We then have:

$$B^c(y^c_t, z^c_t) = \mathbb{P}(Y^c_t = y^c_t \mid Z^c_t = z^c_t).$$

General approach with an MUHMD: we have now all the elements to build an MUHMD, leading with the general procedure presented in section 5.4.3 to formalize the forward and forward-backward algorithm and to evaluate their complexities. The only difficulty is to chose clear notations. We define from equation [5.35] the matrices A_t^{CI} by:

$$A_t^{\mathrm{CI}}[(\boldsymbol{z}_t, \boldsymbol{r}_t, \boldsymbol{e}_t), (\boldsymbol{z}_{t-1}, \boldsymbol{r}_{t-1}, \boldsymbol{e}_{t-1})] =$$

$$\prod_{c=1}^{\mathrm{C}} B^c[y_t^c, z_t^c]\; A_{\mathrm{CI}}^c[(z_t^c, r_t^c, e_t^c), (\boldsymbol{z}_{t-1}, r_{t-1}^c, e_{t-1}^c)]. \qquad [5.36]$$

So, we define the MUHMD $\mathcal{A}^{\mathrm{CI}}$ with multiplexed indices $(\boldsymbol{z}_t, \boldsymbol{r}_t, \boldsymbol{e}_t)$ by, if $\boldsymbol{x}_t = (\boldsymbol{z}_t, \boldsymbol{r}_t, \boldsymbol{e}_t)$

$$\mathcal{A}^{\mathrm{CI}}[\boldsymbol{x}_{\mathrm{T}}, \ldots, \boldsymbol{x}_0] = A_{\mathrm{T}}^{\mathrm{CI}}[\boldsymbol{x}_{\mathrm{T}}, \boldsymbol{x}_{\mathrm{T}-1}] \ldots A_1^{\mathrm{CI}}[\boldsymbol{x}_1, \boldsymbol{x}_0]\, \pi_0[\boldsymbol{x}_0]. \qquad [5.37]$$

Then, we have the following proposition:

PROPOSITION 5.19.– *Let* $(\boldsymbol{Z}_t, \boldsymbol{R}_t, \boldsymbol{E}_t, \boldsymbol{Y}_t)_t$ *be a 1to1-J-MHSMM-CI. Let* $z \in \mathcal{K}$, $r \in \mathcal{D}$; $e \in \mathcal{D}^*$, *with* $|\mathcal{K}| = K$ *and* $|\mathcal{D}| = |\mathcal{D}^*| = D$. *Let us have* C *interacting chains and* T *time steps. Then, the complexity of the elimination or marginalization step in the forward and forward-backward algorithms is in* $\mathcal{O}(T(K^2D^2)^{\mathrm{C}})$.

PROOF.– The approach is similar to the one presented in Algorithm 5.6 and proposition 13 on MHMM with conditional independence. The algorithm to compute W with the forward algorithm can be organized in two parts: first a preparation phase to compute A^{CI} knowing the matrices A_{CI}^c according to equation [5.35], and second computing the normalization constant W with the elimination algorithm on the UHMD built from A^{CI}. This latter part has a complexity in $\mathcal{O}(T\#A^{\mathrm{CI}})$.

We focus on the evaluation of the complexity of the elimination algorithm, that is, the evaluation of $\#A^{\mathrm{CI}}$. A^{CI} is a $K^{\mathrm{C}}D^{2\mathrm{C}} \times K^{\mathrm{C}}D^{2\mathrm{C}}$ matrix, and so has $K^{2\mathrm{C}}D^{4\mathrm{C}}$ terms. However, it is sparse. The reason is twofold: (i) for any chain c, the choice between remaining in same state or having a jump does not depend on the other chains, but on r_{t-1}^c only, and (ii) the transition matrix for a monochain HSMM is sparse. Let us categorize the transitions $(z'^c, r'^c, e'^c) \longrightarrow (z^c, r^c, e^c)$ for chain c into three categories: jumps $(z^c \neq z'^c, r'^c = 0)$, stability $(z^c = z'^c, r'^c > 0)$, or impossible. Let us have a multiplexed row index $\boldsymbol{i} = (\boldsymbol{z}, \boldsymbol{r}, \boldsymbol{e})$ and a multiplexed column index $\boldsymbol{j} = (\boldsymbol{z}', \boldsymbol{r}', \boldsymbol{e}')$ of A^{CI}. We have to evaluate the number of pairs of multi-indices $[\boldsymbol{i}, \boldsymbol{j}]$ for which the coefficient $A^{\mathrm{CI}}[\boldsymbol{i}, \boldsymbol{j}]$ is non-zero. Therefore, because of [5.35], one must have $A_{\mathrm{CI}}^c[(z^c, r^c, e^c), (\boldsymbol{z}', r'^c, e'^c)] \neq 0$ for any c. This means that the transition for chain c modeled by A_{CI}^c must be either a jump, or a stability. Let us evaluate the number of distinct multi-indices $[(z^c, r^c, e^c), (\boldsymbol{z}', r'^c, e'^c)]$ with $z^c \in \mathcal{K}$; $\boldsymbol{z}' \in \mathcal{K}^{\mathrm{C}}$; $r^c, r'^c \in \mathcal{D}$ and $e^c, e'^c \in \mathcal{D}^*$ for which it is the case.

$\boldsymbol{z}'$ can take K^{c} values. Let us fix one value for $\boldsymbol{z}'$. A jump $(z'^c, r'^c, e'^c) \longrightarrow (z^c, r^c, e^c)$ is such that $z^c \neq z'^c$, $r'^c = 0$ and $e^c = 1$. The indices fulfilling these conditions in A^c_{CI} are $[(z^c \neq z'^c, r^c, 1), (\boldsymbol{z}', 0, e'^c)]$. $\boldsymbol{z}'$ being fixed, that is, z'^c as well, there are $a = (K-1)D^2$ of them ($K-1$ values for z^c, D for r^c and D for e'^c). Stability is modeled by $r'^c > 0$ and indices $[(z^c = z'^c, r^c, e^c), (\boldsymbol{z}', r'^c - 1, e'^c + 1)]$. There are $b = D^2$ of them, z^c being fixed through $\boldsymbol{z}'$. This is however an overestimation of the number of indices associated with stability, because one must have $r^c + e^c \leq D$ as well as $r'^c + e'^c \leq D$. This leads to an overestimation of $\#A^{\text{CI}}$. Let us note that $a = b(K-1)$. The set of multi-indices $[\boldsymbol{i}, \boldsymbol{j}]$ with $\boldsymbol{z}'$ fixed such that $A^{\text{CI}}[\boldsymbol{i}, \boldsymbol{j}]$ is non-zero is denoted $\mathcal{E}|_{\boldsymbol{z}'}$ and its cardinality $\#A^{\text{CI}}|_{\boldsymbol{z}'}$. As there are C chains, and as the occurrence of a jump or of stability is independent between chains, $\mathcal{E}|_{\boldsymbol{z}'}$ can be partitioned into non-overlapping sets of s multi-indices corresponding to stable chains and $C - s$ corresponding to chains with jumps, with $0 \leq s \leq C$. So, we have:

$$\begin{aligned}\#A^{\text{CI}}|_{\boldsymbol{z}'} &= \sum_{s=0}^{C} \binom{C}{s} b^s a^{C-s} \\ &= (a+b)^{\text{C}} = (b(K-1)+b)^{\text{C}} = (bK)^{\text{C}} = (KD^2)^{\text{C}}.\end{aligned}$$

Finally, as there are K^c different values for $\boldsymbol{z}'$, we have:

$$\#A^{\text{CI}} = K^c \, \#A^{\text{CI}}|_{\boldsymbol{z}'} = (K^2D^2)^c.$$

Hence, the complexity of the elimination or marginalization algorithm on $\mathcal{A}^{\text{CI}}$ is in $\mathcal{O}(T(K^2D^2)^c)$. □

5.7.2. *Multichain ED-HMM with conditional independence*

In this section, we sketch this approach for a multichain ED-HMM without giving details (see Chapter 1, section 1.2.4 for the formalization of an ED-HMM). We restrict ourselves to 1to1 multichains ED-HMM (there are as many observed variables as hidden variables and y^c_t depends on z^c_t only), with conditional independences of the z^c_t and r^c_t (on transition and emission, see Chapter 3). The interactions between chains occur at a jump only. The transition matrix of the hidden layer in calendar time is denoted A_{ED}. It is a $(KD)^{\text{C}} \times (KD)^{\text{C}}$ matrix. At a jump at time t for chain c, we have:

$$\left|\begin{aligned} A_{\text{ED}}[(\boldsymbol{z}_t, \boldsymbol{r}_t), (\boldsymbol{z}_{t-1}\boldsymbol{r}_{t-1})] &= \prod_{c=1}^{\text{C}} A^c_{\text{ED}}[(z^c_t, r^c_t), (\boldsymbol{z}_{t-1}, r^c_{t-1})] \\ B_{\text{ED}}[\boldsymbol{y}_t, \boldsymbol{z}_t] &= \prod_{c=1}^{\text{C}} B^c_{\text{ED}}[y^c_t, z^c_t] \\ \Rightarrow A^{\text{ED}}_t[(\boldsymbol{z}_t, \boldsymbol{r}_t), (\boldsymbol{z}_{t-1}\boldsymbol{r}_{t-1})] &= \prod_{c=1}^{\text{C}} B^c_{\text{ED}}[y^c_t, z^c_t] A^c_{\text{ED}}[(z^c_t, r^c_t), (\boldsymbol{z}_{t-1}, r^c_{t-1})]. \end{aligned}\right.$$

This allows us to define an MUHMD $\mathcal{A}_{\text{ED}}$ whose coefficients are:

$$\begin{aligned}&\mathcal{A}_{\text{ED}}[(\boldsymbol{z}_{0:\text{T}}, \boldsymbol{r}_{0:\text{T}})] = \\ &A^{\text{ED}}_{\text{T}}[(\boldsymbol{z}_{\text{T}}, \boldsymbol{r}_{\text{T}}), (\boldsymbol{z}_{\text{T}-1}, \boldsymbol{r}_{\text{T}-1})] \ldots A^{\text{ED}}_1[(\boldsymbol{z}_1, \boldsymbol{r}_1), (\boldsymbol{z}_0, \boldsymbol{r}_0)] \, \pi_0[\boldsymbol{z}_0, \boldsymbol{r}_0].\end{aligned}$$

It can be shown that $\#A^{\text{ED}} = (K^2D)^c$ with a reasoning similar to the proof of proposition 19. So, we have the following proposition:

PROPOSITION 5.20.– *The complexity of the elimination (respectively, marginalization) step in the forward (respectively, forward-backward) algorithm for a 1to1 multichain ED-HMM with conditional independences and where interactions between chains occur at jumps only is* $\mathcal{O}(T(K^2D)^{\text{C}})$.

5.7.3. *Different geometries of coupling*

Let us mention without further details that the geometries of couplings between chains proposed in Chapter 3 (section 3.3) for 1to1-MHMM-CI can be transposed in a straightforward way to similar geometries of coupling between 1to1 J-multichain ED-HMM or 1to1-J-MHSMM-CI. It suffices to replace the hidden variable $\boldsymbol{z}_t$ by the pair $(\boldsymbol{z}_t, \boldsymbol{r}_t)$. There is an arc from variable $z_{t-1}^{c'}$ to z_t^c in the graphical representation of dependencies of the semi-Markov chains if, at a jump at time t for chain c, $z_{t-1}^{c'}$ is part of the conditioning of z_t^c. More generally, we can as in section 5.5.3 distinguish the cases where an arc emitted by z_{t-1}^c is toward a hidden variable or an observation in same chain c, or toward another chain $c' \neq c$. When the arcs are emitted toward the same chain, hidden chains are independent conditionally to the observations, and the complexity is dramatically decreased. A table similar to table in Chapter 3 (section 3.5) for complexity of EM algorithm for 1to1-MHMM-CI can be built, and this is left to the reader (complexities are in $\mathcal{O}(TCK^2D)$ for cases (a) and (b), and in $\mathcal{O}(T(K^2D)^c)$ for cases (c) or (d)).

5.8. Conclusions and perspectives

In this chapter, we have elaborated on some models with hidden variables presented and studied in Chapters 1, 3 and 4 to propose a unified framework for designing forward and forward-backward algorithm for each of them, and calculate their complexities, as well as the complexity of EM algorithm. We have defined therefore an algebraic structure, called an UHMD, which is a tensor with coefficients built from the joint law of hidden and observed variables once the observed variables have been fixed. We have shown that the forward algorithm is the elimination algorithm on the UHMD, and the forward-backward algorithm is the marginalizatoin algorithm on it. The unified approach consists of two steps: (i) a preparation phase which consists of writing the UHMD, and (ii) running elimination or marginalization algorithm on it. This allows us to show that the complexity of the forward and forward-backward algorithm for a given family of models are the same because the complexity of the elimination and marginalization algorithm on the associated UHMD are the same. They may however differ through the complexity of the preparation phase. This can be extended to the complexity of EM. This has been extended to multichain models and applied to HSMM using their generative HMM models in calendar time to which one can associate an UHMD.

This chapter develops the UHMD approach on a diversity of models. The main results (some classical, some not published to the best of our knowledge) for the diversity of studied models are presented in Table 5.1 (T is omitted in this table).

Model	**Complexity**
HMM	K^2
multichain HMM	$(K^2)^C$
independent chains	CK^2
General HSMM	$(KD)^2$
ED-HMM	K^2D
1to1 CI J-multichain HSMM	$(K^2D^2)^C$
1to1 J-multichain ED-HMM	$(K^2D)^C$

Table 5.1. *Complexity of main models*

One can identify some regularities when comparing the complexity of monochain or multichain models:

– For a HMM, the complexity is in K^2 for a monochain, and in $(K^2)^C$ for a multichain.

– For a general monochain HSMM, the complexity is in $(KD)^2$, and in $((KD)^2)^c$ for 1to1 multichain models with conditional independence.

– For a ED-HMM, the complexity is in K^2D, and in $(K^2D)^c$ for a multichain ED-HMM.

This is well understood: it is in $\#A$ where A is the Markovian transition matrix, as a general property of algorithm 5.1 on the associated UHMD.

We can observe that the complexity of the general HSMM follows a similar pattern, that is, $\#A_J$, where A_J is the transition matrix of the hidden Markov chain $(J_n, X_{n+1})_n$ of the semi-Markov model. Indeed, KD is the cardinality of state space $\mathcal{K} \times \mathcal{D}^*$ of the hidden Markov chain. So, the complexity of a HSMM with transition matrix A_J is in $\#A_J$. We have shown this result for some models by working model by model with the associated generative Markov model in calendar time. A perspective is to show that it as a general result. If this hypothesis is true, it is expected that the complexity of the multichain general HSMM with interactions being at jumps only is in $\mathcal{O}(T(K^2D^2)^C)$. A well-identified difficulty (see Chapter 4) is that it is hopeless to formalize an MHSMM as a Markov model with variables J_n^c and X_n^c: at a jump in chain c, it is impossible to know the states of the other chains in the absence of a reference calendar time. However, it is possible to write a model for each chain c with Markovian variables $(J_n^c, X_{n+1}^c)_n$ where the new state (J_n^c, X_{n+1}^c)

at jump n depends on all states $\boldsymbol{J}_t$ of all chains at the calendar time t of jump in chain c, as a generative model (knowing X^c_{n+1} at each jump allows us to connect the indices by jump with indices with calendar time). This deserves further studies.

5.9. Notations

Definition	Notation	Domain
Space of the hidden variables	$\mathcal{K}$	$\subset \mathbb{N}$
Cardinality of the space of hidden variables	K	$\mathbb{N}^*$
Space of hidden variables monochain, all times	Ω_{Z}	$\mathcal{K}^{\mathrm{T}+1}$
Space of the observations	$\mathcal{S}$	$\subset \mathbb{N}$
Cardinality of the space of observations	S	$\mathbb{N}^*$
Space of observations, monochain, all times	Ω_{Y}	$\mathcal{S}^{\mathrm{T}+1}$
Number of chains in the hidden layer	C	$\mathbb{N}^*$
Number of distinct observations, multichain	O	$\mathbb{N}^*$
Maximum sojourn time	D	$\mathbb{N}^*$
Hidden state, monochain, one time		
random variable	Z_t	$\mathcal{K}$
realization	z_t	$\mathcal{K}$
Hidden states, monochain, all times	$\boldsymbol{z}$	Ω_{Z}
Hidden state, one chain, one time	z^c_t	$\mathcal{K}$
Hidden state, all chains, one time	$\boldsymbol{z}_t$	$\mathcal{K}^{\mathrm{C}}$
Hidden state, all chains, all times	$\boldsymbol{z}_{0:\mathrm{T}}$	$\mathcal{K}^{\mathrm{C}(\mathrm{T}+1)}$
Remaining time	r_t	$\mathcal{D} = \{0, \dots, D-1\}$
Elapsed time	e_t	$\mathcal{D}^* = \{1, \dots, D\}$
Observations, monochain, one time	y_t	$\mathcal{S}$
Observations, monochain, all times	$\boldsymbol{y}$	Ω_{Y}
Observations, one chain, one time	y^c_t	$\mathcal{S}$
Observations, all chains, one time	$\boldsymbol{y}_t$	$\mathcal{S}^{\mathrm{O}}$
Observations, all chains, all times	$\boldsymbol{y}_{0:\mathrm{T}}$	$\mathcal{S}^{\mathrm{O}(\mathrm{T}+1)}$
UHMD	$\mathcal{A}$	$(\mathbb{R}^{\mathrm{K}})^{\otimes \mathrm{T}+1}$
Transition matrix, HMM, hidden layer	A	$\mathbb{R}^{\mathrm{K}\times\mathrm{K}}$
Factor t in an UHMD	A_t	$\mathbb{R}^{\mathrm{K}\times\mathrm{K}}$
Emission matrix	B	$\mathbb{R}^{\mathrm{S}\times\mathrm{K}}$
Transition matrix of a General MHMM	A_{G}	$\mathbb{R}^{\mathrm{K}^{\mathrm{C}}\times\mathrm{K}^{\mathrm{C}}}$
Transition matrix of a general HSMM		
Markov chain $(J_n, X_{n+1})_n$	A_{J}	$\mathbb{R}^{\mathrm{KD}\times\mathrm{KD}}$
Markov chain $(Z_t, R_t, E_t)_t$	A_{ZRE}	$\mathbb{R}^{\mathrm{KD}^2\times\mathrm{KD}^2}$

Table 5.2. *General notations*

Table 5.2 recalls the main notations of the chapter. Below are some remarks about slight differences in notations between this chapter and Chapters 1, 3 and 4.

1) If a is a vector, the coordinates of a will be denoted either a_i or $a[i]$. If A is a matrix, they will be denoted a_{ij} or $A[i, j]$.

2) The transition matrices $A, A_t \dots$ in this chapter are columnwise stochastic, that is, $A[z, z'] = \mathbb{P}(Z_{t+1} = z \mid Z_t = z')$, whereas in Chapters 1 and 3 they are row-wise stochastic ($A[z, z'] = \mathbb{P}(Z_{t+1} = z' \mid Z_t = z)$).

3) The state space of (hidden) variables is denoted $\mathcal{K}$ here, and Ω_Z in Chapters 1, 3 and 4. Here, Ω_Z denotes the state space of the whole time series, that is, $\Omega_Z = \mathcal{K}^{T+1}$.

4) The initial hidden state of a HMM is denoted π_0, and not π as in Chapter 1. The reason is that the approach in this chapter is based on the recursion formula $\pi_t = A\pi_{t-1}$ starting with $\pi_1 = A\pi_0$.

5.10. Acknowledgments

The author would like to thank Nathalie Peyrard, whose many kind and constructive exchanges throughout this project significantly contributed to clarify and simplify the notions and approaches presented in this chapter.

5.11. Appendix: Viterbi algorithm and most likely state

The Viterbi algorithm is presented here as an appendix, because its first step is a forward algorithm, which can be read as an elimination algorithm on the UHMD associated with the transition matrix, but in a semi-ring, and often referred to as *max-product* algorithm (see Peyrard et al. 2019; Murphy 2012, section 17.4.4). Knowing that allows us to incorporate it into the UHMD approach, to derive it for all models for which such an approach is relevant.

The Viterbi algorithm is an algorithm for HMM that computes the most likely trajectory of hidden variables knowing the observations and the parameters of the model. Let $\boldsymbol{z} = z_{0:T}$ and $\boldsymbol{y} = y_{0:T}$. We are looking for:

$$\boldsymbol{z}^* = \underset{\boldsymbol{z}}{\operatorname{argmax}}\, p(\boldsymbol{z} \mid \boldsymbol{y}).$$

As $p(\boldsymbol{z}, \boldsymbol{y}) = p(\boldsymbol{z}|\boldsymbol{y})\, p(\boldsymbol{y})$, we have:

$$\boldsymbol{z}^* = \underset{\boldsymbol{z}}{\operatorname{argmax}}\, p(\boldsymbol{z}, \boldsymbol{y}).$$

which we will evaluate now.

Let (A, B, π_0) be a HMM with transition matrix A between the hidden states, emission matrix B and initial hidden state π_0. We have seen (equation [5.15] in section 5.4) that the law of the hidden variables $\boldsymbol{z}$, the observations $\boldsymbol{y}$ being fixed, can be written as an UHMD:

$$p(\boldsymbol{z}, \boldsymbol{y}) = \prod_{t=1}^{\mathrm{T}} A_{\mathrm{T}}[z_{\mathrm{T}}, z_{\mathrm{T}-1}] \dots A_t[z_t, z_{t-1}] \dots A_1[z_1, z_0]\pi'_0[z_0], \qquad [5.38]$$

with $\pi'_0[z_0] = B[y_0, z_0]\pi_0[z_0]$, where:

$$A_t[z_t, z_{t-1}] = B[y_t, z_t]\, A[z_t, z_{t-1}], \qquad \pi'[z_0] = B[y_0, z_0]\, \pi_0[z_0]. \qquad [5.39]$$

Let $\mathcal{A}$ be the UHMD with coefficients $\mathcal{A}[\boldsymbol{z}]$ being the RHS of [5.38]. Viterbi algorithm amounts to computing the largest term in $\mathcal{A}$. To do this, a second UHMD, denoted as $\mathcal{A}_{\mathrm{V}}$, will be defined, with operations + and × having been changed.

5.11.1. *UHMD in a commutative semi-ring*

With this change, computing the largest term in $\mathcal{A}$ can be solved by running an elimination algorithm on $\mathcal{A}_{\mathrm{V}}$. The UHMD $\mathcal{A}_{\mathrm{V}}$ is built on a semi-ring that we specify here, following Peyrard et al. (2019). Let us note that the inverse operations (substraction and division) in the field $\mathbb{R}$ appear neither in the definition of an UHMD nor in the matrix × vector products of the elimination algorithm. Operations + and × only are involved. This means that the algorithm can be implemented in any commutative semi-ring. A commutative semi-ring is a set R endowed with two commutative operations $\oplus$ and $\odot$, such that:

- → $\oplus$ is associative $\quad a \oplus (b \oplus c) = (a \oplus b) \oplus c = a \oplus b \oplus c$
- → $\oplus$ has an identity element 0 $\quad a \oplus 0 = 0 \oplus a = a$
- → $\odot$ is associative $\quad a \odot (b \odot c) = (a \odot b) \odot c = a \odot b \odot c$
- → $\odot$ has an identity element 1 $\quad a \odot 1 = 1 \odot a = a$
- → $\odot$ is distributive over $\oplus$ $\quad a \odot (b \oplus c) = (a \odot b) \oplus (a \odot c)$.

It is denoted $(R, \oplus, \odot)$.

Let A be a matrix in $R^{\mathrm{K}\times\mathrm{K}}$ and π a vector in R^{K}. The matrix × vector product $\pi' = A \odot \pi$ in R is defined by, componentwise:

$$\forall\, i, j, \quad \pi'[i] = \bigoplus_{j=1}^{\mathrm{K}} (A[i, j] \odot \pi[j]) \,.$$

Let $\mathcal{A}_{\mathrm{V}}$ be an UHMD, defined by:

$$\mathcal{A}_{\mathrm{V}}[z_0, \dots, z_{\mathrm{T}}] = A_{\mathrm{T}}[z_{\mathrm{T}}, z_{\mathrm{T}-1}] \odot \dots \odot A_t[z_t, z_{t-1}] \odot \dots \odot A_1[z_1, z_0] \odot \pi_0[z_0] \quad [5.40]$$

Then, we can observe the following proposition.

PROPOSITION 5.21.– *The elimination algorithm presented as Algorithm 5.1 can be used in a commutative semi-ring $(R, \oplus, \odot)$ to compute the normalizing constant of UHMD $\mathcal{A}_{\text{V}}$:*

$$W(\mathcal{A}_{\text{V}}) = \bigoplus_{z \in \Omega_{\text{Z}}} \mathcal{A}_{\text{V}}[z].$$

PROOF.– It relies on the distributivity of $\odot$ over $\oplus$, and can be written as follows:

Algorithm 5.8 Computation of the normalization constant W in $(R, \oplus, \odot)$

input T, K
input $(A_t)_{1 \leq t \leq \text{T}}$ with $A_t \in \mathbb{R}^{\text{K} \times \text{K}}$
input $\pi_0 \in R^{\text{K}}$
for $t \in \{1, \ldots, T\}$ **do**
  compute $\pi_t = A_t \odot \pi_{t-1}$ s.t. $\pi_t[z] = \bigoplus_{z'=1}^{\text{K}} \Big(A_t[z, z'] \odot \pi_{t-1}[z'] \Big)$
end for
compute $W = \bigoplus_{z \in \mathcal{K}} \pi_{\text{T}}[z]$
return W

□

5.11.2. *Setting the problem*

Let $\mathcal{A}$ be the UHMD associated with $p(z, y)$ (the observations y being fixed) whose terms are:

$$\mathcal{A}[z] = \prod_{t=1}^{\text{T}} A_{\text{T}}[z_{\text{T}}, z_{\text{T}-1}] \ldots A_t[z_t, z_{t-1}] \ldots A_1[z_1, z_0] \pi_0'[z_0].$$

We are looking for the state z^* for which $\mathcal{A}[z]$ is maximal:

$$z^* = \underset{z \in \Omega_{\text{Z}}}{\operatorname{argmax}}\, Z[z]. \qquad [5.41]$$

This will be done in two steps:

1) find $M = \max_{z \in \Omega_{\text{Z}}} \mathcal{A}[z]$;

2) find z^* such that $\mathcal{A}[z^*] = M$.

Step 1 is done with an elimination algorithm, and step 2 is called the traceback.

5.11.3. *Computing the probability of the most likely state*

A simple solution can be found by remarking that the operations $\oplus = \max$ and $\odot = \times$ confer to $R = \mathbb{R} \cup \{-\infty\}$ a structure of commutative semi-ring. Indeed, we have:

$$\begin{cases} \max(a, \max(b,c)) & = \max(\max(a,b),c) \ = \max(a,b,c) \\ \max(a, -\infty) & = a \\ a \max(b,c) & = \max(ab, ac) \end{cases}$$

(the fact that $\odot$ is associative and has an identity is here obvious). So, $(\mathbb{R} \cup \{-\infty\}, \max, \times)$ is a semi-ring ($-\infty$ has been introduced for this purpose as an identity for $\oplus = \max$). Note that we still need to use $\odot$ and not $\times$ in the product matrix $\odot$ vector because it involves $\oplus$ in $y_i = \bigoplus_{j=1}^{\mathrm{K}} a_{ij} x_j$ in calculation of $y = Ax$.

Let us explicit the matrix $\odot$ vector product in this semi-ring. Let $R = \mathbb{R} \cup \{-\infty\}$, $A \in R^{\mathrm{K} \times \mathrm{K}}$ and $\pi \in R^{\mathrm{K}}$. Let $\pi' = A \odot \pi \in R^{\mathrm{K}}$. Then:

$$\begin{aligned} \pi'[i] &= \bigoplus_{j=1}^{\mathrm{K}} (A[i,j] \odot \pi[j]) \\ &= \max_j \{(A[i,j]\, \pi[j] \mid 1 \leq j \leq K\}. \end{aligned} \qquad [5.42]$$

We have the following proposition:

PROPOSITION 5.22.– *The elimination algorithm in* $(\mathbb{R} \cup \{-\infty\}, \max, \times)$ *can be used for computing* $M = \max_{\boldsymbol{z} \in \Omega_{\mathrm{Z}}} \mathcal{A}[\boldsymbol{z}]$. *This is a forward algorithm, and is the first step of* Viterbi algorithm.

PROOF.– The algorithm can be written as:

Algorithm 5.9 Computation of the highest likelihood M with Viterbi algorithm

input T, K
input $(A_t)_{1 \leq t \leq \mathrm{T}}$ with $A_t \in \mathbb{R}^{\mathrm{K} \times \mathrm{K}}$
input $\pi_0 \in \mathbb{R}^{\mathrm{K}}$
for $t \in \{1, \ldots, T\}$ **do**
  compute $\pi_t = A_t\, \pi_{t-1}$ s.t. $\forall z, \quad \pi_t[z] = \max_{1 \leq z' \leq \mathrm{K}} \left\{A_t[z, z']\, \pi_{t-1}[z']\right\}$
end for
compute $M = \max_{1 \leq z \leq \mathrm{K}} \left\{\pi_{\mathrm{T}}[z]\right\}$
return M

□

5.11.4. *Recovering the most likely state*

This gives M the probability of the most likely state. Here, we show how to recover the most likely state itself (i.e. the state $\boldsymbol{z}^* \in \Omega_Z$ such that $\mathcal{A}[\boldsymbol{z}^*] = M$). The idea is to trace the index yielding the maximum in the calculation presented in line 5 in algorithm 5.9, and then trace back all choices of indices at each step t. Let us present it in a simple example, with $T = 2$, hence $0 \leq t \leq 2$. We have:

$$\mathcal{A}[z_0, z_1, z_2] = A_2[z_2, z_1]\, A_1[z_1, z_0]\, \pi_0[z_0].$$

We are looking for (z_0, z_1, z_2) such that $\mathcal{A}[z_0, z_1, z_2]$ is maximal. The elimination algorithm can be written as:

$$\begin{aligned}
&\max_{z_0, z_1, z_2} \left\{ A_2[z_2, z_1]\, A_1[z_1, Z_0]\, \pi_0[z_0] \right\} \\
&= \max_{z_2, z_1} \left\{ A_2[z_2, z_1] \left[\max_{z_0} \left[A_1[z_1, z_0]\, \pi_0[z_0] \right] \right] \right\} \\
&= \max_{z_2} \left\{ \max_{z_1} \left[A_2[z_2, z_1] \left(\max_{z_0} \left[A_1[z_1, z_0]\, \pi_0[z_0] \right) \right] \right] \right\},
\end{aligned}$$

which leads to the following algorithm, associating the elimination step and the traceback, and is Viterbi algorithm on this example:

Algorithm 5.10 Viterbi algorithm: elimination and traceback

input A_0, A_1, A_2
input π_0
—— *elimination algorithm to compute M*
for $z_1 \in \{1, \ldots, K\}$ **do**
  compute $\pi_1[z_1] = \max_{z_0} \left\{ A_1[z_1, z_0]\, \pi_0[z_0] \right\}$
end for
for $z_2 \in \{1, \ldots, K\}$ **do**
  compute $\pi_2[z_2] = \max_{z_1} \left\{ A_2[z_2, z_1]\, \pi_1[z_1] \right\}$
end for
$M = \max_{z_2} \pi_2[z_2]$
—— *traceback to compute $\boldsymbol{z}$*
select $z_2^* = \operatorname{argmax}_{z_2} \pi_2[z_2]$
select $z_1^* = \operatorname{argmax}_{z_1} A_2[z_2^*, z_1]\, \pi_1[z_1]$
select $z_0^* = \operatorname{argmax}_{z_0} A_1[z_1^*, z_0]\, \pi_0[z_0]$
set $\boldsymbol{z}^* = (z_0^*, z_1^*, z_2^*)$
return $M, \boldsymbol{z}^*$

This concrete example can be extended to any UHMD with $T \geq 2$. This allows us to derive the Viterbi algorithm for all models presented in this chapter, and some others, such as the HMM, MHMM (different guises), HSMM (different guises) and J-MHSMM, for which the general approach presented in section 5.4.3 can be deployed.

5.12. References

Barbu, V.S. and Limnios, N. (2008). *Semi-Markov Chains and Hidden Semi-Markov Models toward Applications*. Springer, New York.

Bishop, C.M. (2006). *Pattern Recognition and Machine Learning*. Springer, Berlin.

Murphy, K.P. (2012). *Machine Learning: A Probabilistic Perspective*. MIT Press, Cambridge, MA.

Oseledets, I.V. and Dolgov, S.V. (2012). Solution of linear systems and matrix inversion in the TT-format. *SIAM J. Sci. Comput.*, 34, A2718–A2739.

Peyrard, N., Cros, M.-J., de Givry, S., Franc, A., Robin, S., Sabbadin, R., Schiex, T., Vignes, M. (2019). Exact or approximate inference in graphical models: Why the choice is dictated by the treewidth, and how variable elimination can be exploited. *Australian and New Zealand Journal of Statistics*, 61(3). doi: 10.1111/anzs.12257.

Rabiner, L.R. (1989). A tutorial on hidden Markov models and selected applications in speech recognition. *Proceedings of the IEEE*, 77, 257–286.

6

Controlled Hidden Semi-Markov Models

This chapter is dedicated to controlled hidden semi-Markov models (HSMMs). The objective is no longer to reconstruct hidden states or to estimate the model parameters but to design the best sequence of decisions in order to minimize some cost function depending on the whole trajectory of the process. In particular, we consider that all model parameters are fully known. We first provide the mathematical framework for sequential decision-making for fully observed Markov chains, called Markov decision processes (MDPs), in the simple case of discrete time and discrete state spaces. Then we will turn to extensions of this basic model in the case of hidden information (or partially observed MDPs), and to the more general class of controlled hidden piecewise deterministic Markov processes (PDMPs) that encompasses both hidden information and Markov or semi-Markov dynamics in continuous time and in continuous state spaces. The main concepts are illustrated using the Squirrel toy models with codes available at https://github.com/orlross/phdnotebook/tree/main. This chapter is a shortened version of Cleynen et al. (2025) that focuses on the links between HSMMs, MDPs and PDMPs.

6.1. Introduction

HSMMs have emerged as powerful tools for modeling systems with latent structures and discrete dynamics. They offer an elegant framework for analyzing temporal processes where transitions between states are governed by probabilistic rules, with sojourn duration following general distributions. While much of this book has focused on HSMMs in discrete time and their extensions, including the coupling of multiple chains, the extension to continuous time opens up a broader spectrum of possibilities and challenges.

Chapter written by Alice CLEYNEN, Benoîte DE SAPORTA, Orlane ROSSINI, Régis SABBADIN and Amélie VERNAY.

To set the stage, we begin with a foundational overview of MDPs. We are interested in problems where, at each time step, a decision-maker acts on the Markov chain by selecting actions that influence its dynamics and generate some reward or cost. MDPs provide a formal structure for such sequential decision-making in stochastic environments. We then present extensions of MDPs, such as partially observed MDPs in the case of hidden information.

Building upon this foundation, we introduce PDMPs as a natural extension of HSMMs into continuous time. PDMPs are characterized by deterministic trajectories punctuated by random jumps, allowing for a hybrid model that seamlessly integrates deterministic evolution with probabilistic events. These models are particularly well suited to describe systems where some Euclidean variables evolve according to deterministic dynamics until a random event triggers a transition to another behavior. In many applications, transitions occur between some hidden discrete states, while the Euclidean variables are observed through noise and in discrete time. Examples include systems in physics, biology and engineering, where abrupt changes are superimposed on continuous evolution. We also discuss how PDMPs can very naturally encompass semi-Markov dynamics and generalize HSMMs to continuous time and space.

The chapter then focuses on the impulse control of PDMPs, which involves making punctual decisions at decision-maker-chosen intervention times to optimize a given objective function. It is the continuous-time counterpart of MDPs. To address the impulse control problem for PDMPs, we demonstrate how these models can still be formalized within the MDP context. This connection provides a powerful analytical and computational toolkit for solving optimization problems associated with PDMPs. Throughout the chapter, the notions are illustrated on an adapted version of the Squirrel toy model introduced in Chapter 1.

The chapter concludes by discussing how the interplay between discrete and continuous decision-making frameworks opens new pathways for research and applications. By bridging the discrete-time HSMM framework and the continuous-time PDMP framework, we not only enrich the theoretical landscape but also enhance the practical applicability of these models to real-world systems requiring control strategies.

For convenience, tables of notation are provided at the end of the chapter in section 6.6.

6.2. Markov decision processes

MDPs provide a robust framework for modeling scenarios where sequential decisions are made in the presence of uncertainty. The objective is to determine an optimal sequence of decisions that maximizes rewards or minimizes costs, while accounting for the probabilistic evolution of the system. At each step, a decision-maker observes the current state, selects an action and receives feedback based on that action, with the system transitioning probabilistically to the next state.

MDPs belong to a class of stochastic control problems that originated in the 1950s (Bellman 1958; Howard 1962). Comprehensive treatments of the subject can be found in foundational texts such as Puterman (1994) and Boutilier et al. (1999). Their ability to model complex, dynamic systems has made MDPs indispensable across a wide range of applications, from robotics and finance to healthcare and artificial intelligence.

The versatility of MDPs in capturing stochastic dynamics and optimizing sequential decisions has led to their adoption across diverse fields, such as engineering, computer science, economics and other social sciences, making them a valuable tool for researchers and practitioners alike. In robotics, MDPs have been employed for tasks such as navigation and control (Chanel et al. 2013). Conversational agents and dialogue systems have leveraged MDPs to model and optimize human–machine interactions (Levin et al. 1998; Young et al. 2013). The medical field has witnessed numerous applications of MDPs, including modeling disease progression and treatment planning for conditions like Alzheimer's disease (Hoey et al. 2010), HIV (Keneally et al. 2016) and diabetes (Roy et al. 2018). MDPs have also been used in clinical decision-support systems for critical care (Schaefer et al. 2004).

In this section, we delve into the formal components of MDPs, including states, actions and cost functions (section 6.2.1), while also introducing the concept of policies, which determine the action selection process (section 6.2.2). Unless otherwise stated, we consider finite spaces for both actions and states. We then introduce the partially observed case (section 6.2.3) and discuss the mathematical and numerical resolution of MDPs (section 6.2.4).

6.2.1. *MDP definition*

Informally, a MDP can be thought of as a Markov chain over a state space Ω_Z that evolves over a horizon of T time steps, where transitions are influenced by actions chosen in a set A by a decision-maker. The decision process can span a finite or infinite number of time steps T, depending on the problem at hand.

The stochastic transitions between states are governed by a transition probability matrix P. The associated costs of these transitions are captured in a (possibly time-dependent) cost function or table c, which quantifies the immediate cost associated with each state-action pair

The formal definition of a MDP is as follows.

DEFINITION 6.1 (MDP).– *A MDP is defined by a tuple* $\langle \Omega_Z, A, T, P, c\rangle$ *where*

– *the* state space Ω_Z *is the finite or countably infinite set of all possible states in the environment;*

– *the* action space A *is the finite set of actions that the decision-maker can take;*

– *the* horizon T *is a positive integer (or positive infinity) counting the number of time steps;*

– *the* transition matrix P *describes the probabilities of transitioning, at any time step* $0 \le t < T$ *from a state* $z_t \in \Omega_Z$ *to another* $z_{t+1} \in \Omega_Z$ *after taking a specific action* $a_t \in A$*. Transition probabilities satisfy the Markov property:*

$$\begin{aligned} &\mathbb{P}(Z_{t+1} = z_{t+1} | Z_0 = z_0, A_0 = a_0, \cdots, Z_t = z_t, A_t = a_t) \\ &= \mathbb{P}(Z_{t+1} = z_{t+1} | Z_t = z_t, A_t = a_t) \ = P(z_{t+1} | z_t, a_t); \end{aligned}$$

– *the* cost function $c : \Omega_Z \times A \times \Omega_Z \to \mathbb{R}$ *is the function that assigns a numerical cost to each state–action–state transition.*

Let us consider a finite-horizon MDP $\langle \Omega_Z, A, T, P, c\rangle$ and an arbitrary initial state $z_0 \in \Omega_Z$. At any time $0 \le t < T$, the decision-maker chooses an action $a_t \in A$ and applies it in the current state z_t. This moves the process to the next state z_{t+1}, following probability $P(\cdot|z_t, a_t)$ and yields a cost $c(z_t, a_t, z_{t+1})$. This process is iterated until termination when the horizon T is reached. A terminal cost $C : \Omega_Z \to \mathbb{R}$ can be defined at the end of the process in the case of a finite horizon. This is illustrated in Figure 6.1, where we follow the same conventions for directed probabilistic graphical models as in Chapter 3.

6.2.2. *Control for MDPs*

We now turn to the control of MDPs, which involves defining how actions may be selected at each time step through policies, a cost criterion associated with each policy and the formal definition of the optimization problem.

6.2.2.1. *Policies*

To construct controlled trajectories for a MDP, it is essential to determine how the decision-maker selects the actions to be taken at each decision step based on the past *history* of the process. This is made through control *policies* that are defined as collections of *decision rules*.

DEFINITION 6.2 (MDP history).– *Consider a given MDP* $\langle \Omega_Z, A, T, P, c\rangle$*. For any* $0 < t < T$*, a sequence of states and actions* $(z_0, a_0, ..., z_{t-1}, a_{t-1}, z_t)$ *is called a* *(length t)* history *of the MDP. A length 0 history is a singleton z_0. The set of all length t histories is denoted $\mathcal{H}_t$.*

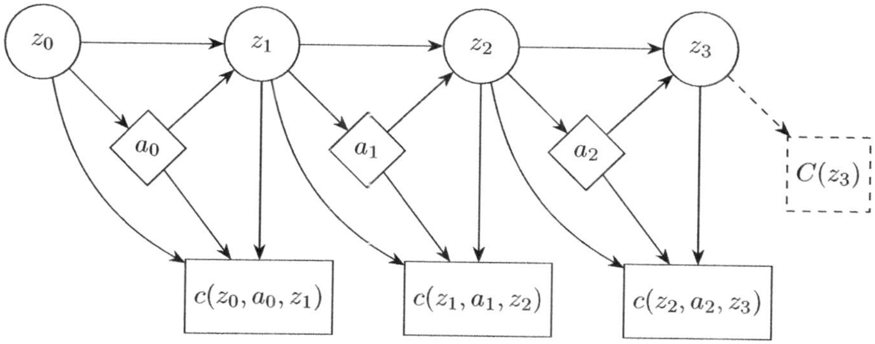

Figure 6.1. *Graphical representation of a Markov decision process with finite horizon $T = 3$. Circles denote states, diamonds represent actions and rectangles indicate costs associated with state-action-state transitions. The dashed rectangle depicts the terminal cost*

DEFINITION 6.3 (History-dependent decision rule).– *A history-dependent decision rule at time t is a mapping $\pi_t : \mathcal{H}_t \to A$ that assigns an action to each history. Let Π_t denote the set of history-dependent decision rules at time t.*

An interesting subclass of decision rules is that of history-independent decision rules.

DEFINITION 6.4 (History-independent decision rule).– *A decision rule at time t, $\pi_t : \mathcal{H}_t \to A$ is history-independent if $\pi_t(h_t) = \pi_t(h'_t)$ for any pair $h_t = (z_0, a_0, ..., z_{t-1}, a_{t-1}, z_t), h'_t = (z'_0, a'_0, ..., z'_{t-1}, a'_{t-1}, z'_t) \in \mathcal{H}_t$ such that $z_t = z'_t$.*

Then, MDP *policies* are time-indexed lists of decision rules that specify the behavior of a decision-maker in a MDP.

DEFINITION 6.5 (MDP policy).– *Let a MDP $\langle \Omega_Z, A, T, P, c \rangle$ be given. A policy for this MDP is a list of decision rules $\pi = (\pi_t)_{0 \leq t < T} = \pi_{0:T-1}$. Let Π denote the full set of policies.*

Again, interesting specific subclasses of policies can be defined.

– Π^M is the subset of *Markovian policies*, that is, policies composed of history-independent decision rules only.

– Π^S is the subset of *stationary policies*, that is, Markovian policies which do not depend on t, where the same (history-independent) decision rule is applied at each time step.

Given a policy π, it is straightforward to simulate a trajectory $(Z_t)_{0 \leq t \leq T} = Z_{0:T}$ from a MDP controlled by π and the generated cost using Algorithm 6.1.

Algorithm 6.1 Simulation of a trajectory of a MDP controlled by policy $\pi = \pi_{0:T-1}$ starting from state z_0 up to horizon T

$Z_0 \leftarrow z_0$
for $t \leftarrow 0$ to $T-1$ **do**
    $A_t \leftarrow \pi_t(Z_0, A_0, \dots, Z_{t-1}, A_{t-1}, Z_t)$
    $Z_{t+1} \sim P(\cdot | Z_t, A_t)$
    $C_t \leftarrow c(Z_t, A_t, Z_{t+1})$
end for

6.2.2.2. *Policy evaluation criteria*

The objective in MDPs is to find a policy that minimizes a cost criterion. Up to a change of sign, minimizing a cost or maximizing a reward are equivalent problems. In the following, we consider cost minimization only.

The transition cost $c(z_t, a_t, z_{t+1})$ represents the immediate penalty of taking action a_t in state z_t and transitioning to state z_{t+1}, while the terminal cost $C(z_T)$ captures the penalty upon reaching the final state of the process. Beyond immediate costs, we must define a criterion that captures the entire random dynamics of the process, considering both the short-term effects of an action and its long-term consequences. The classical evaluation criterion involves some expected sum of future costs.

We distinguish between two common cases: finite-horizon and infinite-horizon optimization criteria.

In the infinite horizon setting, the decision-maker makes decisions indefinitely. In this case, the discounted cost criterion represents the expectation of future transition costs, discounted by a factor $\gamma \in (0, 1)$, which balances the immediate cost against future outcomes and ensures the finiteness of the sum as long as the running cost is bounded.

DEFINITION 6.6 (Discounted expected cost).– *Starting from the initial state $z \in \Omega_Z$ and following the policy $\pi \in \Pi$, the average expected cost is:*

$$V_\gamma(\pi, z) = \mathbb{E}_z^\pi \left[\sum_{t=0}^{\infty} \gamma^t c(Z_t, A_t, Z_{t+1}) \right].$$

In the finite horizon setting, the decision-maker is tasked with controlling the system over a finite number of steps, T. The natural objective in this case is to minimize the total expected cost.

DEFINITION 6.7 (Total expected cost).– *Starting from the initial state $z \in \Omega_Z$ and following the policy $\pi \in \Pi$, the total expected cost is the sum of the costs accumulated over each step, up to the horizon T, plus the terminal cost associated with the final state. It is expressed as:*

$$V_T(\pi, z) = \mathbb{E}_z^\pi \left[\sum_{t=0}^{T-1} c(Z_t, A_t, Z_{t+1}) + C(Z_T) \right].$$

Note that the above evaluation criteria apply to general history-dependent policies as well as all sub-classes of policies.

6.2.2.3. *Optimization*

The optimization problem associated with a cost criterion involves finding a policy π minimizing the chosen evaluation criterion over all possible policies $\pi \in \Pi$. Solving a MDP consists of finding the minimum value of the evaluation criterion called the *value function* and identifying an *optimal policy*, that is a policy with minimal value in every starting state.

DEFINITION 6.8 (Optimal policy).– *Let $\langle \Omega_Z, A, T, P, c \rangle$ be a given MDP. An optimal policy π^* of the MDP, with respect to criterion $V \in \{V_\gamma, V_T\}$ is a policy such that:*

$$V(\pi^*, z) \leq V(\pi, z), \forall \pi \in \Pi, \forall z \in \Omega_Z.$$

Optimal policies are not necessarily unique for a given problem. However, it is well known (see, e.g. Puterman (1994) and Th 1 below) that an optimal policy exists for any finite state space MDP and criterion $V \in \{V_\gamma, V_T\}$. In addition:

– an optimal stationary policy exists for V_γ, in the infinite horizon case;

– an optimal Markovian policy exists for V_T, in the finite-horizon case.

Even though optimal policies need not be unique, they all share the same value function $V^* : \Omega_Z \to \mathbb{R}$, which is unique.

DEFINITION 6.9 (Value function).– *Let $\langle \Omega_Z, A, T, P, c \rangle$ be a given MDP. Let any optimal policy π^* of the MDP, with respect to criterion $V \in \{V_\gamma, V_T\}$ be also given. The optimal value function is then defined as:*

$$V^\star(z) = V(\pi^*, z), \forall z \in \Omega_Z.$$

The optimal value functions and optimal policies of a MDP with respect to the different criteria can be characterized as the solution of sets of nonlinear equations called *Bellman equations*. For the infinite-horizon discounted criterion and for the finite-horizon total cost criterion, the corresponding systems of equations are the following (see, e.g. (Bellman 1958)).

THEOREM 6.1.– *Bellman equations.*
Infinite horizon: the optimal value function $V_\gamma^\star$ is the unique solution of the following nonlinear equation:

$$V_\gamma(z) = \min_{a\in A} \sum_{z'\in\Omega_Z} P(z'|z,a)\left(c(z,a,z') + \gamma V_\gamma(z')\right), \forall z \in \Omega_Z.$$

Any optimal policy is such that:

$$\pi^\star(z) \in \arg\min_{a\in A} \sum_{z'\in\Omega_Z} P(z'|z,a)\left(c(z,a,z') + \gamma V_\gamma^\star(z')\right), \forall z \in \Omega_Z.$$

Finite horizon: the optimal value function $V^\star$ equals $V_0^\star$, where $\{V_t^\star\}_{0\le t\le T}$ is the unique solution to the following nonlinear system:

$$V_t(z) = \min_{a\in A} \sum_{z'\in S} P(z'|z,a)\left(c(z,a,z') + V_{t+1}(z')\right),$$

for all $t \in \{0,\dots,T-1\}$ and $z \in \Omega_Z$, where $V_T(z) = C(z)$ defines the terminal values. Any optimal policy $\pi^\star = \pi^\star_{0:T-1}$ is such that:

$$\pi_t^\star(z) \in \arg\min_{a\in A} \sum_{z'\in\Omega_Z} P(z'|z,a)\left(c(z,a,z') + V_{t+1}(z')\right),$$

for all $t \in \{0,\dots,T-1\}$ and $z \in \Omega_Z$.

We describe in section 6.2.4 a few classical algorithms for computing optimal policies in finite and infinite horizon.

6.2.2.4. *Squirrel toy model as a finite state and action MDP*

We consider a variant of the Squirrel toy model introduced in section 1.2.1. Over the course of a week, we observed a squirrel moving from hiding place to hiding place. Let $m \in \{1,\cdots,4\}$ be the hiding places and $r \in \{0,\cdots 7\}$ be the time spent by the squirrel in its current hiding place. We only have one camera and each day we can decide to place it in front of a specific hiding place. The aim is to take as many photos of the squirrel as possible over the week.

The MDP associated with this example is described by:

$$\begin{aligned}
&\Omega_Z = \{1,\cdots,4\} \times \{0,\cdots 7\},\\
&A = \{1,\cdots,4\},\\
&T = 7,\\
&P(z' = (m',r')|z = (m,r),a)\\
&= (1-e^{-(7-r)})\mathbb{1}_{\{m'=m,r'=r+1)\}} + \frac{e^{-(7-r)}}{3}\mathbb{1}_{\{m'\neq m,r'=0\}},\\
&c(z = (m,r),a) = \mathbb{1}_{a\neq m}.
\end{aligned}$$

The transition probabilities are chosen so that, in essence, the more time the squirrel spends in a given hiding place, the more likely it is to move, and when it does the following hiding place is chosen with uniform probability. Note that in this example, the transition matrix does not depend on the action, the cost does not depend on the next state and the terminal cost is $C = 0$.

Backward finite horizon dynamic programming explicitly provides the value function $V_T = 0$ and the optimal policy which consists of moving the camera in front of the squirrel's hiding place achieves this null total cost.

6.2.2.5. *Extensions*

For simplicity, we defined MDPs with finite or countable state space and with finite action spaces and assumed that all actions are feasible in all states. These assumptions can be relaxed to model more complex scenarios.

Constraints: in many applications, specific actions may be prohibited in certain states. There are two main approaches to handle such constraints. The first one is to assign a prohibitive cost ($+\infty$ for instance) to invalid state–action pairs, preventing them from being chosen by optimization algorithms. The second one is to explicit constraints through the addition of an element K to the MDP definition: $\langle \Omega_Z, A, T, P, c \rangle$ becomes $\langle \Omega_Z, A, T, K, P, c \rangle$, where K is a multifunction from Ω_Z onto the set of subsets of A called the *constraints sets* and $K(z) \subset A$ is the collection of all allowed actions in state z. This approach is explicit and intuitive for finite state and action spaces, requiring only that $K(z)$ be non-empty for all $z \in \Omega_Z$. In the case of countable action spaces, it is usually required that $K(z)$ be non-empty and finite for all z (see, e.g. Hernández-Lerma and Lasserre (1996, Section 1.2) or Hinderer et al. (2016, Section 2.2)). Decision rules π_t are then restricted to select actions only from $K(z)$ for a given state z or history with terminal state z. Such decision rules are called *admissible*, and *admissible policies* are collections of admissible decision rules. The two approaches are equivalent and yield the same value function and optimal policies.

General state or action spaces: MDPs can also be extended to general (uncountable) state and action spaces. While the core definitions remain similar (e.g. transition probabilities replaced by transition kernels), and in most frameworks some version of theorem 1 still holds as well as existence of optimal policies in suitable classes of policies, these extensions introduce significant additional mathematical intricacies.

1) In order to properly define controlled trajectories, Z_t and A_t from algorithm 6.1 must be random variables, that is, measurable with respect to some suitably chosen σ-algebras. This first means that the state Ω_Z and action A spaces must be endowed with σ-algebras and thus have some minimal topological structure (typically Polish spaces). This also means that *admissible* decision rules must also be measurable,

and that the constraints set $\{(z, a); z \in \Omega_Z, a \in K(z)\}$ must contain the graph of measurable mapping from $\Omega_Z \times A$ onto A. Finally, the cost functions also need to be measurable. There exists a wide literature on easy sufficient conditions to ensure all these properties hold, for instance, if Ω_Z is a subset of some $\mathbb{R}^d$ endowed with its Borel σ-algebra, A is a subset of $\mathbb{R}$, and all $K(z)$ are intervals with continuous bounds in z (see, e.g. Hinderer et al. (2016, chapter 16)).

2) Unbounded cost functions in general spaces may lead to undefined expectations. This issue is resolved by assuming cost functions and transition kernels are bounded by a non-negative *bounding function* (see Hinderer et al. (2016, Chapter 16)).

3) Finally, existence of optimal policies requires some minimal regularity on the cost functions and transition kernel (typically lower or upper semi-continuity) and compactness of the set of policies, which often requires the use of relaxed policies that are collections of stochastic (admissible) decision rules. The interested reader is referred to Hernández-Lerma and Lasserre (1996); Bäuerle and Rieder (2011); Hinderer et al. (2016) for further details.

6.2.3. *Partially observed Markov decision processes*

We now turn to the case where the states of the MDP are partially hidden to the decision-maker to take into account controlled HMMs. In this setting, states z are hidden and the decision maker only has access to some information y through a possibly random observation function corresponding to the emission function in HMMs. This encompasses the case of multichains where states z may be multidimensional, with some observed coordinates, and some hidden coordinates, hence the term *partially observed*.

The main difference with standard MDPs is that decisions can now only be taken in view of the *observed* histories.

6.2.3.1. *POMDP definition*

Partially observed MDPs (POMDPs) are an extension of MDPs that were first introduced by Åström (1965) within the context of control theory. In this work, we adopt a more operational perspective of POMDPs, proposed by Kaelbling et al. (1998), which focuses on settings with finite state, observation and action spaces, and finite horizon.

DEFINITION 6.10 (POMDP).– *A finite-horizon POMDP model is a tuple $\langle \Omega_Z, A, T, P, \Omega_Y, O, c, y_0 \rangle$. In addition to the components of a standard MDP introduced in definition 6.1, POMDPs introduce:*

– a finite set Ω_Y representing the possible observations of the system;

– *an observation function:*

$$O : \Omega_Z \times A \times \Omega_Y \to [0, 1],$$

where $O(z', a, y)$ gives the probability of observing $y \in \Omega_Y$, after taking action a, resulting in successor state z';

– *and an initial observation y_0, describing the initial imperfect knowledge we have about the initial state.*

The observation function plays a similar role to the emission function in HMMs, but is now action dependent.

In a POMDP, the objective is still to compute a strategy that minimizes the expected sum of future costs. However, a key difference is that, at each time step t, the true state of the system z_t is not directly observed. Thus, POMDP histories only include actions and observations.

DEFINITION 6.11 (POMDP observed history).– *Consider a given POMDP given by $\langle \Omega_Z, A, T, P, \Omega_Y, O, c, y_0 \rangle$. For any $0 \leq t < T$, a sequence of actions and observations $h_t = (y_0, a_0, y_1, \ldots, a_{t-1}, y_t)$ is called a (length t)* observed history *of the POMDP. The set of all length t observed histories is denoted $\mathcal{H}_t^o$.*

Similarly to solving a MDP, solving a POMDP consists of computing a *policy* assigning an action to every possible state of current knowledge of the POMDP.

DEFINITION 6.12 (POMDP history-dependent decision rule).– *Consider a POMDP given by $\langle \Omega_Z, A, T, P, \Omega_Y, O, c, y_0 \rangle$. A POMDP history-dependent decision rule $\pi_t : \mathcal{H}_t^o \to A$ assigns an action in A to each POMDP observed history $h_t \in \mathcal{H}_t^o$.*

DEFINITION 6.13 (POMDP policies).– *Consider a POMDP given by $\langle \Omega_Z, A, T, P, \Omega_Y, O, c, y_0 \rangle$. A POMDP history-dependent policy is a list of history-dependent decision rules $\pi = \pi_{0:T-1}$. Denote by Π^o the set of all history-dependent policies.*

A POMDP $\langle \Omega_Z, A, T, P, \Omega_Y, O, c, y_0 \rangle$ could also be seen as a MDP $\langle \overline{\Omega}_Z, A, T, \overline{P}, c \rangle$ by setting:

$$\overline{\Omega}_Z = \Omega_Z \times \Omega_Y,$$

$$\overline{P}(\overline{z}' = (z', y') | \overline{z} = (z, y), a) = P(z'|z, a) O(z', a, y'),$$

for all $\overline{z}' = (z', y')$ and $\overline{z} = (z, y)$ in $\overline{\Omega_Z}$. In this setting, the term *partially observed* is more obvious. The main difference between this MPD and the POMDP setting is that in the POMPD, only policies in Π^o are considered, and the initial value Z_0 is randomly drawn from an initial belief b_0, which is a probability distribution over Ω_Z:

$b_0(z) = \mathbb{P}(Z_0 = z|y_0)$ for all $z \in \Omega_Z$. This setting allows us to see Π^o as a subset of Π for the MDP $\langle \overline{\Omega}_Z, A, T, \overline{P}, c\rangle$, and to directly use definition 6.7 to define the total expected cost for a POMPD policy, by setting:

$$V_T(\pi, b_0) = \sum_{z \in \Omega_Z} b_0(z) V_T(\pi, z).$$

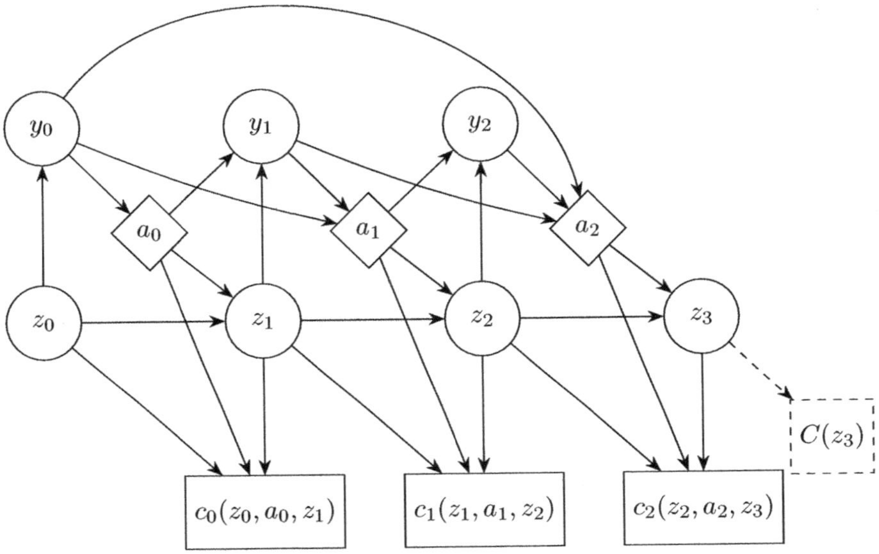

Figure 6.2. *Graphical representation of a partially observable Markov decision process with finite horizon $T = 3$. Circles denote states z_i and observations y_i, diamonds represent actions and rectangles indicate costs associated with state-action-state transitions. The dashed rectangle depicts the terminal cost*

DEFINITION 6.14 (Optimal policy).– *Consider the POMDP $\langle \Omega_Z, A, T, P, \Omega_Y, O, c, y_0\rangle$ with initial belief b_0. An optimal policy π^* for the POMDP, with respect to criterion V_T given in definition 6.7 is a history-dependent policy $\pi^* \in \Pi^o$ such that:*

$$V_T(\pi^*, b_0) \leq V_T(\pi, b_0), \forall \pi \in \Pi^o.$$

Note that here, the expected cost V_T depends on the unobserved states of the process, but must be optimized with respect only to the observed history.

In order to evaluate a POMDP policy and compute an optimal policy, it is useful to notice that a POMDP can be modeled as another particular form of MDP over *belief states*.

6.2.3.2. *Belief MDP*

The knowledge incorporated in the observed history h_t can be encapsulated in a *belief state*, b_t which is a probability distribution on Ω_Z for the unobserved current state given the history:

$$b_t(z) = \mathbb{P}(Z_t = z|h_t).$$

This belief state evolves from an initial belief b_0 based on the POMDP history.

When h_t is completed with a new pair of action/observation (a_t, y_{t+1}), the belief state b_t is updated as follows. Considering that in a given belief state b_t, an action a_t is applied and leads to a new observation y_{t+1} then:

$$b_{t+1} = \tau\left(b_t, a_t, y_{t+1}\right),$$

where τ is a deterministic belief transition function. This function is computed through Bayesian updating:

$$\begin{aligned}
b_{t+1}(z') &= \mathbb{P}(Z_{t+1} = z'|b_t, a_t, y_{t+1}) \\
&= \frac{O(z', a_t, y_{t+1})\mathbb{P}(Z_{t+1} = z'|b_t, a_t)}{\sum_{z"\in\Omega_Z} O(z", a_t, y_{t+1})\mathbb{P}(Z_{t+1} = z"|b_t, a_t)} \\
&= \frac{O(z', a_t, y_{t+1}) \sum_{z\in\Omega_Z} b_t(z)P(z'|z, a_t)}{\sum_{z"\in\Omega_Z} O(z", a_t, y_{t+1}) \sum_{z\in\Omega_Z} b_t(z)P(z"|z, a_t)} \\
&\propto O(z', a_t, y_{t+1}) \times \sum_{z\in\Omega_Z} b_t(z)P(z'|z, a_t).
\end{aligned}$$

A way to deal with partial observability in a POMDP is to consider a corresponding *belief MDP*, which (continuous, multi-dimensional) state space is made of the set of possible belief states of the MDP. The correspondence between POMDPs and belief MDPs is made explicit in the following definition.

DEFINITION 6.15 (Belief MDP).– *A finite-horizon POMDP $\langle\Omega_Z, A, T, P, \Omega_Y, O, c, y_0\rangle$ can be viewed as a belief MDP that is a MDP $\langle\mathcal{B}, A, T, \mathcal{P}, \rho\rangle$ on belief states, where:*

– $\mathcal{B}$ is the continuous set of belief states, it is a subset of the set of probability distributions on Ω_Z;

– A is the action space, unchanged;

– $\mathcal{P}$ is the transition function:

$$\mathcal{P}(b_{t+1}|b_t, a_t) = \sum_{y'\in\Omega_Y} \mathbb{1}_{b_{t+1}=\tau(b_t,a_t,y')}O(z', a_t, y') \sum_{z\in\Omega_Z} b_t(z)P(z'|z, a_t)$$

– *$\rho(b, a, b')$ is the cost associated with action a applied in belief state b resulting to belief state b':*

$$\rho(b, a, b') = \sum_{z \in \Omega_Z} b(z) \sum_{z' \in \Omega_Z} b'(z')c(z, a, z').$$

In essence, ρ corresponds to the expectation of the running cost over the states following distributions given by the beliefs.

The *value* of a history-dependent policy can be equivalently defined within the belief MDP framework (Åström 1965).

THEOREM 6.2.– *POMDP-belief MDP equivalence. Consider a finite-horizon POMDP $\langle \Omega_Z, A, T, P, \Omega_Y, O, c, b_0 \rangle$ and the corresponding belief MDP $\langle \mathcal{B}, A, T, \mathcal{P}, \rho \rangle$, as well as a history-dependent policy $\pi \in \Pi^o$. The value of π can be computed as:*

$$V_T(\pi, b_0) = \mathbb{E}^{\pi}_{b_0} \left[\sum_{t=0}^{T-1} \rho(b_t, \pi_t(h_t), b_{t+1}) \right].$$

Note that belief MDPs are continuous state space MDPs.

6.2.3.3. *Squirrel toy model as a finite state and action POMDP*

Consider the same setting as in section 6.2.2.4 with the squirrel traveling to different hiding places $m \in \{1, \cdots, 4\}$, but now consider you do not fully observe the trajectory of the squirrel. You only observe the presence of the squirrel when the location of the camera corresponds to the squirrel's location. Even then, the time spent by the squirrel in its hiding place remains hidden to you. Denote by $y_t = \mathbb{1}_{\{a_t = m_t\}}$ the observation at time t, that is, 1 if the squirrel is in the same place as your camera, 0 otherwise. The objective remains the same, that is, maximizing the number of pictures of the squirrel taken over the span of the week.

The POMDP associated with this example is based on section 6.2.2.4 with the additional quantities, assuming the camera has just observed the arrival of the squirrel in site $m_0 \in \{1, ..., 4\}$, so that $y_0 = 1$.

$$\Omega_Y = \{0, 1\},$$
$$O(z' = (m', r'), a, y) = \mathbb{1}_{\{a = m'\}} y + \mathbb{1}_{\{a \neq m'\}} (1 - y).$$

This POMDP can be solved, for example, by computing the corresponding belief MDP, which is a finite-horizon MDP, in this case, and solving it backwards.

6.2.4. *Solution algorithms for MDPs and POMDPs*

For finite-horizon MDPs with a known starting state, solving the MDP involves exploring a *min-expectation* tree. The root of this tree corresponds to the initial state, z_0. Each action $a_0 \in K(z_0)$ available at z_0 generates a successor action node. Each action node a_0 further branches into successor state nodes for all z_1 such that $P(z_1|z_0, a_0) > 0$. This process continues across time steps, with the tree's leaves representing the final states z_T. Solving the MDP using exact dynamic programming (theorem 1) involves fully constructing this tree and applying backward induction. Backward induction labels each state node z_t with its optimal value $V_t^*(z_t)$, which represents the return obtained by following the optimal policy π^* from z_t until the horizon T. However, backward induction is computationally intensive because it requires exploring the entire tree. Even storing the complete tree is infeasible in many cases, though depth-first exploration with pruning of dominated actions can help mitigate this storage issue.

To address the computational challenges, Monte Carlo planning methods, and in particular the family of Monte Carlo tree search (MCTS) algorithms, offer an efficient alternative. These heuristic-free approaches, such as the well-known UCT (Upper Confidence bound applied to Trees) algorithm (Kocsis and Szepesvari 2006), focus on exploring the tree through simulations. MCTS algorithms prioritize exploration of the most promising parts of the tree area that show high potential or require further evaluation to refine their estimated values. By doing so, they avoid exhaustive exploration, making them well-suited for large or complex MDPs. There exist many variants of MCTS (Browne et al. 2012; Świechowski et al. 2023) however most of them iteratively update a search tree (which is not exactly the same as the dynamic programming tree) until a computation budget is exhausted, at which point an action, computed for the root node, is returned.

Reinforcement learning (RL) is a computational framework that enables decision-makers to learn optimal mappings from situations to actions in order to minimize a numerical cost. In RL, the environment represents everything beyond the decision-maker's direct control. In the context of MDPs, the environment corresponds to the mathematical structure that defines state transitions and costs. Unlike traditional learning methods, RL does not require complete knowledge of this structure. Instead, it allows decision-makers to learn directly through interactions with the environment. The process of learning optimal actions in RL is approached from multiple perspectives, each characterized by distinct methodologies and challenges. On the one hand, *model-free methods* do not rely on a model of the environment's dynamics. Within this category, *Value-based* approaches focus on estimating the value function to derive optimal policies, while *Policy-based* methods directly search for the optimal policy within the policy space (Sutton and Barto 2018). On the other hand, *model-based methods* involve constructing an explicit model of the environment to predict future states and associated costs while learning (Atkeson and Santamaria 1997).

Recent advancements in deep learning (LeCun et al. 2015; Goodfellow et al. 2016) have significantly enhanced RL capabilities, particularly in addressing challenges posed by high-dimensional state and action spaces. These developments have given rise to deep reinforcement learning algorithms. These algorithms show considerable promise for advancing the field.

For practitioners, implementation can begin by encoding their environment using the Gymnasium python library introduced by Brockman et al. (2016). This library integrates seamlessly with the RLlib library (Liang et al. 2018), which offers a suite of pre-implemented deep RL algorithms ready for use.

6.3. Piecewise deterministic Markov processes

PDMPs are a general class of non-diffusion processes introduced by M. Davis in the 1980s (Davis 1984) covering a wide range of applications, from workshop optimization, queuing theory, internet networks, to reliability, insurance and finance or biology for instance. These continuous-time processes extend both the class of Markov point processes (Davis 1993; Jacobsen 2006; Cocozza-Thivent 2021) and semi-Markov point processes and are characterized by deterministic motions punctuated by random jumps. They are especially suitable to model complex systems with interacting random phenomena, or large population asymptotics.

This presentation of PDMPs is divided into two main topics. In section 6.3.1, we give the definition and examples of use of PDMPs and provide their key features such as simulation properties. In section 6.3.2, we provide a survey on the impulse control problem for this class of processes.

6.3.1. *PDMP definition*

In section 6.3.1.1, we provide a concise formal definition of PDMPs. The special case of semi-Markov dynamics is explored in section 6.3.1.2. Representative examples of PDMPs from their main application domains are presented in section 6.3.1.3. A PDMP version of the squirrel toy model is presented in section 6.3.1.4. We then delve into tools related to PDMPs, beginning with simulation techniques in section 6.3.1.5. The identification and analysis of specific sub-chains of interest are addressed in section 6.3.1.6.

6.3.1.1. *Generic definition*

We start with a formal definition of a PDMP, its state space and its local characteristics.

DEFINITION 6.16 (PDMP).– *A piecewise deterministic Markov process* $Z = (Z_t)_{t\geq 0}$ *is defined by a tuple* $\langle \Omega, \Phi, \lambda, Q \rangle$, *where*

– *the* state space Ω *is a finite union of Borel subsets of* $\mathbb{R}^d$, *for some* $d \geq 1$. *It is the set of all possible states of the process. Let* $\mathcal{B}(\Omega)$ *be its Borel* σ*-field. We denote* $\bar{\Omega}$ *as the closure of* Ω *and* $\mathcal{B}(\bar{\Omega})$ *as the Borel* σ*-field on* $\bar{\Omega}$;

– *the* flow Φ *is a continuous function from* $\Omega \times \mathbb{R}_+$ *onto* $\bar{\Omega}$ *satisfying a semi-group property, that is,* $\Phi(\cdot, t+s) = \Phi(\Phi(\cdot, t), s)$ *for all* $s, t \in \mathbb{R}_+$. *The flow prescribes the deterministic motion between jumps. It may be defined by an explicit function or described as the solution to a system of ordinary differential equations (ODEs);*

– *the* jump intensity, *or hazard rate, or risk function* λ *is a measurable function from* Ω *onto* $\mathbb{R}_+$ *that determines the occurrence of random jumps and is such that for any* z *in* Ω, *there exists* $\epsilon > 0$ *such that:*

$$\int_0^{\epsilon} \lambda(\Phi(z,t))dt < +\infty,$$

forbidding instantaneous jumps;

– *the* jump kernel Q *is a Markov kernel on* $(\mathcal{B}(\Omega), \bar{\Omega})$ *that selects the new location of the process after each jump. It satisfies* $Q(\{z\}|z) = 0$ *so that a jump has to change the state of the process.*

The PDMP dynamics can be described informally as follows: starting from some initial point $z \in \Omega$, the motion of the process follows the deterministic flow $t \mapsto \Phi(z,t)$ until the first jump time S_1. Jumps may occur via two means: *random jumps* occur from the realization of the random clock with intensity λ, while *deterministic jumps* occur when the process reaches the boundary $\partial\Omega$ of the state space. Thus, starting from z at time 0, the first jump time S_1 has the following distribution:

$$\mathbb{P}_z(S_1 > t) = \mathbb{P}(S_1 > t | Z_0 = z) = \exp\left(-\int_0^t \lambda(\Phi(z,s))ds\right) \mathbb{1}_{\{t<t^*(z)\}}, \quad t \geq 0,$$

where $t^*(z)$ is the deterministic time the flow takes to reach the boundary $\partial\Omega$ of Ω when it starts from z:

$$t^*(z) = \inf\{t > 0 : \Phi(z,t) \in \partial\Omega\}.$$

For convenience, we also define:

$$\Lambda(z,t) = \int_0^t \lambda(\Phi(z,s))ds.$$

It is the cumulative risk up to t starting from z. At S_1, the process jumps to a new point $z' = Z_{S_1}$ selected with probability $Q(dz'|\Phi(z, S_1))$ (conditional on S_1), and the motion restarts from this new point as before. A generic representation of a

PDMP is given in Figure 6.3. The flow Φ, the jump rate λ and the Markov kernel Q are called the *local characteristics* of the PDMP. After the first jump time S_1, the process restarts from a new state selected by Q and follows the flow until the next jump time S_2. The sojourn duration $X_2 = S_2 - S_1$ until the next jump is drawn from the jump intensity and time to reach the boundary as before. At the jump time S_2, the post-jump location is selected by the Markov kernel Q and so on iteratively, as described in Algorithm 6.2. Note that Algorithm 6.2 can only simulate the process up to time $S_\infty = \lim_{n\to\infty} S_n$ that may be finite. This phenomenon is called *explosion.* To avoid explosion, we usually also requires that for all $z \in \Omega$, we have:

$$\mathbb{E}_z[\lim_{n\to\infty} S_n] = +\infty.$$

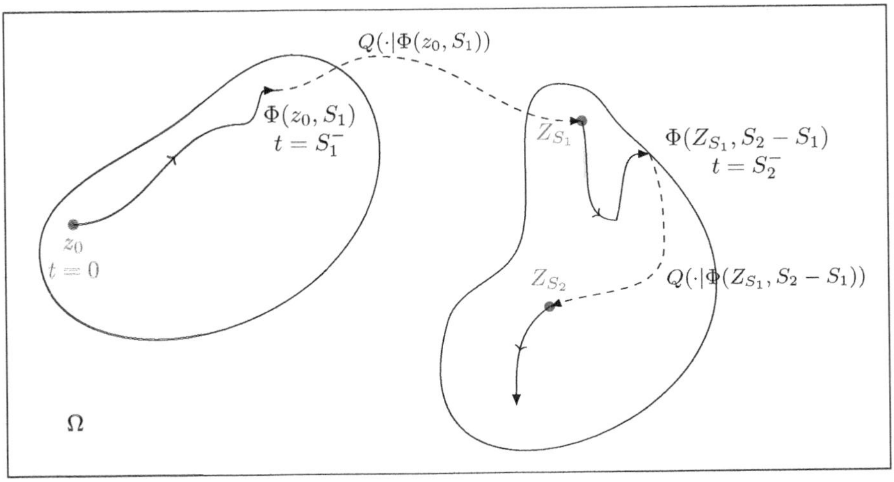

Figure 6.3. *Trajectory of a generic PDMP. Starting from an initial value z_0 at time $t = 0$, the process follows a deterministic trajectory until a jump occurs, either at a random time (as at S_1) or because the process reaches the state boundary (as at S_2). At jump times, the process jumps to a new location drawn from kernel Q*

A process Z defined by a flow Φ, an intensity function λ and a kernel Q satisfying the above properties is well defined and is strong Markov, as is shown, for instance, in Davis (1993) or Jacobsen (2006).

It is particularly convenient, when dealing with PDMPs, to separate the state space Ω and states z into discrete and Euclidean components. In this case, the state space is called *hybrid.* Let M be a finite set, and for all $\mathsf{m} \in \mathsf{M}$, let Ω_m be some Borel subset of $\mathbb{R}^{d_\mathsf{m}}$, which dimension d_m is allowed to change with m. The state space is thus:

$$\Omega = \bigcup_{\mathsf{m}\in\mathsf{M}} \{\mathsf{m}\} \times \Omega_\mathsf{m}.$$

Algorithm 6.2 Iterative construction of a trajectory of a PDMP starting from state z_0 at time 0 and up to the N-th jump time

$z \leftarrow z_0$
$S \leftarrow 0$
for $n \leftarrow 0$ to $N-1$ **do**
  $X \sim \lambda(z)$
  $X \leftarrow \min\{X, t^*(z)\}$
  $Z_t \leftarrow \Phi(z, t-S)$ for $S \leq t < S+X$
  $z' \sim Q(\cdot|\Phi(z, X))$
  $z \leftarrow z'$
  $S \leftarrow S + X$
end for

Any state $z \in \Omega$ can be written as $z = (\mathsf{m}, \mathsf{z})$ with $\mathsf{m} \in \mathsf{M}$ and $\mathsf{z} \in \Omega_\mathsf{m}$. The discrete component m is called the mode or regime and z is called the Euclidean component. In between jumps, mode m remains constant and only z evolves through the flow. At jump times, both the mode and the Euclidean component may be allowed to change through the Markov kernel Q.

In the hybrid setting, we set:

$$\begin{aligned}
\partial\Omega &= \bigcup_{\mathsf{m}\in\mathsf{M}} \{\mathsf{m}\} \times \partial\Omega_\mathsf{m}, \\
\Phi(z,t) &= (\mathsf{m}, \Phi_\mathsf{m}(\mathsf{z}, t)), \\
\lambda(z) &= \lambda_\mathsf{m}(\mathsf{z}), \\
t^*(z) &= t^*_\mathsf{m}(\mathsf{z}), \\
\Lambda(z,t) &= \Lambda_\mathsf{m}(\mathsf{z}, t) \\
Q(\cdot|z) &= Q_\mathsf{m}(\cdot|\mathsf{z}),
\end{aligned}$$

for all $z = (\mathsf{m}, \mathsf{z}) \in \Omega$ and $t \in \mathbb{R}_+$.

The particular case where there are modes but no Euclidean variables corresponds to continuous-time Markov chains, which generalize Markov chains to continuous time. We will see in the next section that the PDMP framework can also encompass semi-Markov dynamics, and therefore it can also generalize semi-Markov chains.

6.3.1.2. *Semi-Markov dynamics*

PDMPs allow for state-dependent intensities, therefore they offer more flexibility for sojourn durations than the exponential distribution, while remaining Markovian. However, this flexibility may still be too limited. For instance, in modes where the flow is constant, the jump intensity also has to be constant, that is, it corresponds to a memory-less exponential distribution. In many practical examples (see, e.g. (Barbu and Limnios 2008)), the exponential distribution may not be realistic to model the underlying phenomenon. However, allowing for more general distributions breaks the Markov property and falls into the class of *semi-Markov* processes.

It is especially easy to encompass semi-Markov dynamics into the PDMP framework by enlarging the state space. This new process is sometimes called the *time-augmented process*; see Davis (1993, Section 31.5) for a slightly different definition in the same spirit.

DEFINITION 6.17 (Time-augmented PDMP).– *A time-augmented piecewise deterministic Markov process is a PDMP defined by the tuple $\langle \Omega, \Phi, \lambda, Q \rangle$, where:*

– The state space Ω has the specific form:

$$\Omega = \bigcup_{\mathsf{m} \in \mathsf{M}} \{\mathsf{m}\} \times \Omega_{\mathsf{m}} \times \mathbb{R}_{+},$$

for some finite mode set M*, where for all* $\mathsf{m} \in \mathsf{M}$*,* Ω_{m} *is some Borel subset of* $\mathbb{R}^{d_{\mathsf{m}}}$.

– The flow Φ has the specific form:

$$\Phi(z, t) = (\mathsf{m}, \Phi_{\mathsf{m}}(\mathsf{z}), u + t),$$

for $z = (\mathsf{m}, \mathsf{z}, u) \in E$, where Φ_{m} are flows from $\Omega_{\mathsf{m}} \times \mathbb{R}_{+}$ onto Ω_{m} satisfying the semi-group property.

– The jump intensity just satisfies the conditions from definition 6.16 and is allowed to depend on all coordinates of $z \in E$, including the new time variable u.

– The jump kernel Q has the specific form:

$$Q(\{m'\} \times A \times B | z) = \mathbb{1}_{0 \in B} Q_{\mathsf{m}}(\{m'\} \times A | \mathsf{z}, u),$$

for $z = (\mathsf{m}, \mathsf{z}, u) \in E$, A any Borel subset of $\mathsf{E}_{\mathsf{m}'}$, B any Borel subset of $\mathbb{R}_{+}$, and where Q_{m} is a Markov kernel on $(\mathcal{B}(\cup\{\mathsf{m}\} \times \Omega_{\mathsf{m}}), \overline{\Omega}_{\mathsf{m}})$.

The additional component u can be interpreted as the time elapsed since the last jump. It moves along the flow at speed 1 and is reset to 0 at every jump.

Figure 6.4 illustrates an example, inspired from the repair workshop model of (Davis 1993, Sec. 2.1), of a time-augmented PDMP that includes boundaries. Consider a machine in a factory. It can be working ($m = 1$) in the factory, under repair ($m = 2$) in the workshop, or under maintenance ($m = 3$) in the workshop. Random jumps from mode 1 to mode 2 are triggered by random failures, with an intensity increasing with the working-time of the machine. Deterministic jumps from mode 1 to mode 3 are scheduled every 30 days to perform maintenance if there is no failure, and deterministic jumps from 2 or 3 to 1 correspond to the deterministic time to repair the machine or realize its maintenance. Here, the PDMP has a mode variable that corresponds to the state of the semi-Markov chain, and an Euclidean variable that corresponds to the time spent thus far in its current state. The later corresponds to the minimum of a random variable with a given intensity and the deterministic time to reach the given boundaries.

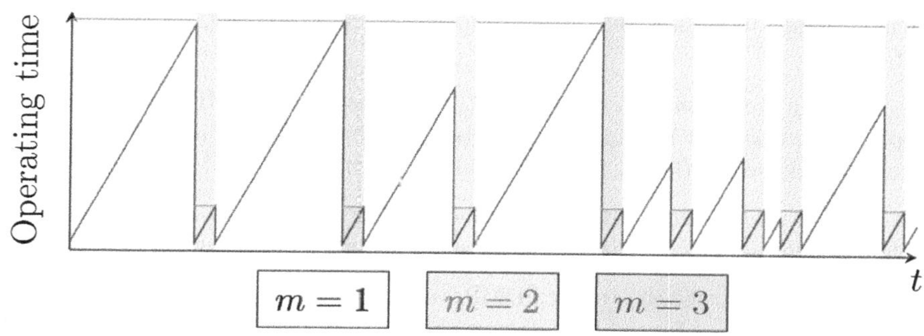

Figure 6.4. *Sample trajectory a repair workshop PDMP model. The PDMP has three modes corresponding to the state of a machine: $m = 1$ working in the factory, $m = 2$, $m = 3$ under repair or maintenance in the workshop. The only Euclidean variable u is the time spent in the factory or workshop. It is deterministic equal to 5 days in the workshop. In the factory, the machine fails with intensity $\lambda_1(u) = \alpha u$, or is sent to the workshop for maintenance if it has been working for 30 days. For a color version of this figure, see www.iste.co.uk/peyrard/guidetohsmm.zip*

One could define more sophisticated time-augmented processes where either u is never reset to 0 (it is thus the time since the beginning, as in (Davis (1993), section 31.5) or resets less frequently than at each jump; see, for example, (de Saporta et al. 2016, Sec. 1.8) for such examples. It is also possible to have multiple time counters, if relevant.

6.3.1.3. *PDMPs as versatile models*

PDMPs are constructed using a small set of intuitive and easily interpretable local characteristics, making them a highly versatile and powerful modeling tool, enabling the representation of a broad spectrum of complex phenomena across various domains.

PDMPs encompass and generalize many classes of Markov processes, including:

– discrete-time and space Markov chains (corresponding to a time-augmented PDMP with no Euclidean variables, null intensity, jumps triggered at boundaries at each time unit);

– continuous-time and discrete state space Markov processes (corresponding to a PDMP with no Euclidean variables and constant intensity for each mode);

– semi-Markov discrete state space processes (corresponding to a time-augmented PDMP with no Euclidean variables and time-dependent intensity), as seen in section 6.3.1.2.

Examples from the PDMPs literature cover a wide range of applications:

– *Workshop optimization and reliability*: the state of the PDMP is the state of a machine subject to failures and repairs or maintenance (Davis 1993; de Saporta et al. 2016).

– *Queuing theory and internet networks*: the state of the PDMP is the number of people in the queue or the load of the network, subject to random or deterministic increases or decreases (Davis 1993; Chafaï et al. 2010; Bardet et al. 2013).

– *Finance and insurance*: the state of the PDMP is the level of an asset subject to random or deterministic increases or decreases (Bäuerle and Rieder 2011).

– *Neurosciences*: the state of the PDMP is the potential of a neuron or of a network of interacting neurons for the Euclidean variables, with modes corresponding to the number of open ionic channels, which opening and closing dynamics depend on the potential and vice-versa (Riedler et al. 2012; Riedler and Thieullen 2015).

– *Genomics*: the state of the PDMP is the state of a gene promoter, level of transcribed RNA and translated protein, either for a single gene or an interacting network of genes (Herbach et al. 2017).

– *Population dynamics*: the state of the PDMP is some trait of an individual or interacting individuals subject to birth, death, division, migration, mutations or other phenomena (Fritsch 2014; Doumic et al. 2015; Costa 2016).

– *Medical treatment*: the state of the PDMP is the state of some marker for a disease and the treatment choice for the individual (Pasin et al. 2018; Cleynen and de Saporta 2018; de Saporta et al. 2024).

6.3.1.4. *Squirrel toy model as a PDMP*

The setting builds on the examples of sections 6.2.2.4 and 6.2.3.3. The squirrel now changes its location in continuous time, and the camera is in a fixed location ℓ. We thus add the camera location to the state space. The state of the PDMP is $z = (m, \ell, u)$, where m is the hiding place, ℓ is the (fixed) location of the camera and u is the time spent by the squirrel in its current hiding place (it corresponds to variable r in sections 6.2.2.4 and 6.2.3.3). The PDMP associated with this example is described by:

$$\begin{aligned}\Omega &= \{1, \cdots, 4\} \times \{1, \cdots, 4\} \times [0, 7], \\ \Phi(z, t) &= (m, \ell, u + t), \\ \lambda(m, \ell, u) &= \lambda(m, u) = \sum_{m'=1}^{4} f(m, m', u), \\ Q(\{m'\} \times \{\ell'\} \times A|z) &= \mathbb{1}_{\zeta \in A} \mathbb{1}_{\ell'=\ell} \frac{f(m, m', u)}{\lambda(m, u)}.\end{aligned}$$

Here, $f(m, m', u)$ corresponds to the intensity to jump from hiding place m to hiding place m' if the squirrel has spent a time u in hiding place m. A trajectory of the associated PDMP is illustrated in Figure 6.5.

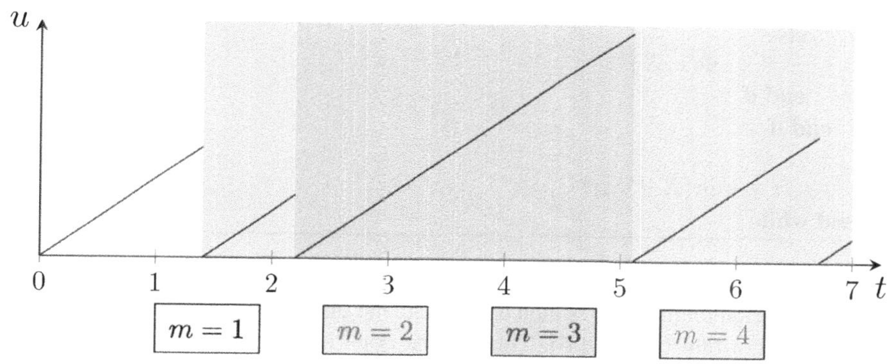

Figure 6.5. *Sample trajectory of the squirrel toy example. The PDMP has a two-dimensional mode corresponding to the squirrel hiding place m and the camera location ℓ (not shown on the figure). The Euclidean variable corresponds to the time spent by the squirrel in its current hiding place. For a color version of this figure, see www.iste.co.uk/peyrard/guidetohsmm.zip*

6.3.1.5. *Simulation*

A very interesting property of PDMPs is that they are particularly easy to simulate due to their intrinsic iterative construction as seen in Algorithm 6.2. Computing the trajectory along the flow (Algorithm 6.2, line 6) is straightforward if the flow is explicit. If it is only known as the solution to a system of ODEs, one just needs to use any suitable ODE solver. Computing the post-jump location from the kernel Q (Algorithm 6.2, line 7) is also usually straightforward. In most applications, the kernel only changes the mode and leaves the Euclidean variable unchanged, or changes the Euclidean variable through some known mapping depending only on its value just before the jump; see examples in section 6.3.1.3.

Algorithm 6.3 Stochastic Simulation Algorithm for a PDMP starting from state $z_0 = (\mathsf{m}_0, \mathsf{z}_0)$ at time 0 and up to time horizon T

$z \leftarrow z_0$
$t \leftarrow 0$
while $t < T$ **do**
  $X \sim \text{Exp}(\overline{\lambda})$
  $Z_{t+s} \leftarrow \Phi(z, s)$ for $0 \leq s < \min\{X, t^*(z)\}$
  if $X > t^*(z)$ **then**
    $z' \sim Q(\cdot|\Phi(z, t^*(z)))$
  else
    $U \sim U[0, 1]$
    if $U \leq \frac{\lambda(\Phi(z,X))}{\overline{\lambda}}$ **then**
      $z' \sim Q(\cdot|\Phi(z, X))$
    else
      $z' \leftarrow \Phi(z, X)$
    end if
  end if
  $z \leftarrow z'$
  $t \leftarrow t + \min\{X, t^*(z)\}$
end while

The only challenging step is then to simulate the jump dates from the non-constant intensity (Algorithm 6.2, line 4). This can be done by an exact simulation method called the stochastic simulation algorithm (SSA), or Gillespie algorithm (Gillespie 1977) that is especially suitable for PDMPs.

The principle underlying the SSA algorithm is to sample from an exponential distribution which (fixed) parameter is an upper bound of the intensity, and accept or reject the proposed date for a jump with a well chosen probability in order to retrieve the target distribution. Even if the date is rejected as a jump, time moves forward at each iteration as detailed in Algorithm 6.3.

The choice of the upper bound $\overline{\lambda}$ can have a strong impact on the speed of the algorithm. If it is too high, it will keep proposing very short times and move forward extremely slowly. If at all possible, one should choose a tight upper bound for the intensity.

6.3.1.6. *Embedded chains*

As for any continuous-time Markov process, one can define many different discrete-time Markov chains embedded in a PDMP. Some of them are particularly interesting for decision-making problems and are specified here.

The *canonical chain* is the most natural Markov chain embedded in a PDMP. It corresponds to the post-jump-locations and inter-jump-times (or sojourn durations) chain. Let us define $J_0 = Z_0$ and $X_0 = 0$, and then iteratively, for all $n \in \mathbb{N}^*$, $J_n = Z_{S_n}$ the value of the PDMP at the *n*th jump time and $X_n = S_n - S_{n-1}$ the inter-jump sojourn time between the $n - 1$th and nth jumps. Then $(J_n, X_n)_{n\in\mathbb{N}}$ is a Markov chain called the *canonical embedded chain* associated with the PDMP process Z. Its transition kernel $P^{J,X}$ is defined by, for any Borel subsets A of Ω and B of $\mathbb{R}_+$:

$$\begin{aligned} P^{J,X}(A \times B|z,t) &= P^{J,X}(A \times B|z) \\ &= \int_0^{t^*(z)} \lambda(\Phi(z,s))e^{-\Lambda(z,s)}Q(A|\Phi(z,s))\mathbb{1}_B(s)ds \\ &\quad + e^{-\Lambda(z,t^*(z))}Q(A|\Phi(z,t^*(z)))\mathbb{1}_B(t^*(z)), \end{aligned}$$

where the first part of the equation corresponds to random jumps, while the second part corresponds to deterministic jumps.

As the motion of the PDMP is deterministic between jumps, it is straightforward to reconstruct the full trajectory of the continuous-time PDMP from the knowledge of $(J_n, X_n)_{n\in\mathbb{N}}$ as:

$$Z_t = \Phi\left(J_n, t - \sum_{i=0}^{n} X_i\right) \quad \text{if } t \in \left[S_n = \sum_{i=0}^{n} X_i, S_{n+1} = \sum_{i=0}^{n+1} X_i\right),$$

with the convention $S_0 = X_0 = 0$. Thus, the canonical chain contains all the information of the process and is especially useful to obtain recursive formulations based on the Markov property. It is the counterpart of the standard (H)SMM introduced in section 1.2.3. It is also at the heart of several numerical approximations for PDMPs (see, e.g. de Saporta et al. (2016)).

Other common embedded chains are the *skeleton* chains defined by sampling the PDMP on a deterministic (but not necessarily homogeneous) grid $(t_1, \ldots, t_n)$. We simply denote $Z_n = Z_{t_n}$, which once again is a (time-inhomogeneous) Markov chain with kernels $P_{t_{n+1}-t_n}$. While kernels $P_{t_{n+1}-t_n}$ might still have explicit forms, unless constraints are added on the PDMP there is no a priori limit in the number of jumps occurring between two time-points t_n and t_{n+1}, which prevents a generic formulation of the kernels. Moreover, unless jump-points are observed, it is not possible to reconstruct the original PDMP process from the observation of $(Z_n)_{n\in\mathbb{N}}$. However, in many application settings, Z_n might be the only available information from the process; see, for instance, the squirrel toy model developed in section 6.4.2.2.

6.3.2. *Impulse control for PDMPs*

In this section, we delve into the control of PDMP. As in the MDP framework, a stochastic control problem involves influencing the dynamics of the process through control variables or decisions to either maximize a reward or minimize a cost. The primary questions of interest include the characterization and regularity properties of the *value function* – representing the optimal reward or cost achievable across all possible decisions or decision sequences – and the existence and properties of optimal decision sequences, called *strategies*.

In this chapter, we focus only on *impulse control*, where at *discrete* decision-maker-chosen times, the process is sent to a new decision-maker-chosen point in the state space. The simplest form of impulse control is *optimal stopping* where the decision-maker selects a single date at which the process is stopped. Impulse control problems for PDMPs have been studied from the theoretical point of view: characterization of the value function and optimal policies through dynamic programming in Gugerui (1986); Costa and Davis (1988); Cloez et al. (2020) for optimal stopping and Costa and Davis (1989); de Saporta et al. (2017) for the general impulse control problem. Numerical schemes to approximate the value function and compute explicit policies close to optimality have also been proposed in Costa and Davis (1989) and de Saporta et al. (2016).

We define the impulse control problem for PDMPs in section 6.3.2.1, and discuss resolution strategies in section 6.3.2.2. We conclude this part with a discussion on the special case of control problems involving hidden information in section 6.3.2.4.

6.3.2.1. *Definition of the impulse control problem*

The formal probabilistic apparatus necessary to rigorously define the impulse control problem for PDMPs is rather technical and will not be used in the following; therefore, for the sake of simplicity, we only present an informal description of the problem. The interested reader is referred to Costa and Davis (1989) or Dufour et al. (2016) for a proper definition.

DEFINITION 6.18 (Impulse strategy).– *A general impulse strategy* $\mathcal{S} = (\tau_n, \chi_n)_{n\geq 1}$ *is a sequence of non-anticipative* $\mathbb{R}_+$*-valued intervention times* τ_n *and* $\mathbb{U} \subset \Omega$*-valued non-anticipative random variables* χ_n.

The trajectory of the PDMP controlled by strategy $\mathcal{S}$ *is described in Algorithm 6.4. Between intervention dates, the PDMP follows its natural trajectory. At the nth impulse (line 12), the process is instantaneously moved to the new starting point* χ_n *and restarts from there.*

The set of admissible strategies is denoted by $\mathbb{S}$.

Algorithm 6.4 Simulation of a trajectory of a PDMP controlled by strategy $\mathcal{S}$ starting from state z_0 up to the n-th intervention

$z \leftarrow z_0$
$S \leftarrow 0$
for $n \leftarrow 1$ to N **do**
  while $S < \tau_n$ **do**
    $X \sim \lambda(z)$
    $X \leftarrow \min\{X, t^*(z)\tau_n - S\}$
    $Z_t \leftarrow \Phi(z, t - S)$ for $S \leq t < S + X$
    if $X < \tau_n - S$ **then**
      $z' \sim Q(\cdot|\Phi(z, X))$
      $z \leftarrow z'$
      $S \leftarrow S + X$
    else
      $z' \leftarrow \chi_n$
      $z \leftarrow z'$
      $S \leftarrow \tau_n$
    end if
  end while
end for

For a strategy $\mathcal{S} = (\tau_n, \chi_n)_{n\geq 1}$. τ_n and χ_n are random variables and must be measurable with respect to the past trajectories of the controlled process up to the nth impulse. Hence, they may depend on the whole past of the process; see Costa and Davis (1989) for details.

DEFINITION 6.19 (Cost of an impulse strategy).– *The total discounted cost of strategy $\mathcal{S}$ for a PDMP starting at z at time 0 and up to infinity is defined by:*

$$\mathcal{V}(\mathcal{S}, z) = \mathbb{E}_z^{\mathcal{S}}\left[\int_0^{+\infty} e^{-\gamma t} c_R(Z_t)dt + \sum_{n=1}^{\infty} e^{-\gamma\tau_n} c_I\left(Z_{\tau_n}, Z_{\tau_n^+}\right)\right],$$

where $\gamma \geq 0$ is a discount factor, c_R is the running cost and c_I is the impulse cost.

The specific assumptions that the cost functions must satisfy are omitted here and can be found in detail in Costa and Davis (1989). For simplicity, we present an infinite-horizon criterion. Alternative criteria are also possible, such as those with a fixed horizon T or a random horizon defined by a prescribed number of jumps or impulses, as briefly discussed after theorem 3.

DEFINITION 6.20 (Value function).– *The value function of strategy $\mathcal{S}$ for a PDMP starting at z at time 0 is defined by:*

$$\mathcal{V}^\star(z) = \inf_{\mathcal{S}\in\mathbb{S}} \mathcal{V}(\mathcal{S}, z).$$

In most impulse control problems for PDMPs, there is no optimal strategy as the infimum may not be reached. We define instead ϵ-optimal strategies.

DEFINITION 6.21 (ϵ-Optimal strategy).– *An ϵ-optimal strategy $\mathcal{S}_\epsilon$ is an admissible strategy satisfying:*

$$\mathcal{V}(\mathcal{S}_\epsilon, z) \leq \mathcal{V}^\star(z) + \epsilon,$$

for all $z \in \Omega$.

Under suitable assumptions (see again Costa and Davis (1989) for details), the value function can be characterized as the unique solution of some dynamic programming equations.

THEOREM 6.3.– *Dynamic programming. The optimal value function $\mathcal{V}^\star$ is the limit of the following recursion. Set $\mathcal{V}_0^\star(z) = \mathcal{V}(\mathcal{S}_\emptyset, z)$ and $\mathcal{V}_{n+1}^\star(z) = \mathcal{L}\mathcal{V}_n^\star(z)$ for all $n \geq 0$, then for all z in Ω, one has:*

$$\mathcal{V}^\star(z) = \lim_{n\to+\infty} \mathcal{V}_n^\star(z),$$

where $\mathcal{V}(\mathcal{S}_\emptyset, z)$ is the cost associated with the no-impulse strategy $\mathcal{S}_\emptyset = (\tau_1 = +\infty)$ starting from z:

$$\mathcal{V}(\mathcal{S}_\emptyset, z) = \mathbb{E}_x^{\mathcal{S}_\emptyset}\left[\int_0^\infty e^{-\gamma t} c_R(Z_t)dt\right],$$

and $\mathcal{L}$ is the single-jump-or-intervention operator defined as:

$$\mathcal{L}V(z) = \inf_{t\in\mathbb{R}+} J(MV, V)(z,t) \wedge KV(z),$$

with

$$MV(z) = \inf_{\chi\in\mathbb{U}}\{c_I(z,\chi) + V(\chi)\},$$

$$J(V,W)(z,t) = \int_0^{t\wedge t^*(z)} e^{-\gamma s - \Lambda(z,s)} \left(c_R(\Phi(z,s)) + \lambda(\Phi(z,s))QW(\Phi(z,s))\right) ds$$

$$+ e^{-\gamma(t\wedge t^*(z)) - \Lambda(z,t\wedge t^*(z))} V(\Phi(x, t\wedge t^*(z))),$$

$$KV(z) = \int_0^{t^*(z)} e^{-\gamma s - \Lambda(z,s)} \left(c_R(\Phi(z,s)) + \lambda(\Phi(z,s))QV(\Phi(z,s))\right) ds$$

$$+ e^{-\gamma t^*(z) - \Lambda(z,t^*(z))} QV(\Phi(z, t^*(z))).$$

Operator M selects the best starting point χ in the control set $\mathbb{U}$ after an intervention. Note that the control set could also depend on the position z before the intervention. Operator $\inf_t J$ selects the best intervention date along the flow, while operator K accounts for cases when it is best to wait for the next natural jump of the process before triggering an intervention. The iterations in the dynamic programming equation thus correspond either to natural jumps of the process or to interventions, hence the name *single-jump-or-intervention* operator. It is based on the Markov property for the canonical embedded chain, and operators can be reformulated as expectations involving the canonical embedded chain (see de Saporta and Dufour 2012).

The iterate $\mathcal{V}_n^*$ can be interpreted as the optimal value function of the impulse control problem where at most n jumps or interventions are allowed, and then strategy $\mathcal{S}_\emptyset$ is applied. Computing the cost of the no-impulse strategy on the whole space may be demanding, as an analytical form is often out of reach and one must turn to Monte Carlo simulations. As the expectation depends on the starting point $z \in \Omega$, one should compute infinitely many Monte Carlo approximations to cover the whole space. Theorem 3 is also valid when replacing $\mathcal{V}(\mathcal{S}_\emptyset, z)$ by any function g such that $\mathcal{V}(\mathcal{S}_\emptyset, z) \leq g(z)$ for all $z \in \Omega$.

We have stated the results for an infinite horizon only. If one is interested in controlling the process up to a finite time horizon T, one just needs to kill the process, or send it to a cemetery state, when time T is reached, and define a null cost on this state to fall back to the infinite horizon framework. The discount factor γ can be set to 0 in this context as the interval of integration is finite. Hence, the cost of strategy $\mathcal{S}$ can typically be defined, in the fixed horizon T, by:

$$\mathcal{V}_{\mathsf{T}}(\mathcal{S}, z) = \mathbb{E}_z^{\mathcal{S}}\left[\int_0^{\mathsf{T}} c_R(Z_t)dt + \sum_{\tau_n < \mathsf{T}} c_I\left(Z_{\tau_n}, Z_{\tau_n^+}\right) + c_T(Z_{\mathsf{T}})\right],$$

where c_T corresponds to some terminal cost function.

6.3.2.2. *Resolution of impulse control problems for PDMPs*

Optimal impulse control of PDMPs attracted attention almost as soon as the concept of PDMP was formalized by Davis. Contributions focusing only on the estimation of the value function are usually based on variational inequalities or value improvement and do not provide explicit strategies (see, for instance, Lenhart (1989); Dempster (1991); Gatarek (1992); Dempster and Ye (1995); Bandini (2018)).

Explicit resolution strategies have also been proposed. In the simpler context of optimal stopping – where the decision-maker can only decide to stop the process without restarting it from a new location – the literature exhibits strategies based on either the discretization of the PDMP Markov jump kernel Q (Costa and Davis 1988); or on the discretization of the Markov kernel of the canonical chain (Costa and Dufour

2008; de Saporta et al. 2010). After discretization, the value function can be estimated via dynamic programming (Gugerui 1986) using the computable discretized kernels.

These approaches have been extended to the more general case of impulse control problems. Costa (1993) considers a discretization of the state space and time to obtain uniform convergence of the discretized problem to the original one. de Saporta and Dufour (2012) consider quantization of the canonical chain to propose an ϵ-optimal strategy. Other approaches, including iteration of a single-jump-or-intervention operator associated with the PDMP, have also successfully been proposed to construct explicit solutions in the infinite horizon context (see, for instance, Costa and Davis (1989); de Saporta et al. (2017)).

Theoretical results on the existence of optimal strategies not relying on the discretization of kernels have also been derived in the case of infinite-horizon impulse control, but they do not provide explicit strategies (see, for instance, Costa and Dufour (2010, 2013)).

All the approaches described above are numerically demanding and work for low dimensional PDMPs with limited complexity. They also require a perfect knowledge of the model and its parameters, together with perfect observation of the process at all times, or equivalently at jumps times.

6.3.2.3. *Squirrel toy model as controlled PDMP*

The setting builds on the PDMP example of section 6.3.1.4. We can now change the location of our camera. A strategy $\mathcal{S} = (\tau_n, \chi_n)_{n \geq 1}$ is then a sequence of (continuous) dates at which we change the camera location and the new location of the process after the change. Here, only the camera location can change: an impulsion moves the process from $z = (m, \ell, u)$ to $\chi = (m, \ell', u)$.

The cost functions are defined by:

$$c_R(z = (m, \ell, u)) = \mathbb{1}_{\{\ell \neq m\}},$$
$$c_I(z, z') = c,$$

and the terminal cost is still 0.

If the impulse cost is null (i.e. $c = 0$), an ϵ-optimal strategy is to move the camera to the new hiding place of the squirrel as soon as possible (with an infinitesimally small delay ϵ) after the jump. If the impulse cost is not zero, the optimal delay before moving the camera may be longer and may depend on the remaining time until the horizon.

6.3.2.4. *Partially observed controlled PDMPs*

A very natural extension to the classical impulse control problem for PDMPs is to allow for hidden information. Suppose that the decision-maker does not have access to the true value of the process, but only to some noisy observation of some of its components. Now decisions can only be taken based on the observations, and not on the hidden state of the process.

The literature is much scarcer for this problem. Constructive resolution approaches have been proposed in the case where jumps are still perfectly observed, but the post-jump locations are only measured through noise. In Brandejsky et al. (2012), the authors consider the equivalent fully observed process on the belief space and apply the quantization approach of de Saporta and Dufour (2012) on the filtered process. Bäuerle and Lange (2018) reduce the problem to a discrete-time MDP and prove the existence of optimal policies, but provide no numerical approximation of the value function or optimal strategies.

In the most challenging case where jump dates are unobserved, the authors have proposed a framework to construct an explicit strategy close to optimality in the optimal stopping context and for more general impulse problems (Cleynen and de Saporta 2021; de Saporta et al. 2024). Those approaches are based on a conversion of the impulse problem into a partially observed MDP, as detailed in section 6.4.2.

6.4. Controlled PDMPs as members of the MDP family

In this section, we demonstrate how impulse control problems for PDMPs can be reformulated within the MDP framework. We also outline how this shift in perspective opens up new avenues for addressing problems that have so far remained unresolved within the PDMP framework.

For example, as discussed in section 6.3.2.4, formulating a rigorous approach to impulse control problems for PDMPs involving hidden jumps and noisy observations presents significant challenges, particularly in defining admissible strategies (see, for instance, Almudevar (2001) or (Costa and Dufour 2013, Section 1.1)). The MDP framework offers a robust structure for overcoming these challenges and for properly formalizing such problems.

Here, we first show how a fully observed impulse control problem for PDMPs can be formulated as a MDP (section 6.4.1), and then turn to the partially observed version (section 6.4.2).

6.4.1. *Controlled PDMPs as MDPs*

We start by encompassing the notion of impulse for PDMPs in the notion of action for MDPs in section 6.4.1.1, then formally state a sub-class of impulse control problems for PDMPs as MDPs in section 6.4.1.2.

6.4.1.1. *Impulse control strategies and MDP policies*

As detailed in sections 6.2.1 and 6.3.2.1, the notions of *actions* for MDPs and *impulses* for PDMPs differ. On the one hand, in the MDP framework, the decision-maker takes decisions at each stage, which act on the dynamics of the process by changing the probability of observing new states through the transition kernel P. On the other hand, in impulse control, the decision-maker typically has two tasks:

1) choosing the next impulse date;

2) at an impulse date, selecting the point from which to restart the process.

To unify both frameworks, we choose to restrict the choice for the impulse strategies of the PDMP decision-maker. Other less restrictive approaches involving more general and complex state spaces are possible and will be briefly discussed at the end of this section. Our main aim here is, on the one hand, to keep the exposition as simple as possible and, on the other hand, to use MDPs with finite-action spaces in view of their numerical resolution. We also consider a finite horizon only for the same reasons.

The first restriction concerns the control set. In the impulse control framework for PDMPs, the decision-maker chooses the new starting point of the process after an impulse, which means they may choose to change either the mode or the location of the process, or both. In the following, we only allow the decision-maker to change the mode of the process.

In this view, we introduce a *mode-augmented* PDMP framework in which we consider two types of modes, one corresponding to the decision-maker-chosen modes and the other to the modes the decision-maker cannot act on. Formally, we decompose the discrete state space M of a PDMP into a two-dimensional finite set $L \times M$, where modes in L can be chosen by the decision-maker while modes in M cannot.

DEFINITION 6.22 (Mode-augmented controlled PDMP).– *A mode-augmented controlled piecewise deterministic Markov process is a PDMP defined by the tuple* $\langle \Omega, \Phi, \lambda, Q \rangle$, *where:*

– *the state space* Ω *has the specific form*

$$\Omega = \{(\ell, m, \mathrm{z}), \ell \in L, m \in M, \mathrm{z} \in \Omega_m^\ell\},$$

for some two-dimensional finite mode set $\mathsf{M} = L \times M$, *and where* $\Omega^{\ell}_{m} = \Omega_{\mathsf{m}=(\ell,m)}$ *is some Borel subset of* $\mathbb{R}^{d_{\mathsf{m}}}$*;*

– *the flow* Φ *just satisfies the conditions from definition 6.16, and we write:*

$$\Phi(z,t) = (\ell, m, \Phi^{\ell}_{m}(\mathsf{z})),$$

for all $z = (\ell, m, \mathsf{z}) \in \Omega$*;*

– *the jump intensity just satisfies the conditions from definition 6.16 and we write:*

$$\lambda_{\mathsf{m}}(z) = \lambda^{\ell}_{m}(\mathsf{z}),$$

for all $z = (\mathsf{m} = (\ell, m), \mathsf{z}) \in \Omega$*;*

– *the jump kernel* Q *satisfies the conditions from definition 6.16 with the additional constraint that* Q *cannot change the value of* ℓ*, as* ℓ *is intended to be decision-maker-chosen. We also set:*

$$Q(\cdot|z) = Q^{\ell}_{m}(\cdot|\mathsf{z}),$$

for all $z = (\ell, m, \mathsf{z}) \in \Omega$*.*

From this rigorously defined mode-augmented PDMP, we can further drop the regime $\ell \in L$ from the state $z \in \Omega$ and include it in the action instead to better fit the standard MDP notation. We denote:

$$\Omega_m = \bigcup_{\ell \in L} \Omega^{\ell}_{m}, \quad \Omega_M = \{(m, \mathsf{z}), m \in M, \mathsf{z} \in \Omega_m\}.$$

We also adapt notations for the kernels $P_{t_{n+1}-t_n}$ associated with the skeleton chain defined in section 6.3.1.6 as follows:

$$\begin{aligned} P^{\ell}_{t_{n+1}-t_n}(B|m,\mathsf{z}) &= P_{t_{n+1}-t_n}(\{\ell\} \times B|\ell, m, \mathsf{z}) \\ &= \mathbb{P}(Z_{t_{n+1}} \in \{\ell\} \times B | Z_{t_n} = (\ell, m, \mathsf{z})), \end{aligned} \qquad [6.1]$$

for any Borel set $B \in \Omega_M$ and $z = (\ell, m, \mathsf{z}) \in \Omega$.

The second restriction concerns the impulse dates. In the MDP framework, decision dates are deterministic and predefined. In contrast, within the impulse control framework for PDMPs, determining the timing of the next impulse is typically part of the decision-making process and is allowed to depend on the current state of the process. In the following, the decision-maker has to make impulses at fixed, regular dates $0, \delta, 2\delta, \ldots, T\delta$, where $T = \mathsf{T}/\delta$ is supposed to be an integer.

In summary, an action a for the MDP corresponding to the controlled PDMP will consist of a decision-maker-chosen mode $\ell \in L$. As mentioned above, other formalisms closer to the original impulse problem for PDMPs are possible. They involve more complex MDPs, with possibly continuous action spaces. In our approach, the strongest restriction is that the delay between consecutive impulses belongs to a fixed grid $\mathbb{T}$, and thus is not allowed to depend on the current state of the process, whereas in the original PDMP framework, decision dates are stopping-times, which allows much more flexibility. This flexibility may be retrieved by defining larger action and constraints spaces (see, e.g. Dufour et al. (2016); de Saporta et al. (2024)). The MDP framework can also increase the flexibility of the impulse control problem for PDMPs as MDP actions are allowed to have a random effect, whereas a PDMP impulse fixes the new starting point of the process.

6.4.1.2. *Formalism of the equivalent MDP*

We can now formally consider impulse control of PDMPs in the MDPs framework.

DEFINITION 6.23 (Controlled PDMP as a MDP).– *Let $\langle \Omega, \Phi, \lambda, Q\rangle$ be a mode-augmented controlled PDMP. The corresponding MDP is defined by the MDP $\langle \Omega_Z, A, T, P, c, C\rangle$ with the following characteristics.*

– The state space is $\Omega_Z = \Omega_M$.

– The horizon T is finite and equals T/δ (which is supposed to be an integer).

– The action space is $A = L$.

– The transition kernel P is the transition kernel P_δ associated with the skeleton chain of the mode-augmented PDMP. More precisely, if $z = (m_{n\delta}, \mathsf{z}_{n\delta})$, and $a = \ell$, then for any Borel subset B of Ω_M, one has:

$$P(z' \in B|z, a) = P_\delta^\ell(B|z),$$

where P_δ^ℓ is defined in equation [6.1].

– The cost function c should in principle be defined by:

$$c(z, a = \ell, z') = c(z, a = \ell) = \mathbb{E}_z\left[\int_0^\delta c_R(Z_t)dt\big|\, Z_0 = (\ell, z)\right] + \tilde{c}_I(z, a = \ell),$$

where $\tilde{c}_I(s, a = \ell) = c_I(Z_{n\delta}, Z_{n\delta^+} = (\ell, z))$ and at time $n\delta$, the decision is taken to move $Z_{n\delta} = (\ell_{n\delta}, z)$ to $Z_{n\delta^+} = (\ell, z)$. In practice, such a cost function is generally intractable, and alternative cost functions close to c, depending only on the values of z, a, z' are considered instead. The terminal cost C is equal to c_T.

One of the main advantages of the MDP framework, is that it is now very easy to rigorously define decision rules π_t as measurable functions from Ω_Z onto A, and to define admissible policies, as mentioned in section 6.2.2.1.

The controlled trajectory of the MDP following policy $\pi = (\pi_t)_{0\leq t<T} \in \Pi$ is defined recursively in Algorithm 6.5.

Algorithm 6.5 Simulation of a trajectory of a controlled PDMP seen as a MDP controlled by policy π starting from state z_0 up to the horizon $\mathrm{T} = \delta T$

1: $Z_0 \leftarrow z_0$
2: **for** $t \leftarrow 0$ to $T - 1$ **do**
3: $\quad A_t = \ell_t = \pi(Z_t)$
4: $\quad Z_{t+1} \sim P(\cdot|Z_t, A_t) = P_\delta^{\ell_t}(\cdot|Z_t)$
5: **end for**

Line 4 corresponds to Algorithm 6.3 for the mode-augmented PDMP starting from state $z = (\ell_n, Z_n)$ at time 0 and up to time horizon δ.

Then the total expected cost of policy $\pi \in \Pi$ starting at $z_0 \in S$:

$$V_T(\pi, z_0) = \mathbb{E}_{z_0}^{\pi}\left[\sum_{t=0}^{T-1} c(Z_t, A_t, Z_{t+1}) + C(Z_T)\right],$$

defined in section 6.2.2 is equivalent to $\mathcal{V}_T(\mathcal{S}, z)$ defined in section 6.3.2.1, and the impulse control problem *corresponds* to the optimization problem:

$$V(z = z_0) = \inf_{\pi \in \Pi} V_T(\pi, z_0).$$

To obtain a formal equality $\mathcal{V}(z_0) = V(z_0)$ is possible but requires a perfect match between the cost functions and the sets of admissible policies. This will not be further discussed here as we chose the MDP framework specifically to avoid delving into the formal definition of admissible policies or cost functions for PDMPs.

6.4.1.3. *Squirrel toy model: from controlled PDMPs to MDPs*

The example is a simplified version of that of section 6.3.2.3. Here, the impulse dates are restricted to correspond to fixed daily times. To fit MDP notations, we now denote $z = (m, u) \in \{1, \cdots, 4\} \times [0, 7]$, and $\ell = a$ is now considered as an action.

In order to properly define its equivalent MDP formalism, we need to compute the transition kernel of the skeleton chain of span 1, as well as the expectation of the running cost over a 1-time increment. Unless we impose a minimum time spent in each hiding place (by for instance choosing a null jump intensity over a span of time), the number of jumps in between two decision dates is not bounded, which prevents writing an explicit transition kernel. Even in the case of bounded number of jumps, the explicit formulation is rather technical and is omitted here. Recall that transitions are easy to simulate in all cases.

In theory, to match the impulse problem set in section 6.3.1.4 the cost function should be:

$$c(z,a) = \mathbb{E}_z \left[\int_0^1 \mathbb{1}_{m_t \neq a} dt \right] + c_I = \int_0^1 \mathbb{P}(m_t \neq a) dt + c_I.$$

This quantity may not have an analytical expression. In practice, one may choose simpler forms for the cost function.

6.4.2. *Partially observed controlled PDMPs as POMDPs*

While properly defining an impulse control problem for a hidden PDMP is very challenging, the framework of POMDPs allows us to deal with partially hidden MDPs. Transforming an impulse control problem for partially observed PDMPs into a POMDP is done in a similar manner as in the perfect observation case, and again greatly simplifies the problem statement.

Here, we will assume that observations are only available at the decision dates $n\delta$, and are given by $Y_n = F(Z_{n\delta}) + \epsilon_n$ with (ϵ_n) real-valued independent and identically distributed random variables with density f independent from the PDMP and the actions, and F a real-valued link function (to simplify the exposition, both the noise and link function could be multi-dimensional). Only partial information about the process Z is obtained through Y_n, and we hence consider $Z_{n\delta}$ as hidden. The objective is now to control the PDMP through policies that may only rely on the observations available up to each decision date.

We first state this problem as a POMDP in section 6.4.2.1, then extend our squirrel toy model example to this more realistic context in section 6.4.2.2, and finally discuss the computation of *belief states* originating from PDMPs in section 6.4.2.3.

6.4.2.1. *Formulation as a POMDP*

We keep the restrictions on the impulses from section 6.4.1.1, and now give a POMDP formulation of the partially observed impulse control problem for a PDMP.

DEFINITION 6.24 (Partially observed controlled PDMP as a POMDP).– *Let* $\langle \Omega, \Phi, \lambda, Q \rangle$ *be a mode-augmented controlled PDMP under partial observations. The corresponding POMDP is defined by* $\langle \Omega_Z, A, T, P, \Omega_Y, O, c, C, y_0 \rangle$ *with the following characteristics:*

– The state space Ω_Z*, the action space* A*, the transition kernel* P*, the cost functions* c *and* C *and the horizon* T *are the same as those of definition 6.23.*

– The observation space is $\Omega_Y = \mathbb{R}$.

– The observation function O corresponds to the observation function of the PDMP, which may or may not depend on decision a. Here, the observation function has density (in the variable y):

$$O(z', a, y) = f(y - F(z')).$$

– The initial observation y_0 is example specific.

The initial belief b_0 is the filter at time 0 associated with the partially observed PDMP and will be discussed more thoroughly in section 6.4.2.3. Here again, the impulse control for the partially observed PDMP is equivalent to its POMDP counterpart, where both the state space Ω_Z and the observation space Ω_Y are continuous. It is then straightforward to consider the associated belief MDP, which can take a particular meaning in the PDMP context.

6.4.2.2. *Squirrel toy model as a partially observed controlled PDMP formalized as a POMDP*

We now combine the features of sections 6.2.3.3 and 6.4.1.3. We now suppose that we do not observe the squirrel, but only have access to a presence/absence indicator at decision times. More precisely, at each decision date n we observe $y = F(z) = \mathbb{1}_{\{a=m\}}$. The observation space, observation function and initial observation are the same as in section 6.2.3.3.

This POMDP can be solved by computing the corresponding belief MDP, which is a finite-horizon MDP in this case, and solving it backwards.

6.4.2.3. *Belief states for PDMPs*

While in the practice of POMDPs the belief is often updated through simulations, as in the context of *particle filtering* (Del Moral 1997; Silver and Veness 2010), the PDMP context offers a particularly convenient setting for the update of the filters, thanks to the Markov property of skeleton chains. Indeed, in the framework of PDMPs, it is very common to introduce the filter of the process, which we will use for the belief state in the context of POMDPs. General considerations can be found in Bäuerle and Rieder (2011) and Brandejsky et al. (2012). Here, we briefly expose the steps in the partially observed controlled PDMP framework.

For $n \leq T$, set $\mathcal{F}_n^O = \sigma(Y_t, 0 \leq t \leq n)$ the σ-field generated by the observations up to n, corresponding to time $n\delta$. Let:

$$b_n(B) = \mathbb{P}(Z_{n\delta} \in B | \mathcal{F}_n^O),$$

denote the filter or belief process for the unobserved part of the process. The standard prediction-correction approach yields a recursive construction for the filter. For any

$n \geq 0$, conditionally on $Y_{n+1} = y' \in \Omega_Y$, $a = \ell \in L$ and $b_n = \theta$, one has $b_{n+1} = \tau(\theta, a, y')$ with:

$$\tau(\theta, a, y')(B) = \frac{\int_{\Omega_M} \int_{\Omega_M} f(y' - F(z')) \mathbb{1}_B(z') P_\delta^\ell(dz'|z) \theta(dz)}{\int_{\Omega_M} \int_{\Omega_M} f(y' - F(z')) P_\delta^\ell(dz'|z) \theta(dz)},$$

for any Borel subset B of Ω_M.

Conditional on being able to compute the integrals, this provides a theoretical framework for belief updates that does not rely on the quality of simulators. In practice, computing this filter is not easy.

6.5. Concluding remarks and open questions

As shown in the previous sections, the MDP formalism makes it possible to address very challenging problems for controlled PDMPs. Conversely, deriving MDPs from PDMPs opens new questions for continuous-state-space MDPs and their practical resolution. We briefly present some of these challenges in this section.

6.5.1. *Open questions in impulse control of PDMPs that might be tackled from the MDP perspective*

As mentioned in sections 6.3.2 and 6.4.1, the very task of formalizing the definition of an impulse controlled PDMP in the partially observed framework is rather technical (even in the fully observed case). Embedding it in the MDP framework provides a very natural environment for this formalization, since by construction strategies are admissible policies and value functions are natural elements of MDP resolution.

The MDP framework provides a wide range of algorithms for the efficient resolution of controlled PDMPs. As mentioned in section 6.4.1 and 6.4.2, a (partially observed) controlled PDMP can be cast as a MDP with a continuous state-space. From there, several possibilities can be considered: discretizing the initial process to reduce it to a finite state space, for which exact resolution through dynamic programming is possible, or using approximate resolution algorithms designed for MDPs with finite but very large state spaces or continuous state spaces. Those two strategies have been developed for a medical example; see, for instance, Cleynen and de Saporta (2018) for an exact resolution of the discretized process in the framework of optimal stopping, Cleynen and de Saporta (2021) for an exact resolution of the discretized process in the general impulse control framework and (de Saporta et al. 2024) for an approximate resolution based on the adaptation of a Monte Carlo tree search algorithm for PDMPs.

Noting that (partially observed) PDMPs are both easy to simulate and can be cast as POMDPs, one can take advantage of the huge amount of resources (both literature and software) available in the field of deep RL (see, e.g. Mnih et al. (2013); Sutton and Barto (2018)). The field of RL is mature and there exist several RL application programming interfaces (e.g. gymnasium Brockman et al. (2016)), connecting POMDP simulation models, to libraries of (deep) RL algorithms (e.g. RLlib Liang et al. (2018)). The embedding of PDMPs in the MDP paradigm hence allows us to address the scalability issues of standard resolution strategies for PDMPs by providing efficient resolution algorithms, including exact and approximate approaches, that provide excellent performances.

6.5.2. *Interesting questions in MDPs arising from converted PDMPs*

In the MDP/RL community, there is an extensive literature on solving problems with infinite or continuous time and state spaces. Recent advancements in deep RL have also led to practical solutions for these problems. However, solving MDPs with continuous state spaces or in continuous time remains a significant challenge, and practical algorithms often fail to perform well in these scenarios.

This difficulty is particularly obvious when the transition function, $P(\cdot|s, a)$, is a continuous probability distribution. Monte Carlo algorithms, for instance, generally struggle in such settings because they cannot effectively explore the resulting min-expectation tree, which has infinitely many successor nodes at each level. Similarly, reinforcement learning approaches require approximations of the Q-values. While deep RL techniques model Q-values using neural networks and are applicable in these cases, they demand immense computational resources and, when relying on real-world data rather than simulations, require vast quantities of data.

Using MDPs derived from PDMPs offers a way to address complex, realistic problems with models that are both easily parameterized and straightforward to simulate. Additionally, the min-expectation tree resulting from applying Monte Carlo tree search to PDMPs maintains a finite, often small branching factor (see, e.g. de Saporta et al. (2024)).

In other words, controlled PDMPs serve as an intriguing benchmark for evaluating MDP solution algorithms. Their complexity lies between that of finite MDPs, which are relatively easy to solve, and MDPs with continuous time, state, or decision spaces, for which existing solving methods typically lack quality guarantees.

6.6. Notations

We gather here Tables 6.1–6.5 that contain the main notations for this chapter.

Definition	Notation	Domain
State space	Ω_Z	
State at time t	Z_t, z_t	Ω_Z
Action space	A	Finite
Action at time t	A_t, a_t	A
Time horizon	T	$\mathbb{N}$ or $\mathbb{R}_+ \cup \{+\infty\}$
Controlled transition matrices or kernels	P	$\Omega_Z \times A \times \Omega_Z \to [0,1]$ $\Omega_Z \times A \times \mathcal{B}(\Omega_Z) \to [0,1]$
Cost function	c	$\Omega_Z \times A \times \Omega_Z \to \mathbb{R}$
Terminal cost function	C	$\Omega_Z \to \mathbb{R}$
Set of length t histories	$\mathcal{H}_t$	$(\Omega_Z \times A)^t \times \Omega_Z$
Decision rule at time t	π_t	$\mathcal{H}_t \to A$
Policy	$\pi = (\pi_t)_{0 \le t < T} = \pi_{0:T-1}$	Π
Set of policies	Π	
Discounted expected cost of policy π starting from $Z_0 = z$	$V_\gamma(\pi, z)$	$\mathbb{R}$
Total expected cost up to time T of policy π starting from $Z_0 = z$	$V_T(\pi, z)$	$\mathbb{R}$

Table 6.1. *Main notations for MDPs*

Definition	Notation	Domain
Observation space	Ω_Y	
Observation at time t	Y_t, y_t	Ω_Y
Observation probability or kernel	O	$\Omega_Z \times A \times \Omega_Y \to [0,1]$ $\Omega_Z \times A \times \mathcal{B}(\Omega_Y) \to [0,1]$
Set of length t observed histories	$\mathcal{H}_t^o$	$(\Omega_Y \times A)^t \times \Omega_Y$
Set of history-dependent policies	Π^o	
State space of the equivalent belief MDP	$\mathcal{B}$	
Belief at time t	b_t	$\mathcal{B}$
Belief update function	τ	$\mathcal{B} \times A \times \Omega_Y \to \mathcal{B}$
Controlled transition kernels of the belief MDP	$\mathcal{P}$	$\mathcal{B} \times A \times \mathcal{B} \to [0,1]$
Cost function of the belief MDP	ρ	$\mathcal{B} \times A \times \mathcal{B} \to \mathbb{R}$

Table 6.2. *Main additional notations for POMDPs*

Definition	Notation	Domain
State space of the PDMP	Ω	Hybrid
State at time t	Z_t	Ω
Set of modes	M	Finite
Mode	m	M
Euclidean state space in mode m	Ω_{m}	$\subset \mathbb{R}^{d_{\mathrm{m}}}$
Euclidean variable	z	Ω_{m}
Time since the last jump	u	$\mathbb{R}_+$
Flow of the PDMP	Φ	$\Omega \times \mathbb{R}_+ \to \Omega$
Deterministic time to reach the boundary starting from z	$t^*(z)$	$\mathbb{R}_* \cup \{+\infty\}$
Jump intensity of the PDMP	λ	$\Omega \to \mathbb{R}_+$
Cumulative intensity of the PDMP	Λ	$\Omega \times \mathbb{R}_+ \to \mathbb{R}_+$
Jump kernel of the PDMP	Q	$\Omega \times \mathbb{B}(\Omega) \to \Omega$
Date of the nth jump	S_n	$\mathbb{R}_+$
State at the nth jump	J_n	Ω
Sojourn duration between jumps $n-1$ and n	X_n	$\mathbb{R}_+$
Transition kernel of the canonical chain	$P^{J,X}$	$\Omega \times \mathbb{B}(\Omega \times \mathbb{R}_+) \to \Omega$
Transition kernel of the PDMP between times 0 and t	P_t	$\Omega \times \mathbb{B}(\Omega) \to \Omega$

Table 6.3. *Main notations for PDMPs*

Definition	Notation	Domain
Impulse strategy	$\mathcal{S}$	$(\mathbb{U} \times \mathbb{R}_+)^{\mathbb{N}}$
Delay between interventions $n-1$ and n	τ_n	$\mathbb{R}_+$
New starting point after the nth intervention	χ_n	$\mathbb{U}$
Running cost	c_R	$\Omega \to \mathbb{R}$
Impulse cost	c_I	$\Omega \times \mathbb{U} \to \mathbb{R}$
Terminal cost function	C_T	$\Omega \to \mathbb{R}$
Control horizon	T	$\mathbb{R}_+$
Discounted expected cost of strategy S starting from $Z_0 = z$	$\mathcal{V}(\mathcal{S}, z)$	$\mathbb{R}$

Table 6.4. *Main additional notations for controlled PDMPs*

6.7. Acknowledgments

Alice Cleynen was supported by European Union's Horizon 2020 research and innovation program (Marie Sklodowska-Curie grant agreement No. 890462). Amélie Vernay was supported by the French Agence Nationale de la Recherche (ANR), grant ANR-20-CHIA-0001-01 (Chaire IA CaMeLOt).

Definition	Notation	Domain
Set of controller-chosen modes, action set	L	Finite
Set of uncontrolled modes	M	Finite
Controller-chosen mode	ℓ	L
Uncontrolled mode	m	M
State space of the PDMP seen as a MDP	Ω_M	Hybrid
Fixed delay between consecutive impulses	δ	$\mathbb{R}_+^*$

Table 6.5. *Main additional notations for controlled PDMPs formalized as MDPs*

6.8. References

Almudevar, A. (2001). A dynamic programming algorithm for the optimal control of piecewise deterministic Markov processes. *SIAM Journal on Control and Optimization*, 40(2), 525–539.

Åström, K.J. (1965). Optimal control of Markov processes with incomplete state information. *Journal of Mathematical Analysis and Applications*, 10(1), 174–205.

Atkeson, C. and Santamaria, J. (1997). A comparison of direct and model-based reinforcement learning. In *Proceedings of International Conference on Robotics and Automation*. IEEE Press, Albuquerque, NM.

Bandini, E. (2018). Optimal control of piecewise deterministic Markov processes: A BSDE representation of the value function. *ESAIM: Control, Optimisation and Calculus of Variations*, 24(1), 311–354.

Barbu, V.S. and Limnios, N. (2008). *Semi-Markov Chains and Hidden Semi-Markov Models Toward Applications: Their Use in Reliability and DNA Analysis*. Springer, New York.

Bardet, J.B., Christen, A., Guillin, A., Malrieu, F., Zitt, P. (2013). Total variation estimates for the TCP process. *Electronic Journal of Probability*, 18(10), 21.

Bäuerle, N. and Lange, D. (2018). Optimal control of partially observable piecewise deterministic Markov processes. *SIAM Journal on Control and Optimization*, 56(2), 1441–1462.

Bäuerle, N. and Rieder, U. (2011). *Markov Decision Processes with Applications to Finance*. Springer, Berlin, Heidelberg.

Bellman, R. (1958). Dynamic programming and stochastic control processes. *Information and Control*, 1(3), 228–239.

Boutilier, C., Dean, T., Hanks, S. (1999). Decision-theoretic planning: Structural assumptions and computational leverage. *Journal of Artificial Intelligence Research (JAIR)*, 11, 1–94.

Brandejsky, A., de Saporta, B., Dufour, F. (2012). Numerical method for expectations of piecewise deterministic Markov processes. *Communications in Applied Mathematics and Computational Science*, 7(1), 63–104.

Brockman, G., Cheung, V., Pettersson, L., Schneider, J., Schulman, J., Tang, J., Zaremba, W. (2016). OpenAI gym. *arXiv*. doi: 10.48550/arXiv.1606.01540.

Browne, C.B., Powley, E., Whitehouse, D., Lucas, S.M., Cowling, P.I., Rohlfshagen, P., Tavener, S., Perez, D., Samothrakis, S., Colton, S. (2012). A survey of Monte Carlo tree search methods. *IEEE Transactions on Computational Intelligence and AI in Games*, 4(1), 1–43.

Chafaï, D., Malrieu, F., Paroux, K. (2010). On the long time behavior of the TCP window size process. *Stochastic Processes and their Applications*, 120(8), 1518–1534.

Chanel, C.C., Teichteil-Königsbuch, F., Lesire, C. (2013). Multi-target detection and recognition by UAVs using online POMDPs. In *Proceedings of the AAAI Conference on Artificial Intelligence*, 1381–1387.

Cleynen, A. and de Saporta, B. (2018). Change-point detection for piecewise deterministic Markov processes. *Automatica*, 97, 234–247.

Cleynen, A. and de Saporta, B. (2021). Numerical method to solve impulse control problems for partially observed piecewise deterministic Markov processes. *arXiv*. doi: 10.48550/arXiv.2112.09408.

Cleynen, A., de Saporta, B., Rossini, O., Sabbadin, R., Vernay, A. (2025). Bridging impulse control of piecewise deterministic Markov processes and Markov decision processes: Frameworks, extensions, and open challenges. *arXiv*. doi: 10.48550/arXiv.2501.04120.

Cloez, B., de Saporta, B., Joubaud, M. (2020). Optimal stopping for measure-valued piecewise deterministic Markov processes. *Journal of Applied Probability*, 57(2), 497–512.

Cocozza-Thivent, C. (2021). *Markov Renewal and Piecewise Deterministic Processes*. Springer, Cham.

Costa, O.L.V. (1993). Discretizations for the average impulse control of piecewise deterministic processes. *Journal of Applied Probability*, 30(2), 405–420.

Costa, M. (2016). A piecewise deterministic model for a prey-predator community. *The Annals of Applied Probability*, 26(6), 3491–3530.

Costa, O.L.V. and Davis, M.H.A. (1988). Approximations for optimal stopping of a piecewise-deterministic process. *Mathematics of Control, Signals, and Systems*, 1(2), 123–146.

Costa, O.L.V. and Davis, M.H.A. (1989). Impulse control of piecewise-deterministic processes. *Mathematics of Control, Signals, and Systems*, 2(3), 187–206.

Costa, O.L.V. and Dufour, F. (2008). Stability and ergodicity of piecewise deterministic Markov processes. *SIAM Journal on Control and Optimization*, 47(2), 1053–1077.

Costa, O.L.V. and Dufour, F. (2010). Average continuous control of piecewise deterministic Markov processes. *SIAM Journal on Control and Optimization*, 48(7), 4262–4291.

Costa, O.L.V. and Dufour, F. (2013). *Continuous Average Control of Piecewise Deterministic Markov Processes*. Springer, New York.

Davis, M.H.A. (1984). Piecewise-deterministic Markov processes: A general class of non-diffusion stochastic models. *Journal of the Royal Statistical Society Series B: Statistical Methodology*, 46(3), 353–376.

Davis, M.H.A. (1993). *Markov Models and Optimization*. 1st edition. Chapman & Hall, London, New York.

Del Moral, P. (1997). Nonlinear filtering: Interacting particle resolution. *Comptes Rendus de l'Académie des Sciences - Series I - Mathematics*, 325(6), 653–658.

Dempster, M.A.H. (1991). Optimal control of piecewise deterministic Markov processes. *Applied Stochastic Analysis*, 5, 303–325.

Dempster, M.A.H. and Ye, J.J. (1995). Impulse control of piecewise deterministic Markov processes. *The Annals of Applied Probability*, 5(2), 399–423.

Doumic, M., Hoffmann, M., Krell, N., Robert, L. (2015). Statistical estimation of a growth-fragmentation model observed on a genealogical tree. *Bernoulli*, 21(3), 1760–1799.

Dufour, F., Horiguchi, M., Piunovskiy, A.B. (2016). Optimal impulsive control of piecewise deterministic Markov processes. *Stochastics*, 88(7), 1073–1098.

Fritsch, C. (2014). Probabilistic and numerical approaches of individual-based chemostat model. Thesis, Université Montpellier 2.

Gatarek, D. (1992). Optimality conditions for impulsive control of piecewise-deterministic processes. *Mathematics of Control, Signals, and Systems*, 5(2), 217–232.

Gillespie, D.T. (1977). Exact stochastic simulation of coupled chemical reactions. *The Journal of Physical Chemistry*, 81(25), 2340–2361.

Goodfellow, I., Bengio, Y., Courville, A. (2016). *Deep Learning*. MIT Press, Cambridge.

Gugerui, U. (1986). Optimal stopping of a piecewise-deterministic Markov process. *Stochastics*, 19(4), 221–236.

Herbach, U., Bonnaffoux, A., Espinasse, T., Gandrillon, O. (2017). Inferring gene regulatory networks from single-cell data: A mechanistic approach. *BMC Systems Biology*, 11(1), 105.

Hernández-Lerma, O. and Lasserre, J.B. (1996). *Discrete-Time Markov Control Processes*. Springer, New York.

Hinderer, K., Rieder, U., Stieglitz, M. (2016). *Dynamic Optimization*. Springer, Cham.

Hoey, J., Poupart, P., Bertoldi, A.V., Craig, T., Boutilier, C., Mihailidis, A. (2010). Automated handwashing assistance for persons with dementia using video and a partially observable Markov decision process. *Computer Vision Image Understanding*, 114(5), 503–519.

Howard, R.A. (1962). *Dynamic Programming Markov Processes*. MIT Press, Cambridge.

Jacobsen, M. (2006). *Point Process Theory and Applications, : Marked Point and Piecewise Deterministic Processes*. Birkhäuser-Verlag, Boston.

Kaelbling, L., Littman, M., Cassandra, A. (1998). Planning and acting in partially observable stochastic domains. *Artificial Intelligence*, 101(1–2), 99–134.

Keneally, S.K., Robbins, M.J., Lunday, B.J. (2016). A Markov decision process model for the optimal dispatch of military medical evacuation assets. *Health Care Management Science*, 19(2), 111–129.

Kocsis, L. and Szepesvari, C. (2006). Bandit based Monte-Carlo Planning. In *European Conference on Machine Learning*. Springer, Berlin, Heidelberg.

LeCun, Y., Bengio, Y., Hinton, G. (2015). Deep learning. *Nature*, 521(7553), 436–444.

Lenhart, S.M. (1989). Viscosity solutions associated with impulse control problems for piecewise-deterministic processes. *International Journal of Mathematics and Mathematical Sciences*, 12(1), 145–157.

Levin, E., Pieraccini, R., Eckert, W. (1998). Using Markov decision process for learning dialogue strategies. In *Proceedings of the 1998 IEEE International Conference on Acoustics, Speech and Signal Processing, ICASSP'98*, Seattle, WA.

Liang, E., Liaw, R., Nishihara, R., Moritz, P., Fox, R., Goldberg, K., Gonzalez, J.E., Jordan, M.I., Stoica, I. (2018). RLlib: Abstractions for distributed reinforcement learning. In *International Conference on Machine Learning*. PMLR, Stokholm.

Mnih, V., Kavukcuoglu, K., Silver. D., Graves, A., Antonoglou, I., Wierstra, D., Riedmiller, M. (2013). Playing atari with deep reinforcement learning. *arXiv*. doi: 10.48550/arXiv.1312.5602.

Pasin, C., Dufour, F., Villain, L., Zhang, H., Thiébaut, R. (2018). Controlling IL-7 injections in HIV-infected patients. *Bulletin of Mathematical Biology*, 80(9), 2349–2377.

Puterman, M.L. (1994). *Markov Decision Processes: Discrete Stochastic Dynamic Programming*. John Wiley & Sons, Hoboken, NJ.

Riedler, M.G. and Thieullen, M. (2015). Spatio-temporal hybrid (PDMP) models: Central limit theorem and Langevin approximation for global fluctuations. Application to electrophysiology. *Bernoulli*, 21(2), 1–48.

Riedler, M.G., Thieullen, M., Wainrib, G. (2012). Limit theorems for infinite-dimensional piecewise deterministic Markov processes. Applications to stochastic excitable membrane models. *Electronic Journal of Probability*, 17, 1–48.

Roy, M., Chowdhury, C., Aslam, N. (2018). Designing transmission strategies for enhancing communications in medical IoT using Markov decision process. *Sensors*, 18(12), 4450.

de Saporta, B. and Dufour, F. (2012). Numerical method for impulse control of piecewise deterministic Markov processes. *Automatica*, 48(5), 779–793.

de Saporta, B. and Zhang, H. (2013). Predictive maintenance for the heated hold-up tank. *Reliability Engineering & System Safety*, 115, 82–90.

de Saporta, B., Dufour, F., Gonzalez, K. (2010). Numerical method for optimal stopping of piecewise deterministic Markov processes. *The Annals of Applied Probability*, 20(5), 1607–1637.

de Saporta, B., Dufour, F., Zhang, H. (2016). *Numerical Methods of Simulation and Optimization of Piecewise Deterministic Markov Processes: Application to Reliability*. ISTE Ltd, London, and John Wiley & Sons, New York.

de Saporta, B., Dufour, F., Geeraert, A. (2017). Optimal strategies for impulse control of piecewise deterministic Markov processes. *Automatica*, 77, 219–229.

de Saporta, B., Thierry d'Argenlieu, A., Sabbadin, R., Cleynen, A. (2024). A Monte-Carlo planning strategy for medical follow-up optimization: Illustration on multiple myeloma data. *PLOS ONE*, 19(12), e0315661.

Schaefer, A.J., Bailey, M.D., Shechter, S.M., Roberts, M.S. (2004). Modeling medical treatment using Markov decision processes. In *Operations Research and Health Care: A Handbook of Methods and Applications*, Brandeau, M.L., Sainfort, F., Pierskalla, W.P. (eds). Springer, Boston, MA.

Silver, D. and Veness, J. (2010). Monte-Carlo planning in large POMDPs. *Advances in Neural Information Processing Systems*, 23, 2154–2172.

Sutton, R.S. and Barto, A.G. (2018). *Reinforcement Learning: An Introduction*. MIT Press, Cambridge, MA.

Świechowski, M., Godlewski, K., Sawicki, B., Mańdziuk, J. (2023). Monte Carlo tree search: A review of recent modifications and applications. *Artificial Intelligence Review*, 56(3), 2497–2562.

Young, S., GaVsić, M., Thomson, B., Williams, J.D. (2013). POMDP-based statistical spoken dialog systems: A review. *Proceedings of the IEEE*, 5, 1160–1179.

List of Authors

Hanna BACAVE
INRAE
Castanet-Tolosan
France

Caroline BÉRARD
Université Rouen Normandie
France

Alice CLEYNEN
CNRS
Canberra
Australia

Marie-Josée CROS
INRAE
Castanet-Tolosan
France

Jean-Baptiste DURAND
Cirad
Montpellier
France

Alain FRANC
INRAE
Cestas
France

Corentin LOTHODÉ
CNRS
Angers
France

Nathalie PEYRARD
INRAE
Castanet-Tolosan
France

Sandra PLANCADE
INRAE
Castanet-Tolosan
France

Orlane ROSSINI
CNRS
Montpellier
France

Régis SABBADIN
INRAE
Castanet-Tolosan
France

Benoîte DE SAPORTA
Université de Montpellier
France

Ronan TRÉPOS
INRAE
Castanet-Tolosan
France

Nicolas VERGNE
Université de Rouen Normandie
France

Amélie VERNAY
Université de Montpellier
France

Irene VOTSI
Université de Lorraine
Metz
France

Index

A, B

C, D

E, F

G, H

I, L

M, N

O, P

R, S

T, V

Other titles from

iSTE

in

Mathematics and Statistics

2025

DUPUY Jean-François
Generalized Linear Models: Problems with Censored, Missing, and Zero-inflated Data

GAIVORONSKI Alexei A., KNOPOV Pavel S., NORKIN Vladimir I., ZASLAVSKYI Volodymyr A.
Stochastic Modeling and Optimization Methods for Critical Infrastructure Protection 1: Stochastic Modeling
Stochastic Modeling and Optimization Methods for Critical Infrastructure Protection 2: Methods and Tools

PICARD Philippe, LEFÈVRE Claude
Abel-Gontcharoff Pseudopolynomials and Stochastic Applications

2024

BOURLÈS Henri
System Theory – A Modern Approach 1: Linear Ordinary and Functional Differential Equations
(New Mathematical Methods, Systems and Applications Set – Volume 1)

MAKAROV Volodymyr, MAYKO Nataliya
Traditional Functional-Discrete Methods for the Problems of Mathematical Physics: New Aspects

2023

AIT BEN HASSI El Mustapha
The Theory of Distributions: Introduction

KOROLIOUK Dmitri, SAMOILENKO Igor
Asymptotic and Analytic Methods in Stochastic Evolutionary Systems

PARROCHIA Daniel
Graphs, Orders, Infinites and Philosophy

2022

CHAKRAVARTHY Srinivas R.
Introduction to Matrix-Analytic Methods in Queues 1: Analytical and Simulation Approach – Basics
Introduction to Matrix-Analytic Methods in Queues 2: Analytical and Simulation Approach – Queues and Simulation

LESFARI Ahmed
Integrable Systems

RADCHENKO Vadym M.
General Stochastic Measures: Integration, Path Properties and Equations

DE SAPORTA Benoîte, ZILI Mounir
Martingales and Financial Mathematics in Discrete Time

SIMON Jacques
Distributions
(Analysis for PDEs Set – Volume 3)

2021

KOROLIOUK Dmitri, SAMOILENKO Igor
Random Evolutionary Systems: Asymptotic Properties and Large Deviations

MOKLYACHUK Mikhail
Convex Optimization: Introductory Course

POGORUI Anatoliy, SWISHCHUK Anatoliy, RODRÍGUEZ-DAGNINO Ramón M.
Random Motions in Markov and Semi-Markov Random Environments 1: Homogeneous Random Motions and their Applications
Random Motions in Markov and Semi-Markov Random Environments 2: High-dimensional Random Motions and Financial Applications

PROVENZI Edoardo
From Euclidean to Hilbert Spaces: Introduction to Functional Analysis and its Applications

2020

BARBU Vlad Stefan, VERGNE Nicolas
Statistical Topics and Stochastic Models for Dependent Data with Applications

CHABANYUK Yaroslav, NIKITIN Anatolii, KHIMKA Uliana
Asymptotic Analyses for Complex Evolutionary Systems with Markov and Semi-Markov Switching Using Approximation Schemes

KOROLIOUK Dmitri
Dynamics of Statistical Experiments

MANOU-ABI Solym Mawaki, DABO-NIANG Sophie, SALONE Jean-Jacques
Mathematical Modeling of Random and Deterministic Phenomena

SIMON Jacques
Continuous Functions
(Analysis for PDEs Set – Volume 23)

2019

BANNA Oksana, MISHURA Yuliya, RALCHENKO Kostiantyn, SHKLYAR Sergiy
Fractional Brownian Motion: Approximations and Projections

GANA Kamel, BROC Guillaume
Structural Equation Modeling with lavaan

KUKUSH Alexander
Gaussian Measures in Hilbert Space: Construction and Properties

LUZ Maksym, MOKLYACHUK Mikhail
Estimation of Stochastic Processes with Stationary Increments and Cointegrated Sequences

MICHELITSCH Thomas, PÉREZ RIASCOS Alejandro, COLLET Bernard, NOWAKOWSKI Andrzej, NICOLLEAU Franck
Fractional Dynamics on Networks and Lattices

VOTSI Irene, LIMNIOS Nikolaos, PAPADIMITRIOU Eleftheria, TSAKLIDIS George
Earthquake Statistical Analysis through Multi-state Modeling
(Statistical Methods for Earthquakes Set – Volume 2)

2018

AZAÏS Romain, BOUGUET Florian
Statistical Inference for Piecewise-deterministic Markov Processes

IBRAHIMI Mohammed
Mergers & Acquisitions: Theory, Strategy, Finance

PARROCHIA Daniel
Mathematics and Philosophy

2017

CARONI Chysseis
First Hitting Time Regression Models: Lifetime Data Analysis Based on Underlying Stochastic Processes
(Mathematical Models and Methods in Reliability Set – Volume 4)

CELANT Giorgio, BRONIATOWSKI Michel
Interpolation and Extrapolation Optimal Designs 2: Finite Dimensional General Models

CONSOLE Rodolfo, MURRU Maura, FALCONE Giuseppe
Earthquake Occurrence: Short- and Long-term Models and their Validation (Statistical Methods for Earthquakes Set – Volume 1)

D'AMICO Guglielmo, DI BIASE Giuseppe, JANSSEN Jacques, MANCA Raimondo
Semi-Markov Migration Models for Credit Risk (Stochastic Models for Insurance Set – Volume 1)

GONZÁLEZ VELASCO Miguel, del PUERTO GARCÍA Inés, YANEV George P.
Controlled Branching Processes (Branching Processes, Branching Random Walks and Branching Particle Fields Set – Volume 2)

HARLAMOV Boris
Stochastic Analysis of Risk and Management (Stochastic Models in Survival Analysis and Reliability Set – Volume 2)

KERSTING Götz, VATUTIN Vladimir
Discrete Time Branching Processes in Random Environment (Branching Processes, Branching Random Walks and Branching Particle Fields Set – Volume 1)

MISHURA Yuliya, SHEVCHENKO Georgiy
Theory and Statistical Applications of Stochastic Processes

NIKULIN Mikhail, CHIMITOVA Ekaterina
Chi-squared Goodness-of-fit Tests for Censored Data (Stochastic Models in Survival Analysis and Reliability Set – Volume 3)

SIMON Jacques
Banach, Fréchet, Hilbert and Neumann Spaces (Analysis for PDEs Set – Volume 1)

2016

CELANT Giorgio, BRONIATOWSKI Michel
Interpolation and Extrapolation Optimal Designs 1: Polynomial Regression and Approximation Theory

CHIASSERINI Carla Fabiana, GRIBAUDO Marco, MANINI Daniele
Analytical Modeling of Wireless Communication Systems
(Stochastic Models in Computer Science and Telecommunication Networks Set – Volume 1)

GOUDON Thierry
Mathematics for Modeling and Scientific Computing

KAHLE Waltraud, MERCIER Sophie, PAROISSIN Christian
Degradation Processes in Reliability
(Mathematial Models and Methods in Reliability Set – Volume 3)

KERN Michel
Numerical Methods for Inverse Problems

RYKOV Vladimir
Reliability of Engineering Systems and Technological Risks
(Stochastic Models in Survival Analysis and Reliability Set – Volume 1)

2015

DEVOLDER Pierre, JANSSEN Jacques, MANCA Raimondo
Basic Stochastic Processes

LE GAT Yves
Recurrent Event Modeling Based on the Yule Process
(Mathematical Models and Methods in Reliability Set – Volume 2)

DE SAPORTA Benoîte, DUFOUR François, ZHANG Huilong
Numerical Methods for Simulation and Optimization of Piecewise Deterministic Markov Processes

2014

COOKE Roger M., NIEBOER Daan, MISIEWICZ Jolanta
Fat-tailed Distributions: Data, Diagnostics and Dependence
(Mathematical Models and Methods in Reliability Set – Volume 1)

MACKEVIČIUS Vigirdas
Integral and Measure: From Rather Simple to Rather Complex

PASCHOS Vangelis Th
Combinatorial Optimization – 3-volume series – 2nd edition
Concepts of Combinatorial Optimization / Concepts and Fundamentals – Volume 1
Paradigms of Combinatorial Optimization – Volume 2
Applications of Combinatorial Optimization – Volume 3

2013

COUALLIER Vincent, GERVILLE-RÉACHE Léo, HUBER Catherine, LIMNIOS Nikolaos, MESBAH Mounir
Statistical Models and Methods for Reliability and Survival Analysis

JANSSEN Jacques, MANCA Oronzio, MANCA Raimondo
Applied Diffusion Processes from Engineering to Finance

SERICOLA Bruno
Markov Chains: Theory, Algorithms and Applications

2012

BOSQ Denis
Mathematical Statistics and Stochastic Processes

CHRISTENSEN Karl Bang, KREINER Svend, MESBAH Mounir
Rasch Models in Health

DECREUSEFOND Laurent, MOYAL Pascal
Stochastic Modeling and Analysis of Telecom Networks

DEVOLDER Pierre, JANSSEN Jacques, MANCA Raimondo
Stochastic Methods for Pension Funds

2011

MACKEVIČIUS Vigirdas
Introduction to Stochastic Analysis: Integrals and Differential Equations

MAHJOUB Ridha
Recent Progress in Combinatorial Optimization – ISCO2010

RAYNAUD Hervé, ARROW Kenneth
Managerial Logic

2010

BAGDONAVIČIUS Vilijandas, KRUOPIS Julius, NIKULIN Mikhail
Nonparametric Tests for Censored Data
Nonparametric Tests for Complete Data

IOSIFESCU Marius, LIMNIOS Nikoloas, OPRISAN Gheorghe
Introduction to Stochastic Models

VASSILIOU P.C.G.
Discrete-time Asset Pricing Models in Applied Stochastic Finance

2008

ANISIMOV Vladimir
Switching Processes in Queuing Models

FICHE Georges, HÉBUTERNE Gérard
Mathematics for Engineers

HUBER Catherine, LIMNIOS Nikolaos, MESBAH Mounir, NIKULIN Mikhail
Mathematical Methods in Survival Analysis, Reliability and Quality of Life

JANSSEN Jacques, MANCA Raimondo, VOLPE Ernesto
Mathematical Finance

2007

HARLAMOV Boris
Continuous Semi-Markov Processes

2006

CLERC Maurice
Particle Swarm Optimization

Printed and bound by CPI Group (UK) Ltd, Croydon, CR0 4YY

13/09/2026

14961012-0003